From Theoretical Physics to Biology

Institut de la Vie

Proceedings of the Third International Conference
'From Theoretical Physics to Biology'
Versailles, June 21–26, 1971

From Theoretical Physics to Biology

Editor: M. MAROIS, Paris

145 figures, 9 tables, 1973

S. Karger · Basel · München · Paris · London · New York · Sydney

CHEMISTRY

S. Karger · Basel · München · Paris · London · New York · Sydney
Arnold-Böcklin-Strasse 25, CH-4011 Basel (Switzerland)

Copyright 1973 by S. Karger AG, Verlag für Medizin und Naturwissenschaften, Basel
Printed in Switzerland by City-Druck, Glattbrugg
ISBN 3-8055–1578-2

Contents

Physical Aspects of the Order in Biological Systems

The First Steps of Evolution and the Nature of Life

Contents

Systems of Order Recognition and of Recognition

Systems of Sensorial Analysis

Neurophysiological Aspects of Vision

Round Table Session

Perspectives of Theoretical Physics and Biology and Their Social Implications

Conference Data

Third International Conference
'From Theoretical Physics to Biology',
Versailles, June 21–26, 1971

Organizing Committee

P. AUGER, Paris
S. BENNETT, Chapell Hill, N. C.
S. E. BRESLER, Leningrad
G. CARERI, Rome
E. G. D. COHEN, New York, N. Y.
A. COURNAND, New York, N. Y.
M. EIGEN, Gottingen
W. FAIRBANK, Stanford, Calif.
A. FESSARD, Paris
H. FRÖHLICH, Liverpool
D. GLASER, Berkeley, Calif.
D. HODGKIN, Oxford
F. JACOB, Paris
M. KAC, New York, N. Y.
A. KATCHALSKY †, Rehovot
J. C. KENDREW, Cambridge
M. KOTANI, Tokyo
R. KUBO, Tokyo

A. LICHNEROWICZ, Paris
P. O. LÖWDIN, Uppsala
F. LYNEN, Munich
O. MAALØE, Copenhagen
K. MENDELSSOHN, Oxford
R. S. MULLIKEN, Chicago, Ill.
M. F. PERUTZ, Cambridge
I. PRIGOGINE, Brussels
I. RABI, New York, N. Y.
L. ROSENFELD, Copenhagen
E. E. SALPETER, Ithaca, N. Y.
S. L. SOBOLEV, Novosibirsk
A. SZENT-GYÖRGYI, Woods Hole, Mass.
H. THEORELL, Stockholm
A. TISELIUS, Uppsala
V. F. WEISSKOPF, Cambridge, Mass.
E. P. WIGNER, Princeton, N. J.
C. N. YANG, Stony Brook

Sponsors

Monsieur GEORGES POMPIDOU, Président de la République Française, représenté par Monsieur ROBERT POUJADE, Ministre délégué auprès du Premier Ministre, chargé de la protection de la nature et de l'environnement.

International Union of Pure and Applied Physics

Participants

AMASSIAN, V. E., Albert Einstein College of Medicine of Yeshiva University, 1300 Morris Park Avenue, Bronx, NY 10461 (USA)

AUGER, P., 12, rue Emile Faguet, Paris 14e (France)

BENNETT, S., The University of North Carolina, The Laboratories for Reproductive Biology, 111 Swing Building, Chapel Hill, NC 27514 (USA)

BLOCH, C., Commissariat à l'Energie Atomique, Centre d'Etudes Nucléaires de Saclay, 91 Gif-sur-Yvette (France)

BLUNDELL, T. L., Molecular Biophysics Laboratory, South Parks Road, Oxford (Grande Bretagne)

BRONK, D., President Emeritus, The Rockefeller University, New York, NY 10021 (USA)

CAMPBELL, F. W., Physiological Laboratory, Cambridge (Grande Bretagne)

CARERI, G., Istituto de Fisica «Guglielmo Marconi», Universita di Roma, Piazzale delle Scienze 5, Rome (Italie)

COHEN, E. G. D., The Rockefeller University, New York, NY 10021 (USA)

COOPER, A., Department of Zoology, Laboratory of Molecular Biophysics, University of Oxford (Grande Bretagne)

COOPER, L., Brown University, Rhode Island, New York (USA)

CORDONE, L., Istituto di Fisica dell' Universita, Via Archirafi 36, Palerme (Italie)

COURNAND, A., Prix Nobel, College of Physicians and Surgeons, Columbia University, Department of Medicine, 630 West 168th Street, New York NY 10032 (USA)

COWAN, J. D., Chairman of the Committee on Mathematical Biology, University of Chicago, 959 East 57th Street, Chicago, IL 60637 (USA)

DIAMOND, R., MRC Laboratory of Molecular Biology, Hills Road, Cambridge (Grande Bretagne)

DODGE, F., The Rockefeller University, New York, NY 10021 (USA)

DUBOS, R., The Rockefeller University, New York, NY 10021 (USA)

EDELMAN, G. M., The Rockefeller University, New York, NY 10021 (USA)

EIGEN, M., Prix Nobel, Max Planck Institut für Physikalische Chemie, Bunsenstrasse 10, D-2400 Göttingen (RFA)

ELEY, D., Department of Chemistry, University of Nottingham, Nottingham (Grande Bretagne)

FAIRBANK, W., Department of Physics, Stanford University, Stanford,

FASELLA, P., Centro Biologico Moleculare, C.N.R., Istituto di Chimica Biologica, Citta Universitaria, Rome (Italie)

FESSARD, A., Professeur au Collège de France, 4 avenue Gordon Bennett, Paris 16ᵉ (France)

FOX, S., University of Miami, Coral Gables, Florida (USA)

FRÖHLICH, H., University of Liverpool, The Chadwick Laboratory, P.O.Box 147, Liverpool L69 3BX (Grande Bretagne)

DE GENNES, P. G., Service de Physique des Solides, Faculté des Sciences, Bat. 510, 91 Orsay (France)

GLASER, D., Prix Nobel, Department of Molecular Biology, University of California, Berkeley, CA 94720 (USA)

HAHN, H., Institut A für Theoretische Physik, Mendelsohnstrasse 1, Braunschweig (RFA)

HAKEN, H., Institut für Theoretische und Angewandte Physik, Azenbergstrasse 12, Stuttgart (RFA)

HEPP, M. C., Institut für Hirnforschung, Universität, Zürich (Suisse)

HEPP, K., Ecole Polytechnique de Zürich, Zürich (Suisse)

Madame HODGKIN, D., Prix Nobel, Chemical Crystallography Laboratory, South Parks Road, Oxford (Grande Bretagne)

JERNE, N. K., Basel Institute for Immunology, 487 Grenzacherstrasse, Bâle (Suisse)

JULESZ, B., Head of the Sensory and Perceptual Processes, Bell Telephone Laboratories, Mountain Avenue, Murray Hill, NJ 07974 (USA)

KAC, M., The Rockefeller University, New York, NY 10021 (USA)

KASHA, M., Florida State University, Institute for Molecular Biophysics, Tallahassee, Florida 32306 (USA)

KATCHALSKY, A., The Weizmann Institute of Science, Rehovot (Israel)

KATCHALSKI, E., Department of Biophysics, The Weizmann Institute of Science, Rehovot (Israel)

KENDREW, J. C., Prix Nobel, MRC Laboratory of Molecular Biology, University Postgraduate Medical School, Hills Road, Cambridge (Grande Bretagne)

KNIGHT, B. W., Jr., The Rockefeller University, New York, NY 10021 (USA)

KOSHLAND, D. E., Jr., University of California, Department of Biochemistry, Berkeley, CA (USA)

KUBO, R., University of Tokyo, Department of Physics, Faculty of Sciences, 5–1 Hongo, 7-Chome, Bunkyo-Ku, Tokyo (Japon)

LEVINTHAL, C., Columbia University, Department of Biological Sciences New York, NY 10027 (USA)

LICHNEROWICZ, A., Professeur au Collège de France, 6 avenue Paul Appell, Paris 14² (France)

LIFSON, S., Chemical Physics Department, The Weizmann Institute of Science, Rehovot (Israel)

LING, G., Pennsylvania Hospital, Philadelphie, PA (USA)

LIPMANN, F., Prix Nobel, The Rockefeller University, New York, NY 10021 (USA)

LITTLE, W. A., Department of Physics, Stanford University, Stanford, CA 94305 (USA)

LONGUET-HIGGINS, H. C., University of Edinburgh, Department of Machine Intelligence and Perception, Edinburgh (Grande Bretagne)

LÖWDIN, P. O., Department of Quantum Chemistry, Uppsala University, Box 518, Uppsala (Suède)

MAALØE, O., Det Mikrobiologiske Institut, Øster Farimagsgade 2A, DK 1353 Copenhague K (Danemark)

MACKAY, D., Department of Communications, University of Keele, Keele (Grande Bretagne)

MAFFEI, L., Laboratorio di Neurofisiologia del C.N.R., Via S. Zeno 49/A, Pisa (Italie)

MARGOLIASH, E., Department of Molecular Biology, The Abott Laboratories, University of Chicago, North Chicago, IL 60064 (USA)

MAROIS, M., Professeur à la Faculté de Médecine de Paris, Président du Conseil d'Admini-stration de l'Institut de la Vie, 89 bld St-Michel, 75 Paris 5ᵉ (France)

MATTHIAS, B. T., Institute for Pure and Applied Physical Sciences, P.O.B. 109, La Jolla, CA 92307 (USA)

MATSUBARA, T., Department of Physics, Faculty of Science, Kyoto University, Kitashirakawa-Oiwakecho, Sakyo-Ku, Kyoto (Japon)

MAZUR, P., Institut-Lorentz voor Theoretische Natuurkunde, Leiden (Pays Bas)

McCONNELL, J. R., Dublin Institute for Advanced Studies, School of Theoretical Physics, Dublin (Irlande)

MENDELSSOHN, K., University of Oxford, Department of Physics, Clarenton Laboratory, Oxford OXI 3PU (Grande Bretagne)

METCALF, D., Head Cancer Research Unit., The Walter and Eliza Hall Institute of Medical Research, Royal Melbourne Hospital, Victoria (Australie)

MILLER, J., The Walter and Eliza Hall Institute of Medical Research, Royal Melbourne Hospital, Victoria (Australie)

MITCHISON, N. A., Medical Research Council, National Institute of Medical Research, Mill Hill, Londres NW 7 (Grande Bretagne)

MÖLLER, G., Division of Immuno-Biology, Karolinska Institutet, Wallenberg Laboratorium, Stockholm 50 (Suède)

MÜLDNER, H., Max Planck Institut für Physikalische Chemie, Göttingen (RFA)

NICOLIS, G., Faculté des Sciences, Chimie Physique II, Université Libre de Bruxelles, 50 avenue F. D. Roosevelt, Bruxelles 5 (Belgique)

NINIO, J., Centre National de la Recherche Scientifique, Centre de Génétique Moléculaire, 91 Gif-sur-Yvette (France)

NOSSAL, G. J. V., Director, The Walter and Eliza Hall, Institute of Medical Research, Royal Melbourne Hospital, Victoria (Australie)

ONSAGER, L., Prix Nobel, Department of Chemistry, Yale University, New Haven, Connecticut (USA)

ORGEL, L. E., The Salk Institute for Biological Studies, P.O.B. 1809, San Diego, CA (USA)

OOSAWA, F., Institute of Molecular Biology, Faculty of Science, Nagoya University, Chikusa-Ku, Nagoya (Japon)

PALMA, M. U., Istituto di Fisica, Via Archirafi, Palerme (Italie)

PATTEE, H. H., Microwave Laboratory, High Energy Physics Laboratory, Stanford University, W. W. Hasen Laboratory of Physics, Stanford, CA 94305 (USA)

PHILLIPS, D. C., Department of Zoology, University of Oxford (Grande Bretagne)

PRIGOGINE, I., Faculté des Sciences, Chimie-Physique II, Université Libre de Bruxelles, 50 avenue F. D. Roosevelt, Bruxelles 5 (Belgique)

RABI, I. I., Prix Nobel, Department of Physics, Columbia University, 538 West 120th Street, New York, NY 10027 (USA)

RATLIF, F., Professor of Physiological Psychology, Rockefeller University, New York (USA)

RICHARDS, F. M., Chairman, Department of Molecular Biophysics and Biochemistry, Box 1937, Yale Station, Newhaven, CT 06520 (USA)

ROSE, A., R.C.A. David Sarnoff Center, Princeton. NJ 08430 (USA)

ROSENFELD, L., Nordisk Institut for Theoretisk Atomfysic, Blegdamsvej 17, Copenhague (Danemark)

SCHLEIF, R., Graduate Division of Biochemistry, Brandeis University, Waltham, Mass. (USA)

SELA, M., Head of the Department of Chemical Immunology, The Weizmann Institute of Science, Rehovot (Israel)

SALPETER, E. E., Laboratory of Nuclear Studies, University of Cornell, Ithaca, NY (USA)

SOBOLEV, S. L., Académie des Sciences de l'U.R.S.S., Novossibirsk (U.R.S.S.)

SPEKREIJSE, H., Laboratorium voor Medische Fysica, Universiteit van Amsterdam, Amsterdam (Pays Bas)

SZENT-GYÖRGYI, A., Prix Nobel, Laboratory of the Institute for Muscle Research, The Marine Biological Laboratory, Woods Hole, MA 03543 (USA)

TARJANNE, P., University of Helsinki, Department of Theoretical Physics, Brovergester-rassen 20, Helsinki 17 (Finlande)

ULAM, S., University of Colorado, Department of Mathematics, Boulder, Colorado (USA)

VALLEE, B. L., Biophysics Research Laboratory, Peter Bent Brigham Hospital, Boston, Mass. (USA)

WAGNER, M., Institut für Theoretische Physik, Azenbergstrasse 12, Stuttgart (RFA)

WEISSKOPF, V. F., M.I.T., 545 Technology Square, Cambridge, MA 02139 (USA); C.E.R.N., 1211 Genève 23 (Suisse)

VON WEIZSÄCKER, E., Forschungsstätte der Evangelischen Studiengemeinschaft, Schweilweg 5, 69 Heidelberg 1 (RFA)

WIGNER, E. P., Prix Nobel, Palmer Physical Laboratory, Princeton, NJ 08540 (USA)

WILKINS, M., Prix Nobel, 30 St John's Park, Londres SE 3 (Grande Bretagne)

YOMOSA, S., Department of Physics, Faculty of Science, Nagoya University, Foroocho, Chikusa-Ku, Nagoya (Japon)

Preface

The Institut de la Vie establishes in a concerted manner the dialogue between theoretical physics and biology. It is aware of the difficulty and the audacity of the undertaking; but it also realises its importance and necessity; it foresees its long term benefits for the understanding of the mechanisms of life and the oneness of knowledge.

The Third International Conference 'From Theoretical Physics to Biology' was placed under the distinguished sponsorship of Monsieur GEORGES POMPIDOU, President of the French Republic, represented by Monsieur ROBERT POUJADE, Deputy Minister to the Prime Minister, responsible for the protection of nature and environment.

One of its outstanding events was the solemn awarding of the Prize of the Institut de la Vie, on June 22, 1971, in the Hall of Mirrors of the Château de Versailles. This Prize, in the amount of 250.000 french francs, is awarded every two years by an International Jury drawn from outstanding personalities who represent the highest level of human civilisation and aspirations. The object of this Prize is to propose to the appreciation of all people a person or a body that would have acquired merits with regard to life either by defending it or by illustrating it. The simultaneous timing of these two events was deliberate: it was of major symbolic value and thus established a tangible link between the highest level of scientific thought concerning life and the recognition of its value.

Continuity characterizes all efforts of the Institut de la Vie. The Third Conference followed two earlier ones which were held at Versailles, respectively from June 26 to 30, 1967[1] and from June 30 to July 5, 1969[2].

1   From Theoretical Physics to Biology, North-Holland Publishing Company, 1969, 443 pages.
2   Edition du Centre National de Recherches Scientifiques, 1971, 446 pages.

A Fourth Conference will take place in 1973 and the International Organizing Committee established its programme on May 6, 1972 during a meeting organized by Professor EIGEN, at Göttingen.

In the time between two such important international meetings, the Institut can continue to expand its activities. For instance, the sessions of the International Organizing Committee now foster colloquies on precise subjects i.e.: the Göttingen meeting featured a symposium of 'Self-organization of matter and early stages of biological evolution'.

L'Institut de la Vie établit d'une manière concertée le dialogue entre la physique théorique et la biologie; il a conscience de la difficulté et de l'audace de l'entreprise, mais il en mesure aussi l'importance et la nécessité; il en pressent les bénéfices à long terme pour la compréhension des mécanismes de la vie et pour l'unité de la connaissance.

La Troisième Conférence Internationale «De la Physique Théorique à la Biologie» était placée sous le haut patronage de Monsieur GEORGES POMPIDOU, Président de la République Française, représenté par Monsieur ROBERT POUJADE, Ministre délégué auprès du Premier Ministre, chargé de la protection de la nature et de l'environnement.

L'une de ses manifestations fut la remise solennelle du Prix de l'Institut de la Vie, le 22 juin 1971, dans la Galerie des Glaces du Château de Versailles. Ce Prix, d'un montant de 250 000 francs, est attribué, tous les deux ans, par un Jury international représentatif de la plus haute civilisation et de l'espérance humaines. Il a pour objet de proposer à l'estime des hommes une personnalité ou une organisation qui auraient bien mérité de la vie, soit pour l'avoir défendue, soit pour l'avoir illustrée. La coïncidence des deux événements fut voulue; elle avait une grande importance symbolique: ainsi fut établi un lien visible entre la plus haute réflexion scientifique sur la vie et la reconnaissance de sa valeur.

La continuité caractérise l'effort de l'Institut de la Vie. La Troisième Conférence fait suite aux deux premières qui se sont tenues à Versailles, respectivement du 26 au 30 juin 1967[1] et du 30 juin au 5 juillet 1969[2].

Une Quatrième Conférence aura lieu en 1973 et le Comité International d'Organisation en a fixé le programme le 6 mai 1972, au cours d'une réunion organisée par le Professeur EIGEN, à Göttingen.

1   Physique Théorique et Biologie, North-Holland Publishing Company, 1969, 443 pages.
2   Edition du Centre National de Recherches Scientifiques, 1971, 446 pages.

L'initiative de l'Institut de la Vie va s'amplifiant puisqu'entre deux grandes réunions internationales, les sessions du Comité International d'Organisation sont désormais l'occasion de colloques sur des sujets précis: c'est ainsi que la réunion de Göttingen a été marquée par un symposium sur l'auto-organisation de la matière et les premiers stades de l'évolution biologique.

Physical Aspects of the Order
in Biological Systems

Proc. 3rd int. Conf. From Theoretical Physics to Biology, Versailles 1971, pp. 2–11
(Karger, Basel 1973)

General Introduction

Introduction by H. FRÖHLICH

Chadwick Laboratory, University of Liverpool, Liverpool

The International Union of Pure and Applied Physics has sponsored these conferences several times now and it is in the hope that the discussions between biologists and physicists will lead to novel points of view and, in particular, that this time we can have real discussions between physicists and biologists that this sponsorship has been renewed. It is largely on the suggestion of MAALØE that the organisation of this first day is somewhat loose so as not to inhibit the discussions.

In physics many interesting investigations on the properties of materials have made use of the concepts of collective behaviour. In a naive way one might think then that since all matter is composed of atoms it should be desirable to find all possible modes of motion of these atoms. In fact, however, this would be most undesirable because the amount of information thus obtained would be so excessive that sorting out the relevant parts would be a greater task than looking for them in the first place. Thus, near thermal equilibrium 1 cm^3 of a material involves some exp (10^{22}) microscopic states as we know from entropy. A computer working since the beginning of the universe (10^{17} sec) would thus have had quite a negligible time available for each state. Clearly, one should expect concepts of collective behaviour to be also of considerable relevance in biology and much of what has been done during the last four years was in an attempt of finding a basis for the proposal of experiments which might achieve some contact of physical ideas with biology. Now one very serious difference exists here: in physics, one preferably deals with materials in or near their thermal equilibrium. In biology, on the other hand, energy passes continuously through the system. My original proposal was that this energy will lift a few modes of motion far from thermal equilibrium and put them into different states, collective or otherwise. It should be emphasised that this refers to very few degrees of freedom only, for all the individual atoms will carry out their thermal motion practically identical with that in thermal equilibrium.

At first sight one might think that the energy supply required to maintain such excited states is excessively high. There, however, another important feature comes into play, namely, stabilisation of such excited states through non-linear effects transforming these excited states into meta-stable ones.

It has in particular been suggested that in view of the remarkable dielectric properties of biological materials, certain electric polarisation waves should be strongly and coherently excited and it has been shown on model-calculations that this indeed is to be expected if the energy supply exceeds a critical one and if non-linear effects are considered [1, 2].

A frequency range of 10^{11}—10^{12} sec^{-1} was proposed but quite different frequencies might be possible. Additional ionic displacements may then lead to the establishment of ferro-electric or anti-ferroelectric states [3]. This suggestion again was based on a model-calculation in which a field of polarisation waves is coupled in a nonlinear way to a field of elastic waves. Extended to the case of giant molecules this would imply that they have meta-stable excited states with very large dipole moments (rather than ferro-electricity in the strict sense).

It is of interest to note that in thermal equilibrium, establishment of an ordered state (like the ferro-electric one) from a disordered one requires extraction of energy. In our present case, however, continuous supply of energy may impose a similar order on an otherwise disordered system.

According to these ideas the energy produced in biological activity imposes a certain "order" or "organisation" on the otherwise disordered system and stabilises it through various types of deformation. If realised these features can have far-reaching consequences on the dynamic behaviour of a biological system. For the non-linearity implies that properties of the system may depend on its excitation.

Two classes of consequences may be distinguished referring to

(I) Specific forces

(II) Energy storage

Case (I) presents possibilities for recognition and attraction of one system by another and of induction or repression of certain processes. Thus, stabilised coherent excitation of electric vibrations of certain frequencies

ω_A in a system A may lead to very strong long-range attraction of a previously not excited system B, provided it possesses corresponding frequencies ω_B close to the ω_A (resonance). Such long-range recognition and attraction may, of course, be supplemented by short-range features of a different nature (e.g. chemical), once A and B have come close together. The non-stabilised B excited through polarisation by A, will then suffer frictional losses. If the proximity of A and B leads to chemical reactions (case a; B → B_a) then the energy liberated herewith may over-compensate the frictional losses, and may then be used to maintain the vibrational activity in A which, in turn, may help to activate further A-systems, which can then attract more B_a-systems, etc. This idea has been used in a model for the action of enzymes [3]. For as KOSHLAND has pointed out, "... the essential mystery of their [the enzymes] enormous catalytic power remains ...". Estimates of the catalytic power on existing concepts are wrong by orders of magnitude which can be 10^{10} or more. The ideas presented here might help, therefore, to introduce new qualities which seems to be required to overcome such high discrepancies.

In case b no chemical reaction takes place in (B → B_b). The frictional loss may then lead to a collapse of the vibrational activity in A. Very little energy extraction may suffice because the coherent excitation in A requires a state of energy supply only slightly larger (say by ε) than a critical one s_c. Frictional losses need then only exceed the small ε, not the much larger total $s_c + \varepsilon$. Now if this rate of energy supply $s_c + \varepsilon$ is of the type of case a, arising from chemical activity in molecules B_a, then collapse of the activity in A represses this chemical activity.

If the ideas about the sense of smell proposed about 20 years ago by WRIGHT [5] are correct then case b might apply. In this model the property of smell of a molecule S rests in a combination of certain dipolar vibrational frequencies supposed to be in resonance with similar frequencies in the receptor organ. The difficulty that the molecules S cannot transfer any energy and hence not stimulate a nerve impulse, can be overcome by assuming the receptor to be activated as described above for a system A, and S to behave like B_b. Minute energy extraction ε by B_b suffices then to cause a strong change in the activity of A which then might stimulate a nerve impulse.

It is quite conceivable that a single molecule B_b may extract more than the minute energy supply ε which triggers off the collapse in A, thus liberating the *whole* of the energy stored in A; this may well be sufficient to stimulate a nerve impulse.

Similar effects might arise in other highly sensitive biological processes, such as vision, where the action of a single light quantum leads to a change in a single molecule which in turn, however, might release appropriately stored energy required to cause a nerve impulse.

An example of the possible use of the ideas presented here for energy storage in muscle has been suggested by CHAPLAIN [6] of Magdeburg. I quote from his letter:

" ... Now the points where your work became of considerable interest to me. First, all kinds of striated muscles can be activated, if strain is transmitted from one cross-bridge to another along the filament backbone. Would you equally think it possible that the underlying mechanism involves a mechanical displacement of the amino acid dipoles from their working points? Second, we just discovered that muscle can store energy within the cross-bridge. This energy fraction is derived from the hydrolysis of ATP, of which only some 60% is needed to return the cross-bridge to the resting state. Muscle can actually make use of this stored energy in the late phase of the twitch to generate power or redevelop tension after a quick release. Naturally this energy storage dissipates with time.

With refined methods I have been able to measure the exact time course of energy liberation in presence and absence of external tension (see enclosed figures). Tension may slow the dissipative processes ...

As an additional information—we got good estimates of the energy the individual cross-bridge expends and makes use of at the various stages of the cycle. A cross-bridge (about 380 Å long, double-helix) can store maximally $8.7 \, 10^{-21}$ cal as 'energy', which is more than twice it needs to generate work and dissipate some energy during its extension ..."

Within this context it should also be emphasised that SZENT-GYORGYI [7] has, from biological evidence, come to the conclusion that " ... certain propagated electric waves belong to the most fundamental phenomena of life ...". It would be of considerable importance to link these to the coherent waves discussed above.

A final suggestion arose when I remembered LONGUET-HIGGINS' lecture of two years ago on the storage of memory in the brain, referring to ideas of PRIBRAM [8] that this storage might be based on the principles developed for the holograph such that any particular memory storage involves changes in the whole brain rather than in local centres only. PRIBRAM presents much evidence for the latter but a difficulty arose because the analogy with the holograph would require some kind of coherent vibrations. According to the ideas presented above, such coherent vibrations are actually available.

I realise, of course, that memory is a very very complex subject requiring many details. But, as a general feature, if we accept PRIBRAM's ideas of holographic storage then we would need a coherent vibration. This vibration would be available.

References

1 FRÖHLICH, H. in MAROIS Theoretical physics and biology, vol. 13 (North-Holland Publ. Company, Amsterdam, 1969).
2 FRÖHLICH, H.: Int. J. Quantum Chem. 2: 641 (1968).
3 FRÖHLICH, H.: Nature, Lond. 228: 1093 (1970).
4 KOSHLAND, D. E.: KOSHLAND, D. E. and NEET, K. E.: Ann. Rev. Biochem. 37: 359 (1968).
5 WRIGHT, R. H. and BURGESS, R. E.: The chemistry of smell; in HOUBEN-WEYL Methodicum chimicum, vol. XI, chapt. 20.2 (Thieme, Stuttgart 1971).
6 CHAPLAIN, R. A.: Personal commun.
7 SZENT-GYORGYI, A.: Personal commun.;
8 PRIBRAM, K. H.: The neurophysiology of remembering. Sci. Amer. 220: 73 (1969).

Author's address: H. FRÖHLICH, University of Liverpool, The Chadwick Laboratory, P.O. Box 147, *Liverpool L69 3BX* (England)

Introduction by O. MAALØE

I'm taking over a job extremely different from that of FRÖHLICH. At times, the headline, "de la Physique théorique à la Biologie" could properly be modified to read "jusqu'à la Biologie", because the gap os very wide indeed. The way we had in mind when organizing this presentation, was to emphasize this gap by having two co-chairmen; one impersonating theoretical physics and one, myself, far out in the direction of biology. By 'far out' I mean a person who can listen to FRÖHLICH's presentation and, totally without malice, take it in as a piece of science fiction. However, science fiction can be justified and very stimulating. And a biologist cannot help being tickled by the possibility that theories coming straight out of physics might reveal elements characteristic of biological systems, which had been neglected by biologists.

In this situation the innocent biologist must ask himself the following question: how can I know whether it's science fiction I'm confronted with, or whether it contains elements which might come to take a very important place in biology? Our discussion here may help answering this question, since everybody in this room occupies a position somewhere between the extremes represented by the co-chairmen of this morning. What I would like this session to become is in fact a dialogue which might begin with a biologist asking FRÖHLICH, the founder of the theories, to be as specific as he feels he can with respect to the actual signals of one kind or another that one might hope to pick up in order to verify the actual operation of the various mechanisms we have heard about. Having been informed as precisely as possible about the predictions of theory, one would then turn to the physicists who have been seriously occupied with the special instrumentation with which the predicted signals might be picked up. And, finally, we should return to the biologist and ask him what is the most likely biological material that might emit such signals and in what conditions it should be maintained to maximize the chance of picking up the expected signals.

I shall not comment at all on FRÖHLICH's conjectures regarding enzyme behaviour or the nature of the signals that you pick up and register as smell; I shall confine myself to some remarks about bacteria which are small, independently growing cells, known, under suitable conditions, to behave very uniformly. You can have an assembly of bacteria in a dynamic steady state and I for one believe that the bacterial membranes are uniform enough to suggest that a very narrow signal frequency might be generated, and since a few measurements have in fact been tried on bacteria, I hope we shall have some matter-of-fact discussion of the possibilities of testing, and, perhaps, that which to many biologists sounds like science fiction at the moment, may eventually turn out to have a place in real live biology. The best we can hope for right now, I imagine, is that our discussion may lead to a more profound study being initiated by a group with proper expertise in theory, in the technical problems of measurement, and in the handling of the biological material on which the measurements would have to be made.

As FRÖHLICH already mentioned we will now hear prepared contributions relevant to some of these different aspects. Whether this morning or this afternoon we will get as far as to discuss practical steps in detail cannot be said now. We hope to have a lively discussion after lunch, and for this reason no fixed time table has been laid down.

Author's address: O. MAALØE, Det Mikrobiologiske Institut, Øster Farimagsgade 2A, DK-1353 Copenhagen (Denmark)

Discussions

H. Fröhlich: Discussion is now open.

H. C. Longuet Higgins: May I make one or two points? You were kind enough to refer to some ideas of your own about the possibility of information being stored in our brains in a non-localized form and, more ·specifically, as a power spectrum of the incoming information. I think this is still a possibility, but my own thinking has moved on from there and in the report of the last meeting we reprinted a paper in which we examined the possibility that the storage of information was not in the form of a power spectrum of the incoming signal. As most people who have thought about this we believe that the power spectrum is just the Fournier-transform of the autocorrelation function. So from the point of view of the information stored the two are very much the same. It does not seem to be easier to construct a neural model if one adopts the correlation view rather than the power spectrum view. But I think nonetheless the other possibility is quite interesting, and should be kept in mind.

H. Fröhlich: I take it that in your possibility you would not require coherent vibrations.

H.C. Longuet Higgins: No, one would not need coherent vibrations for storing correlation. only for the other purpose.

H. Fröhlich: One would need it?

H.C. Longuet Higgins: Yes, but may I make also a more general remark about your theory, as it rests on assumptions about resonance interaction between the vibrations of molecular systems? I think that one cannot overlook a very important experimental implication here, about the reflects of isotopic substitution. Now it is a fact that one can substitue biological systems, indeed whole organisms, with deuterium in place of hydrogen and they will still live quite satisfactorily. But when one replaces hydrogen by deuterium in large molecules one totally alters the vibrational spectrum. If any of the phenomena that you were discussing, for example enzyme action, actually required a high degree of specificity, it would be totally ruined by isotopic substitution. And that's why I find myself really unable to believe that, for example, the mechanism of smell has molecules, even if one could think of some kind of biological action which would excite the *smelly* molecules! I should have thought that shape was a much more important factor—it is inaltered, of course, by isotopic substitution.

H. Fröhlich: May I just add the fact that in the article [5] on smell which I quoted, there is a remark that they have made isotope substitutions in some molecules and did find indications positive in nature, i.e., a change in smell. But this I can show you later. Obviously, smell investigation is not a very objective one. But they say definitely they have found it.

H.C. Longuet Higgins: Well, I actually did a few small experiments myself some years ago. I managed to get hold of some deuterium substituted hydrocarbons and I couldn't tell the difference between perdeuterobenzene and ordinary benzene by taking a sniff of the bottle: they seemed to be extraordinarily similar, whereas the vibrational spectra are utterly different. Utterly. I think one must take this objection quite seriously.

U. PALMA: As I remarked here already two years ago (and as it will appear in the Proceedings of the 2nd Conference of l'Institut de la Vie), there are indeed sharp threshold effects of deuteration. I showed then a plot I had made, using published data by MARSLAND and ZIMMERMAN on the effect of partially deuterated water on the development of sea urchin eggs. The data evidenced the existence of a sharp threshold at about 70% D_2O in H_2O. It should be appreciated, however, that this is a complex problem, if we want to understand it in a system such as a sea urchin egg. A much simpler system should be selected in the hope of sorting out the individual contributions. I shall discuss in my talk today the effects of deuteration in a simple aqueous system.

Now, may I have, myself, a question for the biologists? I should like to know whether they think there may be in principle a connection between the so called biological clocks and oscillations of the type which has been described by Fröhlich: I mean, could the biological clocks be related in some cases with some kind of oscillations?

O. MAALØE: Would anyone like to answer the question? I am not terribly familiar with biological clocks.

M. EIGEN: At least some simple clocks consisting of chemical reactions have been analysed, and it is quite clear that it is a dissipative type of phenomena as described by Prigogine which makes it possible for chemical systems far from equilibrium to oscillate. They have been established rather well. I don't know of any such case which would have to be traced back to the phenomena described by Fröhlich. Now one little remark with respect to "shape" I take it that you interpret shape in a wider sense. After all benzene and pyridine do not smell at all alike, although they have the same shape but pyridine has a polar group. In other words, it is the distribution of polar, non polar groups, hydrogen bonds, etc. . . which determine "shape".

H. C. LONGUET HIGGINS: One must consider the screening effect of the electrolytes. This implies that any long range collective modes must be of a frequency higher than the diffusion time of the electrolyte, in other words, something of the order of 10^{10} or 10^{11} per second. So in this sense one would not expect electrostatic modes to be of any direct relevance to circadian rhythms; the frequencies are far too low.

H. FRÖHLICH: The predicted one is between 10 to 11th and 10 to the 12th.

H.C. LONGUET HIGGINS: Yes, so you are just in the right spot . . . Could I take up your point about enzymes, because this raises another problem? As I understood it you required for your collective excitations, a continual supply of energy to a system. This is fine for living systems. With enzymes, however, we can isolate the individual molecules the test tube, away from the living system, in an essentially equilibrium state, and yet the enzymes still promote their reactions, at the same rate. There are no indications that the *in vitro* reactions are any different from those in the living system. We are still faced with the problem of why the isolated enzymes are so active as catalysts.

H. FRÖHLICH: Well, the idea is that since chemical reactions take place this supplies the energy . . . and that lifts the enzyme into the excited state which might be very close to the non-ferroelectric ground state.

H.C. LONGUET HIGGINS: Yes, but this energy doesn't become available until the reaction has actually taken place.

H. FRÖHLICH: So it has to have a short period to get started like a chain type of reaction, but this may be a very short period and I am told that an analysis of this first stage might be just on the edge of what is possible and some people try to do this.

L. ORGEL: Prof. FRÖHLICH's theory in my opinion makes a sharp prediction in the case of smell. Molecules of D- and L-enantiomers of optically active substances have identical infra-red spectra, but very different shapes. If Prof. FRÖHLICH's theory is correct, optically enantiomeric molecules should have the same smell; if his theory is incorrect, they should have different smells. I don't know the answer, but I imagine there is someone who can tell us.

H. FRÖHLICH: This is WRIGHT's theory, not mine and he says that he has an enormous amount of material in favour of it.

H.C. LONGUET HIGGINS: But I think this would be absolutely conclusive and could be done in 10 minutes.

M. KASHA: The answer is negative. It is now clearly established that pairs of enantio-morphs have contrasting olfactory properties. This result argues powerfully against N. WRIGHT's olfactory hypothesis.

D. MACKAY: In trying to evaluate Dr. FRÖHLICH's theory I feel we must distinguish between large-scale and small-scale phenomena. I think that PRIBRAM must have been referring to evidence that ablation or interference with the electrical properties of different areas of cortex doesn't necessarily abolish memory. In other words, when he speaks of the storage being 'distributed' this is over a range of centimeters, whereas in talking about molecular recognition we are talking about a totally different order of magnitude. I would suggest that the evidence on 'distributive memory', which goes back to LASHLEY's work in the 1940's, rather tells against the idea that it should depend on vibrations of the sort you have been discussing involving electric fields.

One of the things that LASHLEY did, and others following him, was to embed strips of mica and pieces of gold conductor in the brain so as deliberately to interfere with the electrical properties of cortex, and to show that within limits this didn't have the sort of destructive effects that you would have expected if the electrical properties of the medium were important. This was of interest at the time in connection with the Gestalt theory of form perception; but the point I'm making is that this kind of evidence hardly supports you. If anything, it is negative.

H. FRÖHLICH: The frequencies involved would be of the order 10 to the 11th or 10 to the 12th and I don't think mica would make any difference in this case. This is a very particular frequency range and one would first have to see very carefully what are mica's resonant frequencies. I don't think it has any in this range.

H.C. LONGUET HIGGINS: The problems we face, it seems to me, are about time delays of the order of milliseconds. I mean the sort of times that are involved in the transmission of nervous impulses and I think one has no reason to doubt that the transmission of nervous impulses is the effective thing as far as the actual behavior of the brain is concerned. The time it takes for a neuron to recover from having delivered a volley of impulses is of the order of milliseconds and the time taken to cross a synapse is again of the order of milliseconds. So I think that one ought to be thinking about times in that sort of range when considering the mechanism of memory. Might I just add a remark about a situation to which your Theory obviously does apply, but it is a very macroscopic situation indeed? The emission of sound from my larynx is just the sort of thing your theory accounts for very well; here is a vibrating system which is quite non-thermodynamically activated by the flow of energy in a certain biological system. I feel that that is the kind of area in which your ideas might be most relevant.

H. SPEKREIJSE: Since Prof. MACKAY mentioned cortical areas, it might be appropriate to call your attention to the frequency selectivity in the spontaneous electro-encephalogram. When a subject is at rest, spontaneously occurring potential variations can be measured with electrodes placed on the scalp. Particularly from the occipital region, sometimes strongly mono-rhythmic waves can be recorded with a frequency of about 10 Hz. These are the so-called α-waves or α-rhythm.

A number of theories has been advanced to explain the origin of the α-rhythm. Some of them explicitly mention relaxation oscillators. We were, however, not successful in applying these oscillators to the α-rhythm, although with stimuli weaker than needed for full synchronization, certain pseudoresonance effects could be mimicked. Due to the highly nonlinear character of relaxation oscillators they can, however, also be driven near twice their natural frequency. This means that for a natural frequency of 10 Hz, also frequency selectivity can be found near 20 Hz. If the driving force of 20 Hz becomes so strong that locking occurs, then the relaxation oscillator moves back to the 10 Hz region. Since a selectivity at twice the α-frequency was never demonstrated in normal subjects, I am inclined to reject the relaxation model. The above illustrates how careful one should be with qualitative descriptions based on nonlinear models.

F. LIPMANN: I'm sorry to interrupt the discussion, but I want to make a remark on the use of deuterium in tissue cultures, mentioned earlier. I remember that in the early thirties when I was in ALBERT FISCHER's laboratory in Copenhagen, deuterium had just become available and experiments on tissue cultures were reported from Oslo. Deuterium was found to inhibit strongly the growth of such cultures. However, after purification of the deuterium, it turned out that one could use any amount of it without harming the tissue. So I would like to warn about the effects of using high dosages of deuterium without paying full attention to the absence of inhibitor concomitants.

Proc. 3rd int. Conf. From Theoretical Physics to Biology, Versailles 1971, pp. 12–20
(Karger, Basel 1973)

Organic Ferroelectricity

B. T. MATTHIAS

University of California, San Diego[1], La Jolla, Calif.
and Bell Telephone Laboratories Murray Hill, N.J.

Collective phenomena, and with them long range order effects, are of decisive importance for all areas in the natural sciences. While the mechanisms leading to ferroelectricity are different from those causing ferromagnetism, and while superconductivity is a low temperature phenomenon as compared to melting which usually occurs at high temperatures—all these cooperative phenomena have two things in common. First, none of them, at present, permits computation of the critical temperature below—or above—where the cooperative phenomenon takes place. And second, they are sometimes confused with each other. So, one is reduced to consider mostly the phenomenological aspects of their existence and, at the same time, look at their relationships with respect to one another.

It has become evident during the last few years that ferroelectricity and superconductivity are mutually exclusive [1, 2]. Nevertheless, long range order in biological systems has frequently been mentioned as having its origin in superconductivity. This, in the course of time, led to a great number of speculations about organic superconductors, with critical temperatures far above ambience. Unfortunately in all these speculations, superconductivity had once again been confused completely with ferroelectricity.

Long range ordering in systems with essentially one dimension only, has thus far always led to ferroelectricity, and never to superconductivity. Inversely, ferroelectricity has never yet occurred in more than one direction only. Pseudolinear organic compounds are never superconductors or even

1   Research sponsored by the Air Force Office of Scientific Research, Air Force Systems Command, USAF, under AFOSR grant number AFOSR-631-67A.

metals for that reason. However, during the last decade the number of organic ferroelectrics has increased by orders of magnitude, many of them in the category just mentioned.

One of the most interesting groups is formed by the many glycine compounds discovered by PEPINSKY and MATTHIAS more than a decade ago. They are all quite different from one another and, at present, their sole common feature seems to be the existence of glycine in them. The following are the compounds known at present:

$G_3 \cdot H_2So_4$
$G_3 \cdot H_2SeO_4$
$G_3 \cdot H_2BeF_4$
$G \ \cdot AgNO_3$
$G_2 \cdot MnCl_2 \ \cdot 2H_2O$
$G_2 \cdot HNO_3$
Sarcosine$_3$ CaCl$_2$

Since glycine is involved in so many biological systems, it seems quite likely that many of the biological systems will ultimately also be ferroelectric. At the same time, the Curie points cover precisely the correct temperature range.

There is also another set of observations which lends added credence to this hypothesis. Every ferroelectric crystal has to be piezoelectric, while the reverse is not necessarily so. Just the same, the probability for finding ferroelectricity increases considerably once piezoelectricity has been observed. Over the past ten years, DUCHESNE and his collaborators have succeeded in detecting piezoelectric response in a very large number of organic compounds. In addition, certain types of DNA and RNA have actually been reported as being ferroelectric [3].

The postulate of superconductivity in organic materials and at temperatures above room temperature is totally unsupported by any experimental evidence [4]. It seems much more sensible to consider ferroelectricity as the long range ordering mechanism in organic matter. At least, this has been observed [3].

References

1 MATTHIAS, B.T.: Superconductivity versus ferroelectricity. Mat. Res. Bull. 5: 665 (1970).
2 BURAVOV, L.I.; KHIDEKEL', M.L.; SHCHEGOLEV, I.F., and YAGUBSKII, E.B.: Supercon-

ductivity and dielectric constant of highly conductive complexes of tetracyanoquino-
diemethane (TCQM), JETP Letters *12:* 99 (1970).

3 POLONSKY, J.; DOUZOU, P., and SADRON, C.: Evidence of ferroelectricity in DNA. C. R.
   250: 3414 (1960).
4 Proc. Int. Conf. on Organic Superconductors, Hawaii 1969. J. Polymer Sci. *29:* 1-224
   (1970).

Author's address: B.T. MATTHIAS, Institute for Pure and Applied Physical Sciences,
University of California, P.O. Box 109, *La Jolla, CA 92037* (USA)

Discussion

A. KATCHALSKY: The question of piezo- and ferroelectricity in living systems is very
intriguing. To the best of my knowledge neither of these phenomena was ever found in
aqueous media. Although what you have described is of great interest for crystalline systems,
it does not seem that any hydrated medium exhibits clear cut ferroelectricity. DNA and some
proteins were found to give ferroelectric signals when rather dry. Upon adding water, however,
the phenomenon disappears. An interesting case of piezoelectricity in semi-dry systems is that
of bone. It was found that upon twisting bones piezoelectric charging appears, and it was
even hypothesized that the regeneration of tissue is related to the stimulus of piezoelectric
charges.

B. MATTHIAS: You are quite right. The trouble is, with the solvated system we just cannot
measure it.

A. KATCHALSKY: I would like to mention that although piezoelectricity does not develop
on a microscopic scale, it seems that other domain phenomena may appear in single biopo-
lymer molecules. Thus hysteresis of electrical charging and discharging (produced in a
potentiometric titration) was found in RNA and complexes of polynucleotides. My coworker,
Dr. Eberhard Neumann from Eigen's laboratory, has studied more carefully this molecular
hysteresis from the point of view of memory recording in biopolymeric systems. Recently
we have found that strong electric fields of the order of magnitude of 20 kV/cm—which is
close to the field effects of action potentials in the nerve—may imprint conformational changes
on biopolymeric systems in molecular dispersion. These hysteresis phenomena, which resemble
magnetization-demagnetization in ferromagnetics, seem to be the first observed in aqueous
solution.

H. FRÖHLICH: May I say that this is just exactly what I wanted to have. You see, in
the crystalline state the ground state, the relative ground state would be ferroelectric. In the
disordered state it is not. But as you have energy passing through it, or as you have these
local electric fields, the ferroelectric state which would be close by it, can be excited. The
usual ferroelectricity requires infinite systems. Ferroelectric is here a manner of speaking
of shorter ranges that are polarized.

B. MATTHIAS: All I am trying to say is that if you have a ferroelectric crystal, the energy
intrinsic in this crystal due to the spontaneous polarization can be released in many different
ways.

H.C. LONGUET HIGGINS: Obviously, this is happening in an externally applied field. It was my understanding that the crystal generates its own electric fields, that you don't have to apply external electric fields.

B. MATTHIAS: In the normal state it does not generate it. You have to polarize the crystal by one way or the other. Otherwise, the crystal is just like a piece of iron, non-polarized. You can polarize it in different ways. You don't even need an external field. An external pressure will do too.

D.D. ELEY: It is my understanding that ferroelectricity involves domains, which are polarized in one of two directions, the polarization of this crystal being the vector sum of the polarization of the domains.

H. FRÖHLICH and B. MATTHIAS: That's right.

B. MATTHIAS: Now, to come back to why I thought electricity has something to do with biology. It is due to the fact that so many of the organic materials have become ferroelectric. This involves a large number of glycine, sarcosine and other compounds which you probably know. We have found that as far as glycine is concerned, we can make just about any compound with it and this will turn into a ferroelectric. Now I don't assume that the biological mechanism inside of us is due to any of these glycine compounds, but what one amino acid does, another one will probably do too, especially since we have found so many others which also are ferroelectric. If one is willing to assume that the ferroelectric mechanism has anything to do with biological phenomena then of course one might even try to understand the mechanisms in muscles because the ferroelectricity is characterized by its strong piezoelectric effect—that is to say, the deformation with an implied field. In the same way, if it has anything to do with propagation inside of us, a large pressure on a ferroelectric crystal could also prevent the polarization from being switched—because if you reorient the polarization, you then reorient the mechanical deformation. Thus, you can also suppress ferroelectricity by a sufficiently large mechanical pressure. Now, to the last thing that I particularly had in mind. I don't think that in the memory a holographic process takes place—FRÖHLICH mentioned this—what I would like to think of is strictly a very primitive computer memory where you store information by up or down, by one or zero, i.e., by a bit. This can be done with a ferroelectric very beautifully since if you have a ferroelectric crystal and you apply a field, then according to the field direction, the crystal will be polarized. If the coercive force is sufficiently large, then when the field is taken off, the spontaneous polarization will remain until you reverse the field. So one could build—and clearly nature has done it differently—ferroelectric memories which have electrodes on both surfaces at right angles to one another. The cross section between two intersecting electrode lines will now give a bit. How much can one store in a ferroelectric crystal? How many bits of information can one store? I understand from my biological friends that the average human being requires for his memory capacity at least 2 billion bits. A very bright person has 6 billion bits. Thus, if one is willing to consider the memory in such a form, ferroelectricity could be a mechanism to permit you to store that much information without any great space requirement. I will show you immediately why. Clearly, it's the crystallite size or its material, or even if you have one single crystal, once it shrinks below a certain limit, you no longer will get ferroelectricity. However, the one thing we know for certain, is that a crystallite size must have at least about 100 Å cube edge in order to be ferroelectric. So, that means we can store 1 bit in 10^{-18} cm^3. Therefore, even for a small brain, one will have no problems whatsoever to store as much as one is required in order to explain 6 billion bits. Let me make it quite clear. This is only a "gedanken" experiment, nothing more. It is only to show you that the ferro-electric storage, is at least capable of storing that much information. The access, of course, is another question, but at least you can store that much. Now one would say that these are very small sizes. Do these things really exist in the human body? There is a whole complex of extremely interesting data by DUCHESNE and his collaborators. As I told you before, when you have ferroelectricity you also have very pronounced piezoelectricity. And as DUCHESNE has shown in a large number of materials, the piezoelectricity of most organic materials is abundant and goes up to extremely high frequencies—up to the range of 100 megacycles which is 10^8. Thus, we have a possibility of evaluating how large, if we are talking about ferroelectrics of very small size, the sides would actually have to be in order to give the capacity. Let me show you. When you have a piezoelectric crystal, the resonance frequency in general is one over the length, the elastic modulus; and the density square root. Generally, for a very soft piezoelectric, you get something in the order of 10 kilocycles for 1 cm. In order to get DUCHESNE's frequencies, we have to assume 10^{-4} cm. All this is very rough, of course, but even if we have 10^{-4} cm we still have 10^{-12} cm^3 and again we will be able to at least store enough in larger ranges to get the required number of bits. Obviously, I know as well as you do, that this is not going to be what it is. But at least it gives you a certain picture of how things could have been. During the last 10 years there have been dielectric measurements made on material such as DNA. As I have shown you before—the hysteresis loop which is characteristic of a ferroelectric has also been found for certain types of DNA. It is in a work by POLANSKY. Clearly, we cannot explain anything on the basis of ferroelectri-city, and by explaining I mean that we can predict. The one thing we can do, however, is we can try to make a picture on the basis of ferroelectricity.

L. ONSAGER: You can of course put polar systems into aqueous solution. Any ions present will tend to relieve electric stresses; but under favorable conditions enough may be left over to produce orientation phenomena. The best known example is vanadium pentoxide, which is closely related to some of the commercial ferroelectrics. I might also point out that the alpha helix is automatically polar; the end charge ought to be somewhat greater than one elementary charge.

K. MENDELSSOHN: Well, I think before one goes into this question perhaps I should say something more generally, namely, that after this third conference definite progress has been made. I remember four years ago we were at the stage where our biologist friends— or most of them—would fight tooth and nail against the proposition that any process of life is based on stereo-chemical fitting by nearest-neighbor forces. Now, long-range interaction is generally accepted.

VOICE: By whom?

K. MENDELSSOHN: Well, by the physicists at least . . . as I said, there are still some die-hards among our biologist friends who will stick to stereo-chemical fitting—but let us not bother about that now. I would like to place a note of caution on what was said this morning. At the last conference I tried to advocate a more general approach, saying, let's look at life as an aggregation of matter in the same way we have looked at other aggregations and see how this aggregation differs from the others. It differs a lot. We have to look at these differences to see where we ought to start thinking. Now MATTHIAS just said very confidently that you can have either ferromagnetism, ferroelectricity or superconductivity. Nothing much is ferromagnetic, superconductivity is obsolete so it's going to be ferroelectricity. I agree that there may be many things that might involve ferroelectricity. I don't know, but I think this is an oversimplification because if we have a pattern in states the this order pattern may be much more fundamental than the mechanism producing it. Take the example of a crystal, for instance—that crystal can be made by resonance forces, by ionic forces, by exchange forces and these are all very different things. Nevertheless, the pattern is the crystal. I can imagine that other processes, such as magnetic interaction, electric interaction and superconductivity, let's say, collective electron modes—are all playing their parts in those structures that are most important in the function of life. However, one thing should be clear. Even if these well-known phenomena of non-living matter are involved in the structures of life, it is most unlikely that they will manifest themselves in exactly the same form. Instead, we may possibly expect co-operative effects which in their pattern bear a certain resemblance to phenomena such as ferroelectricity, ferromagnetism and superconductivity. To look for their exact replica is a bit too naive. Even a benzene ring shows certain characteristics of a persistent superconductive current. Interaction between carbon atoms at opposite corners of the ring is entirely vested in the collective electron modes shared equally by all carbon atoms in the ring.

Altogether, it seems to me that at our present state of knowledge physicists can do no better than to learn what the biologists have discovered, translate these discoveries into the language of physics and use them as a basis for their own work.

S. Fox: I wonder if the theoretical requirement of a dry state for ferroelectricity in the living system is met by the observation that living systems are not uniformly aqueous, they are not uniformly solvated, and in fact they have some hydrophobic zones such as lipid layers and certain regions in protein molecules.

D. MACKAY: Just a comment on the alleged biological evidence for memory. I think it's mainly based on what people can recall under hypnosis. The big difficulty there, of course, is to allow for the redundancy. If you simply multiply the number of bits that somebody is supposed to have recalled, in a particular hypnotic exercise, by the number of perceptual moments in his life, then you come up with figures of the order of 10^{10} to 10^{14} or even more. But if you allow for the enormous redundancy, in the information theoretical sense, in the sequence of experiences that we have throughout life, that figure of even 10^{10} is I think a bit suspect. With regard to the claim you were making, Dr. MATTHIAS, I would have thought that this was an encouragement not to bother to look for storage mechanisms on such a micro scale. As someone else has remarked, 10^{10} is of the same order as the number of neurons in the brain. In other words, all we need look for is a functionally significant way of modifying the structure of neurons to the tune of one bit of information per neuron on the average. Since most neurons have numerous dendritic and axonal connections with other neurons, which are presumably modifiable, there is already at a gross structural level far more than enough theoretical storage capacity without having to go to the ferroelectric mechanisms you were postulating. I'm not saying they don't exist. I'm only saying that you can't argue that you *need* them on the score of the alleged figure of 10^{10} bits.

I also wondered whether in terms of your model you envisage anything analogous to the wiping of a ferromagnetic store by applying magnetic fields? In other words, if the brain were storing information ferroelectrically, would you not predict that the application of alternating electric fields at high frequencies would wipe out or at least affect memory?

B. MATTHIAS: If you could get the electric field to it. But you cannot because most people are unwilling to expose themselves or their brains to 100 kV, roughly the coercive force. That would be sort of dangerous . . . In animals yes, but how do you define the memory of an animal?

H.C. LONGUET HIGGINS: I see the chairman is pointing at me. Apologies for intervening again. I would think that perhaps it might be worth it to get quite clear in our minds what counts as ferroelectricity and what doesn't. I mean, if one is simply thinking of a physical system which in its most stable state has a polarization . . . well, everybody knows that the nerve membrane has a resting potential. So, are we just talking about systems which have polarization? Because if we are, we know already that such systems occur in living tissue. The second point: I was intrigued by Prof. MATTHIAS' invention of a ferroelectric brain, but after all, we do know an awful lot about the anatomy of the brain. It is neural network. All one has to do is look at the thing. There has been some very important recent work, in particular by DAVID MARR, in which he makes quite definite and very plausible hypotheses which, to a certain extent, are supported by experimental evidence. On the way, for example, we learn motor skills, with our cerebellum. It would appear—as any neurophysiologist would think entirely reasonable—that it is the synapses which essentially store the bits of information. It can't very well be the axons because once an impulse has started along an axon it goes on until it gets to the end. The only place where the storage could possibly occur is at the switches. These switches are quite well characterized, we know a lot about them from electronmicroscopy, electrophysiology, and so forth. The real problem here is, exactly what is it about the synapse that makes it either conduct an impulse which reaches it, or not conduct it. But then again, we have quite a lot of knowledge and there is a large literature on the subject. To speak in these terms about ferroelectric memory seems to me ignoring all that is already known about the brain and I think that this is scientifically irresponsible.

B. MATTHIAS: Well, we are just trying to understand the biologist, and I am not saying that this is the way of our brain or memory functions. I am saying that this is the simplest picture the primitive experimentalist like myself could visualize of how the memory might have been constructed. Clearly, it isn't. When you talk about ferroelectricity, that everything is ferroelectric, well then we agree. We have come to the conclusion over the last 20 years that everything that shows dialectic anomalies will show saturation and nonlinear behavior and, associated with it, very strong mechanical deformation. That is all I really wanted to say.

U. PALMA: I should like to go back to H-bonded systems. This class of ferroelectrics behaves in quite a singular way, as compared with other types of ferroelectrics. The reason for this, as some work at our laboratory has shown, is that, within the crystal and above the transition temperature, the proton system behaves as a fluid more than as a solid—at least with respect to the appropriate degrees of freedom. This makes it not very appropriate to speak of phonons (in the ordinary sense) above the critical temperature and this is probably the reason why the Fröhlich-Cochran (soft mode) theory of ferroelectrics is not so satisfactory for H-bonded ferroelectrics, while it is so successful, e.g., for the barium titanate class. The different behaviour is due to the exceptionally light mass of protons which allows tunneling and jumping-over effects. Now, the array of potential barriers and wells seen by one single proton depends upon the motional state of the remaining protons. But the same array, in turn, imposes the motional (and tunneling) state of that one proton. Therefore, there is some kind of feedback between the two quantities, and this plays a central role in the transition. Something similar may also happen in an aqueous system: consider a macromolecule with its surface array of potential wells and barriers as seen by one (water) proton. At not too high temperatures, this array may, in principle, be capable of giving a structural imprinting to the water. This imprinting could be propagated by the feedback effect that I have mentioned and this could, in principle, entail a variety of effects from the imprinting of different motional-geometric water structures, to, e.g., enzyme-substrate long-range recognition, etc. All this would not be ferroelectricity, of course, still it would be something conceptually close to it.

B. MATTHIAS: You are quite right. I tried to say this, but apparently I didn't say it clearly enough. The hydrogen bonded ferroelectrics in themselves cover an enormous range. Obviously, potassium dihydrogen phosphate for instance, has very little to do with the ferroelectricity of ice. So I quite agree with you. Only I didn't want to go into details of that kind . . .

VOICE: What is ferroelectricity, then?

B. MATTHIAS: Ferroelectricity is when one has a polar axis that can be switched, by an external field or external pressure, a pyroelectric crystal in which the direction of spontaneous polarization is not frozen.

VOICE: Every possible polarized system which can be switched, will be called ferroelectric?

B. JULESZ: I would like to join Profs. MACKAY and LONGUET HIGGINS in their criticism. Some of us physiological psychologists who came to participate in this interdisciplinary gathering would appreciate it if speakers in other disciplines would not ignore the many findings which have been collected in the life sciences. For instance, with respect to memory there is half a century of work on synaptic systems, and if someone could elucidate the chemical or physical processes that may occur on synaptic membrane it would be most informative. However, if someone descends from this level by three or four orders of magnitude to speculate on how crystals might store memories, I feel disappointed, and do not see the purpose of an interdisciplinary meeting. When psychologists, physiologists, physicists, and chemists are invited to a meeting, we would like to stay at a convenient level in the hierarchy where we can go one step down or one step up, depending on the particular discipline. Only under such circumstances can I imagine useful interdisciplinary collaboration.

With respect to memory itself, I would like to make one comment, that in the central nervous system there are at least 10^{10} neurons (and if we add the cerebellum we might have 10^{10}). Furthermore, each neuron has 100 to 1000 synapses and each of these synapses might interconnect neurons with various strengths. So there is adequate capacity to store vast amounts of memories even if the given state of a neuron or a synapse might describe a memory. On the other hand, a memory state does not have to be stored by one neuron or synapse; it could be the firing state of all the available neurons or synapses. Thus, the central nervous system can store all the memories that might ever be needed in one's lifetime, and there is no need to evoke storage mechanisms as unlikely as crystals.

B. MATTHIAS: Yes, but you didn't listen at all to what I said. I said right in the beginning, if on the basis of cooperative phenomena, such as the ones I named–they are the only ones I know, if we want to explain living things or describe or even talk about them, on the basis of these collected phenomena, then this is what it might look like. We never said, «This is the way it has to be». Undoubtedly you have more information of a different kind, but we were talking about collective, cooperative phenomena and whether it's 10^6, 10^9, 10^{12}, it really doesn't bother us in any way. We can accommodate you. But we were not really talking about the biology we were trying to show you what the point of view of a physicist is.

That is all we were saying. We were not telling you that this is the way it has to be.

V.F. WEISSKOPF: I would like to speak under the impression of what we heard this morning. I'm neither a solid state physicist and certainly not a biologist, but somehow I find it very disappointing. Let me try to apply psychology for a change—in which I am also not an expert. Physics always was lucky enough to deal with simple phenomena. Also, and more importantly and more relevant to what we discussed today, most phenomena exhibit a lot of linearity in those fields of physics which have been well investigated. The great success of quantum mechanics was to show that linear oscillations play a much more important role than the 19th century had made one to believe. When we are faced with nonlinear phenomena very strange things happen, things we are not accustomed to find and things that are very surprising. This is true both for nonlinear phenomena and also for phenomena where many bodies and long-range interactions are involved. These phenomena remind some people of the living world. They don't remind me of it, but some people are reminded of it. These people are inclined to jump to the conclusion that such phenomena have something to do with the process of life. Isn't that the same thing that happened here in this palace when people built those automatic puppets in the 17th century and believed that they had come nearer to the riddle of life? I find it most interesting to look at complicated processes in physics; I think that's what physicists should do, in particular the processes of many-particle-interactions and nonlinearities. They require completely new ways of thinking. But to jump to conclusions that this has anything to do with living phenomena seems to me to be incredibly presumptious and incredibly naive. I am speaking here as a physicist, I feel that we physicists should show a lot more humility facing the problems of the living world, than has been shown here. If one finds a few possibly long-range effects between molecular systems, and says that it has anything to do with smell, well, I feel uncomfortable. I'm sorry . . . but I think that one should really go back to the foundations and see first of all what we can learn from nonlinear processes and from long-range effects. Then we should learn from the biologists what the real life phenomena are and where the complications come in. I believe the whole concept of phenomenon is different in physics and in biology. This is the basis of some of the misunderstandings. We physicists like to see a phenomenon as a well separated group of observations and we were terribly lucky that such a type of approach pays off. In facing biology the whole concept of thinking of a phenomenon must be changed. I may exaggerate this. I do not want to imply that conversation between physicists and biologists is impossible. On the contrary. I think we have a lot to learn from each other. But I don't think the last hour has been the way to do this.

Proc. 3rd int. Conf. From Theoretical Physics to Biology, Versailles 1971, pp. 21–34
(Karger, Basel 1973)

Evidence for Collective Modes[1]

M. U. Palma

Istituto di Fisica dell'Università and Gruppi Nazionali Struttura della Materia del
Consiglio Nazionale delle Ricerche, Palermo, Italy

Preamble

Admittedly, the title of this talk is far-fetched. A somewhat more
conservative title could have been: 'Evidence for collective behaviour in
an aqueous system'. It should, nevertheless, be remarked that these meetings
of the Institut de la Vie differ from ordinary Conferences, inasmuch as we
are encouraged to speak free of inhibitions, to receive (and perhaps to
contribute) inter disciplinary stimuli, rather than a number of rigorous,
cautious and perhaps understated specialistic results.

The Problem

In the spirit of these meetings, I shall discuss the results of some
experiments that we have undertaken on a simple model system. These
experiments were related to the possibility for certain macromolecular solute
species to act as templates, to control and determine the cooperative transition from ordinary liquid water to another phase (different from ice). This
problem is, in turn, closely related to the old and still unsettled question:
is cell water a mere solvent or has it a specific role in a variety of phenomena, as assumed in so-called 'holist' theories, (e.g., a role in the functional
coherence of the protoplasm), in the sense that within the cell, water, ions

---

1   Based on unpublished work by G. AIELLO, P. L. INDOVINA, M. L. MARINO, M. S.
MICCIANCIO-GIAMMARINARO, S. MICCIANCIO, M. B. PALMA-VITTORELLI and E. TETTAMANTI.
Their permission to quote their results is gratefully acknowledged.

and biomolecules constitute an ordered system, or a (limited) set of such systems?

At first sight it would seem rather difficult to distinguish other possible phases between ordinary liquid water and ice, with the physics of ordinary liquid water still far from being understood and the entropy of fusion of ice already being so low and many features of liquid water being close to those of ice. Nevertheless, evidence for the existence of more than one type of water in cells and model systems has become available in the last few years [1, 2, 3, 4]. These data have been often interpreted as supporting the assumption [5, 6] of the existence and the specific functional role of a 'structured' water within the cell.

The meaning of this term (structured water) needs clarification. Let us consider that, although permanent or long-lived arrangements of molecules do not exist in the liquid, such arrangements, or 'structures' are expected to be found on a suitable time scale, perhaps of the order of 10^{-11} sec for ordinary liquid water [7]. By 'structured water' we shall thus intend water having a more solid-like (yet, not necessarily ice-like) structure than ordinary water at the same temperature. This type of structure will be found on longer time scales and it will be characterized by geometries not necessarily identical to those of ordinary liquid water. Different geometries will in principle imply different excitations spectra. A change in the boundary conditions, due to the introduction of solute macromolecules acting as 'templates' or 'pinning points', will be capable of affecting the water structure (i.e., both the geometry and the excitations spectrum), and this will not necessarily result in a situation more complex than in ordinary water. On the contrary, in certain specific instances a simpler situation may result.

A differently structured phase of liquid water would best be studied by starting from its *onset,* i.e., from the relevant phase transition in a simple water-macromolecule system, which could presumably serve as a model system. To be fitt for this purpose, our model system should thus be capable of undergoing repeated temperature cycling (perhaps up to 100 °C) without suffering irreversible changes: this, in fact, would facilitate the observation of the onset of a phase transition. Another requirement would be to deal with fairly simple (and possibly not artificial) macromolecules, capable of acting as templates to imprint a specific geometry on water.

These requirements are satisfied by the choice of the agar-water system. Agar is a natural substance occurring as a cell-wall constituent of a group of red seaweeds. Chemically, it is a family of polysaccharides, having a rather simple molecular structure, which shows and even exaggerates all of the important properties of gelling systems [8]. A well characterized member of the family is agarose, a galactoglycan, in which D-galactopyranose units, connected $\beta - (1 \to 4)$ alternate with 3.6 anydro-L-galacto-pyranose units, connected $\alpha - (1 \to 3)$ [8, 9]. The $O_1 - O_2 \dots O_3 - O_4 \dots$ etc. distances in D-galactopyranose are 2.85 Å, essentially coincident with the hydrogen-bond length in liquid water [8, 9, 10]. This puts agar in the special class of molecules characterized by repeat-distances multiples of a significant distance in the ordinary water structure. These molecules are thus in principle capable of acting as templates for a collective arrangement of water molecules. In point of fact, water has not been found to be 'structured' in a variety of other gels and viscous solutions of molecules similar to agar, but not belonging to the mentioned class [10, 11].

Experimental Results

A suitable tool to study this type of situation is NMR spectroscopy of water protons. Steady-state and impulsive NMR techniques provide information about the geometric and dynamic situation of protons, through lineshapes, linewidths and relaxation times.

A first set of experimental results is shown in figures 1 to 4. The main features to be remarked are: the abrupt and hysteretical temperature-dependence of both the linewidth and the T_2 relaxation time (fig. 1); the

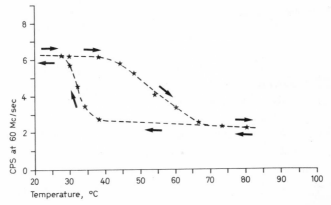

Fig. 1. Steady-state, high resolution proton magnetic resonance of the water protons (60 Mc/sec) in a 2% agar-water system. Linewidth (between half-intensity points) is plotted vs. temperature. Measurements started from a temperature close to 100°C, following ebullition of specimen to allow complete dissolution.

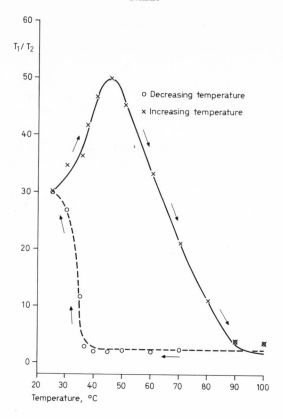

Fig. 2. Results of spin-echo experiments (by Drs. P. L. INDOVINA and E. TETTAMANTI) using the Carr-Purcell method. The proton T_1/T_2 is plotted vs. temperature for a system as for figure 1.

absence of isotopic effects in the *fractional* line broadening (fig. 3–4) and the remarkable departure (fig. 2) of the T_1/T_2 ratio from unity, which is the theoretical value in the present situation, i.e., when both $1/T_1$ and $1/T_2$ *decrease* for increasing temperatures [12].

For higher agar concentrations, sharper transitions and a shift of the transition toward higher temperatures were observed.

Finally, the *proton* resonance linewidth in the lower-temperature phase depends upon the agar concentration, C, according to a law of the type:

$$\Delta \nu = \beta \, C^\gamma \tag{1}$$

and a least-squares fit of the experimental results yields the following values for the parametres:

$\beta = 1.93 \pm 0.05$
$\gamma = 0.63 \pm 0.02$ (index of determination $\rho^2 \sim 0.97$)

These values coincide for both 100% H_2O and 0.2 H_2O in D_2O. In other words, the *proton* resonance $\Delta\nu_{max}$ seems to behave as $C^{2/3}$, i.e., as the square of the inverse agar-agar distance.

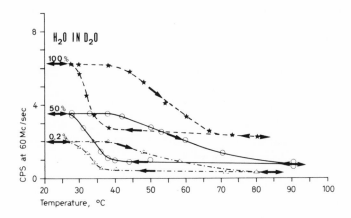

Fig. 3. Same as figure 1, for different concentrations of H_2O in D_2O. Note that data always refer to *proton* (not deuteron) resonance. 100% curve refers to 2% agar in *normal water*.

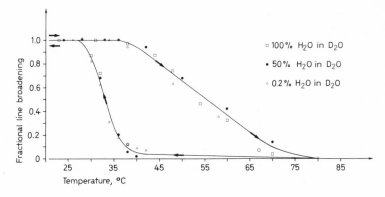

Fig. 4. Same as figure 3. The fractional line broadening $\quad f = \dfrac{\Delta\nu - \Delta\nu_{min}}{\Delta\nu_{max} - \Delta\nu_{min}}$

is plotted vs. temperature for the three different concentrations of H_2O in D_2O.

Discussion

Abrupt changes (or 'transition') in proton magnetic resonance line-widths and/or T_2 relaxation times as a function of temperature have been observed in a variety of systems in which hindered motions of interacting H^+ ions are possible. I shall not even attempt to quote the vast literature in this field.

Linewidths and relaxation times depend in a reasonably smooth way upon a 'triggering' frequency ν_c (the so-called correlation frequency) which, in turn, depends exponentially upon the temperature:

$$\nu_c = \nu_o \exp\left(-\frac{W}{KT}\right) \tag{2}$$

Abrupt changes (or 'transitions') in linewidths and relaxation times are, therefore, indicative of similarly abrupt changes in the correlation frequency. These, in turn, can only be due to changes (or 'transitions') in the geometric structure, and, therefore, in the energy levels involved in motions.

Clearly, these 'transitions' have a motional and geometric character and should be viewed under these two aspects jointly (or in what we physicists call the 'phase space', which is constituted by the two subspaces of coordinates and momenta). The amount and type of interaction among protons will determine the possibly cooperative character of these transitions.

If we want to view these transitions mainly under the motional (or, alternatively, under the geometric) aspect, we must provide a conceptual link between the two. This can be expressed in terms of a feedback which can be visualized as follows. Protons and heavier ions as well, contribute to the potential pattern, as seen by one single proton. The contribution due to protons will depend upon their motional state and hence upon the temperature, in the sense that the more their motion is excited, the more their charges will appear spread and the more their contribution will appear washed out. (This effect shall be much less pronounced for the contribution due to the heavier ions.) Now, it is the potential pattern which, in turn, imposes the motional (including tunneling) state of protons. This inter-dependence can, therefore, be visualized in terms of some kind of feedback between the motional and geometric states of protons. The incidence and the role of this feedback will depend upon how, respectively, the heavier ions' and the protons' contributions to the potential barrier pattern will compare.

A proton motional-geometric transition having a conspicuously co-operative character (i.e., one in which the above-mentioned feedback has

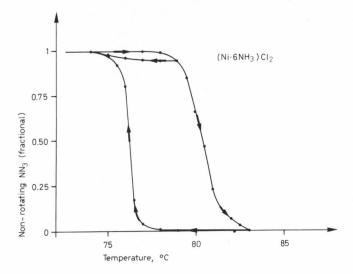

Fig. 5. Fraction of non-rotating NH_3 groups in $NiCl_2 \cdot 6NH_3$ crystals vs. temperature. Note the conspicuous thermal hysteresis and the partial hysteretical loop. Substitution of ND_3 for NH_3 causes an upward shift of several degrees of the whole hysteretical cycle. For cystals grown from a mixture of 50% NH_3 and 50% ND_3, the transition occurs at an intermediate temperature and its sharpness is unaffected [14, 15].

a dominant role), has been studied at our laboratory in the case of a simple system which may serve as a model for more complex ones. This is shown in figure 5.

Let us now discuss more specifically the present experimental results in agar-water systems. The thermal hysteresis shown in figure 1 is indicative of a cooperative behaviour [13]. This conclusion also agrees very well with the results on the dependence of $\Delta \nu$ upon C in the low-temperature phase (Eq. 1). In fact, if two types of water were present in the gel, rapid proton exchange would cause the reciprocal of relaxation times to depend *linearly* upon the agar concentration [3, 4]. Data shown in figure 1 and the best-fit values for parametres in Eq. 1, therefore, concur in suggesting the existence of one collective structure, governed by the agar-agar distances. Data shown in figure 4, on the other hand, might appear (also in view of the close similarity between figure 1 and figure 5) to be in contrast with the case of figure 5, in which [14, 15] an average-mass isotope effect was found. In fact, one might attempt to explain the results of figure 4 by assuming that the correlation frequency ν_c for the individual resonating proton does not depend upon the dynamics or the system as a whole but that it interests, instead the *individual* resonating *proton* itself. However, this would be strongly in contrast with the cooperative behaviour that we have just discussed. We must conclude, therefore, that the dynamics of the water protons does not dominate the phase transition. but that it is collectively modified as a consequence of it. This will be further clarified below.

In the case of figure 5 we have a heavier ions matrix of 'pinning points' (or 'template matrix') constituting a solid crystal both above and below the transition. The potential pattern contributed by this self-consistently arranged system is well defined and does not change across the transition. In addition to this, there will be the proton contribution which has already been shown to be dominant [15, 16]. For this reason the average-mass isotope effect was expected and, indeed, found in this case [14].

In the agar-water system above the transition temperature, on the contrary, macromolecules move at random, so that they do not contribute a well defined and concerted potential pattern. In these conditions, even by assuming that each macromolecule is capable of imprinting a definite structure in water, this will never give rise to a cooperative phase transition (such as illustrated in figure 1), unless the different macromolecules undergo a concerted spatial and motional ordering, giving rise to a self-consistent overall 'template matrix'. In this way only, the different macromolecules will, in a mutually consistent way, contribute to the building up of a structure extended to the agar-water system as a whole.

One might think that no ordering in the above sense is needed to explain the observed transition, if this is simply due to a macromolecular intermeshing which would make the whole system rigid and would, therefore physically trap all the water. This would not be correct, however, because: a) The intermeshing should be hysteretical, i.e., it should occur cooperatively, with bond energies depending upon the existence of other bonds. Now, the agar-agar distance is several hundreds Angstroms and this would hardly be compatible with cooperativity of intermeshing, if not with intermeshing itself; b) similarly, the helix stability has never been explained on these bases [8]; c) the already discussed concentration dependence of NMR linewidths would not be explained.

We must return, therefore, to the idea of a concerted spatial and motional ordering of macromolecules. This ordering will dominate the transition and will occur because the structural information will be carried by the water molecules. The mechanism for this, as we have already seen, can be visualized in terms of a geometric-motional feedback in the proton system, which will *affect* the proton dynamics, rather than being *dominated* by it. It follows that substitution of H_2O by D_2O will not necessarily affect the transition.

A more specific model (if not necessarily the correct one) for the transition could be the following. At high temperature, the motion of both the agar and the water molecules is sufficiently chaotic and no concerted ordering, or interlocking, can occur with (or through) water. For the occurrence of the cooperative phase transition, agar and water must fulfill the following conditions: a) agar molecules must be in the appropriate conformation (e.g., double helix [8]) and their motion (including internal motion) must be slow or concerted, and b) water molecules must be in a motional state such to accept (and to propagate by interlocking) the structure imposed by the agar molecules. Let us call T_a and T_b the temperatures below which conditions a) and b) are respectively fulfilled. If $T_a < T_b$ the onset of the transition will be controlled by the agar molecules. Furthermore, since it will be, of course, $(T_b)_{H_2O} < (T_b)_{D_2O}$ this model will predict (in agreement with the experimental data) the absence of an isotope effect in the transition temperature, upon substitution of heavy water for normal water.

In other words, D_2O will, as a consequence of the heavier mass of deuterium, be capable of 'transmitting' the geometric-dynamical situation at temperatures higher than those of normal water. Therefore, if the structuring is controlled by the agar motion, it will 'a fortiori' occur in heavy water, at the same temperature as in normal water.

Conclusions (and a Few Speculations)

We have thus seen that the cooperative phase transition from ordinary liquid water to a type of water exhibiting a different kind of order in the phase space has probably been observed. This transition appears to be macromolecule-controlled.

This type of ordering may (although not necessarily) be related to that due to a condensation of longitudinal modes in macromolecules, first suggested by FRÖHLICH [17, 18, 19]. In fact, this condensation (occurring in a manner typical of Bose condensation), would provide a long-range dynamical coherence across the macromolecule. In our model, this coherence is one of the possible causes which can play a role in a geometric-dynamical structuring of water.

More recently, FRÖHLICH has proposed a model of nonlinear interaction between the macromolecules' polar modes and compressional modes in the fluid (supposed neutral) in which the molecules are suspended [20]. This leads to very interesting theoretical conclusions. Our proposed mechanism could be one of the reasons for this nonlinear interaction between macromolecules and fluid. It provides a possibility for specific macromolecules in solution or at surfaces (e.g., at cell walls), to impose well defined and cooperative structures to water.

The same species of macromolecules in different dynamical states could, in principle, impose different structures, having different symmetries and long-range correlations. (By the same argument, even very different molecular species dissolved in water need not be incompatible with one and the same water structure, imposed by one or more of these species.) One might venture to think of this as a possibility for obtaining one water structure (or compatible structures) within a cell in some specific conditions, in spite of the variety of solutes present therein.

These possibilities have a direct bearing on a variety of facts of wide biological interest, most of which are centred on the transport properties of membranes and on the alternative view of a role of structured water in these properties [5, 6].

In point of fact, the data I have just discussed are also suggestive of a possibility for membrane theory and colloidal theory to merge together. Very interesting, also, is the observation that a cation selectivity is also found in clays [21]. As is well-known, K+ ions made available in the course of rcck weathering, tend to be embodied in clays much more preferentially than Na+ ions. This also occurs when the genesis of clays takes place in the presence of sodium concentrations that are much higher than potassium concentrations. This has vast (and, again, biological) consequences [21]. It is worth noting here that, as KATCHALSKI has recently shown, synthesis of polypeptide chains in primeval conditions seems to have occurred on montmorillonite clays [22].

Already in the previous conferences of the Institut de la Vie, the possible biological relevance of a study of the properties of biomolecules in the space of momenta has been suggested. The observation of the cooperative onset of a geometric and dynamical order in a water-macromolecule system, which we have just discussed, is in line with this speculation. A great advance may be hoped for in biology, by the joint consideration of the properties of biological systems in geometric space, as well as in the space of momenta, and by the consideration of their interdependence. Such a step would be similar to the one which marked the passage from crystal-lography to solid state physics. An extension in this sense of molecular biophysics towards *system* biophysics may be very rewarding.

References

1 COPE, F. W.: Biophys. J. *9:* 303 (1969).
2 HAZLEWOOD, C. F.; NICHOLS, B. L. and CHAMBERLAIN, N. F.: Nature, Lond. *222:* 747 (1969); we are grateful to Dr. HAZLEWOOD for a letter giving details on this work.
3 DASZKIEWICZ, O. K.; HENNEL, J. W. and LUBAS, B.: Nature, Lond. *200:* 1006 (1963).
4 GLASEL, J. A.: Nature, Lond. *220:* 1124 (1968).
5a LING, G. N.: in ROSE, A. H. Thermobiology (Academic Press, New York 1967).
5b LING, G. N.: A physical theory of the living state (Blaisdell, New York 1962).
6 SZENT-GYÖRGYI, A.: Perspectives in biology and medicine; vol. 14, p. 239 (1971).
7a BERNAL, J. D.: in HADZI, D. Hydrogen bonding (Pergamon Press, Oxford 1959).
7b CLIFFORD, J. and PETHICA, B. A.: in COVINGTON, A. K. and JONES, P. Hydrogen-bonded solvent systems (Taylor & Francis, London 1968).
7c DERSEY, N. E. (Ed.): Properties of ordinary water-substance (Reprinted). (Hafner, New York 1968).
7d DROST-HANSEN, W.: in BROWN, H. D. Chemistry of the cell interface (Academic Press, New York [to be published]). We are grateful to DORST-HANSEN, W. for a preprint of this work.
7e EISENBERG, D. and KAUZMAN, W.: The structure and properties of water (Oxford Univ. Press, London 1969).
7f FRANK, H. S.: Science *169:* 635 (1970).
7g HOLTZER, A. and EMERSON, M. F.: J. phys. Chem. *73:* 26 (1969).
7h KAVANAU, J. L.: Water and solute water interactions (Holden Day, San Francisco 1964).
7i KLOTZ, I. M.: Horizons in biochemistry; p. 523 (Academic Press, New York 1962).
7j NÉMETHY, G.: Angew. Chemie (Int. Ed) *6:* 195 (1967).
8 REES, D. A.: Adv. Carbohydrate Chem. *24:* 267 (1969).
9 O'NEILL, A. N. and STEWART, D. K. R.: Canad. J. Chem. *34:* 1700 (1956).
10 STERLING, C. and MASUZAWA, M.: Makrom. Chemie *116:* 140 (1968).
11 HECHTER, O.; WITTSTRUCK, T.; MCNIVEN, N. and LESTER, G.: Proc. Nat. Acad. Sci., Wash. *46:* 783 (1960).
12 BLOEMBERGEN, N.; PURCELL, B. M. and POUND, R.: Physiol. Rev. *73:* 679 (1948).
13 JUSTI, E. and LAUE, M. VON: Physik. Z. *35:* 945 (1934). (We are grateful to TISZA, L. for having brought this paper to our attention and for a discussion on this subject.
14 AIELLO, G.; PALMA, M. U. and PERISCO, F.: Phys. Ltrs *11:* 117 (1964).
15 AIELLO, G. and PALMA-VITTORELLI, M. B.: Phys. Rev. Ltrs *21:* 137 (1968).
16 AIELLO, G. and PALMA-VITTORELLI, M. B.: To be published.
17 FRÖHLICH, H.: in MAROIS, M. Theoretical physics and biology, vol. I, Proc. 1st Int. Conf. on Theor. Physics and Biology, Versailles 1967, p. 13 (Noth-Holland Publ. Co., Amsterdam 1969).
18 FRÖHLICH, H.: Phys. Ltrs *26A:* 402 (1968); see also [20].
19 FRÖHLICH, H.: Int. J. Quantum Chem. *2:* 641 (1968).

20 FRÖHLICH, H.: Personal commun. We wish to thank FRÖHLICH, H. for correspondence
    and for a number of vigorous discussions and for having provided the initial stimulus
    to the present work.
21 We are grateful to CARAPEZZA, M. for having drawn our attention to this specific fact
    and also to TONANI, F. and LEONE, M. for a number of discussions and suggestions on
    the subject of clays.
22 PAECHT-HOROWITZ, M.; BERGER, J. and KATCHALSKI, A.: Nature, Lond. *228:* 636 (1970).

Author's address: M. U. PALMA, Istituto di Fisica, Via Archirafi 36, *Palermo* (Italy)

Discussions

H. FRÖHLICH: Well, now we shall have some discussion on the question of bound water in cells and anything referring to it. Perhaps Dr. LING will comment?

G. LING: Dr. PALMA, you have kindly sent to me—through Dr. FRÖHLICH, a preprint of the work which you have just communicated. I am glad to tell you that Mr. CHRISTOPHER MILLER in our laboratory has confirmed the gist of your findings.

As is typical of biology, that which appears to be a simple problem at first, often becomes more and more complex, as more and more data are gathered. Thus, WOESSNER and SNOWDEN [J. Coll. Interf. Sci. *31:* 290, 1970] in a recent study of the relaxation times of water in agar gel, came to a conclusion quite different from yours. That is, they concluded that only a minor fraction of the water strongly interacts with the agar macromolecules. The line width increase they attributed to rapid exchange between this small fraction and the bulk of normal water. However, as you pointed out, the linear relation between line width and the cube root of the agar concentrations certainly is in favor of your interpretation.

Our own studies of living cells and model systems have led us to think that the problem is not a simple question of normal liquid water state versus another abnormal bound water state. Instead, there appears to be more than one state of organized water. The major reason behind this idea is that conclusions drawn from different criteria for 'abnormal' water sometimes match and at other times they do not.

The criteria that we have used frequently are: 1. reduced NMR relaxation times; 2. lowered solubility or equilibrium distribution coefficient (i.e., q-value) of certain solutes like sugars and Na^+ ion; and 3. altered freezing temperature and pattern of ice formation.

Thus, water in montmorillonite clay has been shown to exhibit split NMR peaks due to spatial orientation in deep layers between clay platelets [WOESSNER and SNOWDEN, J. Coll. Interf. Sci. *30:* 54, 1969]. Yet, our own preliminary studies indicate that clay water (or water in agar gel) does not show pronounced ability to accomodate less sucrose than normal water.

Water in a concentrated gelatin gel has long been known to refuse freezing even at liquid nitrogen temperature [MORAN, Proc. Roy. Soc. A *112:* 30, 1926]. Yet, the spin-spin relaxation time, T_2, appears not greatly reduced [C. MILLER, unpublised].

Man-made sulfonate ion exchange resins contain 50% water. The T_2 of this water is only moderately reduced [GORDON, Chem. and Ind. *1962:* 267]. Yet, the equilibrium distribution coefficient (q-value) of sucrose between the resin water and an external normal aqueous solution is only about 20% [LING, Ann. N.Y. Acad. Sci. *195:* 401, 1965].

From the purely physical point of view each of these phenomena deserves studies on its own right. However, for a conference of this nature, it will be more important to emphasize the crucial roles which organized water may play in the biology of living cells [LING, Int. J. Neurosci. *1:* 129, 1970].

It has long been thought that the ability of living cells to exclude Na^+ ion in the resting state and the ability to admit Na^+ ion during functional activity reflects the properties of a microscopic cell membrane containing a variety of pumps, including the Na^+ pump. This view became seriously doubted when it was shown that under carefully controlled conditions, the postulated Na^+ pump would have to consume energy at the rate of 150 to 300 times that available to the cells [LING, A physical theory of the living state, Blaisdell, Waltham, Mass., 1962].

More recent experiments have shown evidence that the low q-value for Na^+ ion and sugars in the cell water reflects organization of the bulk of cell water by the proteins in the interior and on the surface of the living cells. Furthermore, the maintenance of this physical state of cell water is under the control of the ultimate product of metabolism, ATP, which interacts with the proteins. Perhaps just as noteworthy is our finding that when cells become cancer, the ability to exclude sugars vanishes [unpublished] and with this the NMR relaxation time T_1, becomes lengthened to a value closer to that of normal liquid water [DAMADIAN, Science *171:* 1151, 1971].

L. ONSAGER: Agar rearranges much more rapidly at the final high temperature than it did while the temperature was rising. Afterwards, a substantial undercooling was needed to restore a tightly bound system.

U. PALMA: You are thinking along the line which is essentially at the basis of the thermodynamic theory of the sol-gel transitions. This type of phenomenological theory has in fact been developed and it closely resembles the Devonshire theory of ferroelectrics. To let it work, however, one must include phenomenologically it the existence of cooperativity. One must subsequently provide a microscopic explanation for the occurrence of this cooperativity. The model I have described provides one possible microscopic explanation. Well, I do not claim that this is *the* model, but if you don't have a microscopic mode, the thermodynamic theory does not take you any further and you are left at a stage similar to that of ferroelectricity prior to the advent of microscopic models and theories. More specifically, I agree that there may be a network formation and an interaction between polysaccharide chains (especially at the higher concentrations). However, we must account for cooperativity in this intermeshing. Now, the water proton magnetic resonance is strongly affected and the water dielectric constant is also strongly affected, so we conclude that water is the channel through which cooperativity goes. The specific model is more debatable, of course.

D. D. ELEY: I just wondered whether the T_1's and the T_2's referred to the protons in the water or in the agar?

U. PALMA: The agar is present in concentrations up to 5% by weight, so that the T_2 data refer essentially to the water protons. In fact, the hysteretical temperature dependence of T_2 and that of the high-resolution linewidth are essentially identical, in spite of the fact that the impulsive (wide-line) technique brings all the proton signals into your measurement, while in the steady-state (high resolution) experiments, you specifically look at the water protons resonance.

D. D. ELEY: I can't remember the state of 2% of agar. Is it a sol or a gel?

U. PALMA: It is a firm gel. Let me show another slide: here you see the influence of the macromolecule concentration. In the 5% system, as compared with the 2%, the transition temperature is higher and the transition itself is steeper. This again confirms that it is the macromolecules which control the whole system.

D. D. ELEY: Were small crystallites formed below the transition? Is not the situation a little like a glass transition temperature Tg in a simple polymer system? I wonder whether there is any tendency for the agar molecules to form crystallites or other linking points (entanglements)?

U. PALMA: We have looked at the system in crossed polaroids and it appears absolutely uniform. At least up to a certain concentration.

L. ROSENFELD: I just wanted to comment on what WEISSKOPF said a moment ago, largely to express agreement with his sentiment. In particular, I was somewhat frightened by the appearance of this myth of ferroelectricity. Since the time of LORENTZ, one knows that there is an internal field, which we now call the Lorentz field, in any substance electrically polarized and there is nothing more to ferroelectricity than just the effect of this field. Now I come to WEISSKOPF's main point: he rightly emphasized that there is a question of method we have to think about in order to make the collaboration between physicists and biologists fruitful. I feel that we have now reached the stage—thanks to the biologists—in which we are presented with clear evidence evincing various fundamental biological structures with more and more detail; the task of the physicist, therefore, should not be to invent models, but to study these structures and try to apply the known laws of physics to them as well as he can and see what happens. WEISSKOPF further raised the point of nonlinearity, a property which, as we know already from the experience we have of it in physics, gives rise to phenomena of a new kind, not familiar to the physicist, because the physicist (being as lazy as any other person) prefers to concentrate on the simplest systems, which are the linear ones. More recently, however, precisely with a view to comparing biological with physical phenomena, some physicists have started to investigate what happens when you allow for nonlinear interactions and also for another circumstance which is very important, namely the fact that biological systems are not anywhere near thermodynamic equilibrium. Those points have been emphasized by FRÖHLICH this morning. When you combine these two things, some results, obtained by PRIGOGINE and his people and also by HAKEN, indicate that you do indeed find phenomena of a quite unfamiliar kind, not comparable with those occurring in the neighbourhood of thermodynamic equilibrium, and even belonging to a family of solutions of the equations which is not connected by any continuous transition with the thermodynamic branch. One aspect that has resulted from this purely theoretical investigation is the appearance of structures, that is, of a critical parameter to which you may give the dimensions of a length and which characterizes certain structures. Such structures have actually been observed: Benard cells in hydrodynamics, and chemical reactions in which you see a sedimentation of the products under certain conditions. I am not suggesting that these particular chemical and physical facts have any direct relationship with biological phenomena, but they show at least that there are more things in physics than physicists dreamt of.

Proc. 3rd int. Conf. From Theoretical Physics to Biology, Versailles 1971, pp. 35–49
(Karger, Basel 1973)

Cooperative Phenomena in Systems far from Thermal Equilibrium

H. HAKEN

Institute for Theoretical Physics, University of Stuttgart, Stuttgart

It is a great pleasure for me to at tend this conference for the third time. One of the comments we have heard each time at these conferences is that there is an enormous gap between theoretical physics and biology. It appears to me, however, that this gap is becoming narrower and narrower, at least in certain disciplines, and I want to substantiate this point of view in my paper. One of the most striking features of life is the transition from disorder to order or, in other words, the phenomenon of self-organization. The first obvious question that a biologist may well ask a physicist is this: are there in physics any phenomena which are good examples for the transition from disorder to order in analogy to processes in biology?

There are, in fact, a number of phenomena in physics that are well understood and which clearly show a transition from disorder to order. An example, for instance, is crystallization. In the disordered phase, the matter is a liquid in which the positions of the atoms are more or less at random. After crystallization, the atoms are in a spatially completely fixed and regular order. In this example, the ordering occurs with respect to the spatial position. In physics also other quantities may undergo an ordering process ('phase transition') [1], for instance, in ferromagnets. Here, above a certain temperature the elementary magnets (spins) may point in random directions, whereas below a certain temperature, the so-called Curie-temperature, the magnets become aligned, showing a high degree of order. There are quite a number of other examples but in the eyes of a biologist all these examples have a serious drawback: the new, ordered state occurs if the temperature is lowered, i.e., if energy is taken out of the system. This is in striking contrast with what we observe in biology. Here, living, ordered states are those which are maintained by a flux of energy through them and certain functions can be achieved only when the energy flux is increased. Of course, there are also systems in physics that are far from thermal equilibrium and function only if there is a certain flux of energy through them (actually, any machine has this property). It came as a large surprise to many of us, however, that in spite of this, these systems showed transitions from order to disorder which are strongly reminiscent of the old disorder-order transition in equilibrium systems, e.g. in ferromagnets. One of these new systems is the laser, which, by the way, I proposed as a good model of biological processes at the first Versailles conference [2]. Let me briefly remind you of the laser [3]. It is an optical device by which incoherent light (the so-called pumplight) or other forms of energy are converted into a highly directional, very intense and very coherent beam of light. A typical experimental set-up is shown in figure 1. The laser consists of a rod of a so-called 'active' material, which is excited from the outside. Because of the special geometry (e.g., mirrors at the endfaces) laserlight is emitted in the axial direction.

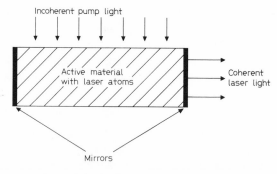

Fig. 1. Scheme of a laser.

The analogy between a ferromagnet and a laser may roughly established as follows [4] (fig. 2): the ferromagnet may be visualized as consisting of many individual elementary magnets (spins) which, in the disordered phase, show in different directions. When we operate the laser as a usual lamp, i.e., for small pump energy, the atoms of the active material emit light waves independently of each other. The optical phase is completely at random, analogous to the complete randomness of the spin directions. Now consider the ordered phase: in the ferromagnet the elementary magnets show all in one direction. Similarly, we know that in the laser (at a sufficiently high pump energy) the atoms emit light in a completely coherent fashion, i.e.,

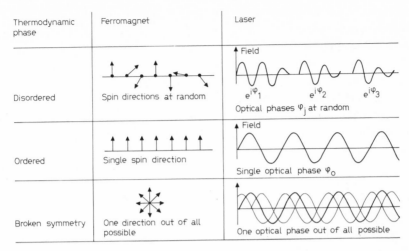

Thermodynamic phase	Ferromagnet	Laser
Disordered	Spin directions at random	Field $e^{i\varphi_1}$ $e^{i\varphi_2}$ $e^{i\varphi_3}$ Optical phases φ_j at random
Ordered	Single spin direction	Field Single optical phase φ_0
Broken symmetry	One direction out of all possible	One optical phase out of all possible

Fig. 2. Analogy between laser and ferromagnet.

no atom is allowed any longer to emit light with a random optical phase. All the atoms emit just one wave. Thus, analogous to the fixed direction of the spins, we have now a fixed optical phase.

In both cases we observe the phenomenon of 'broken symmetry' which plays an eminent role in modern physics. What is meant by this concept is the following: whereas the spins may show 'in principle' in every direction, the point in only one direction in the ordered phase. Similarly, whereas the optical phase is 'in principle' undetermined, one definite phase is selected in the ordered state.

I should like to mention a second concept which plays an important role in the theory of phase-transitions, namely, that of the order parameter. In view of the enormous number of spins in a ferromagnet, it would be rather hopeless to treat each individual spin. Instead, physicists use the magnetization M, i.e., the total field produced by the spins as a quantity ('order-parameter') to describe the degree of order. It turns out that in the disordered phase M $\neq$ 0, but in the ordered phase M $\neq$ 0. In an analogous manner, it wouldn't make much sense to describe the actual state of all the different atoms in a laser. Instead, one uses the lightfield E as order-parameter. Again, one finds E = 0 in the disordered and E $\neq$ 0 in the ordered phase. The choice of the proper order-parameter is often crucial in the theory of phase-transitions and sometimes it is quite difficult to detect the right quantity. I have a strong feeling that the concept of order-parameter will play an eminent role in biological processes, but that we are still far removed from having found it, except in simple cases (see below).

As I will show below, there are further analogies between the two systems (e.g., ferromagnet and laser) which can be expressed in a completely quantitative manner. Most important in our present context, however, is the fact that in the ferromagnet the ordered state is reached by lowering the temperature, whereas in the case of a laser the energy pumped into the system must be increased in order to obtain the ordered state.

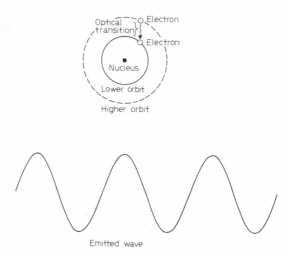

Fig. 3. Schematic description of optical emission.

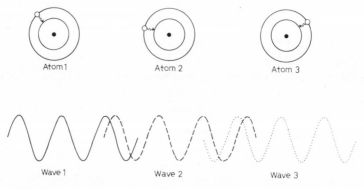

Fig. 4. 'Disordered' state: incoherent emission of the excited atoms.

Let me now say a few words about how the laser emits coherent light, in terms that are nowadays often used in biology. Let us consider a set of atoms of the active material. In a naive picture we describe the atoms by the Bohr-model (fig. 3). In the ground state, the atom is running in the smaller orbit; in the excited state, it runs in the larger orbit. We consider a number of atoms which are excited by the pumplight. According to the different positions, the electrons may fall down emitting light with random phases (fig. 4). This occurs in a usual lamp and represents the disordered state. What happens in a laser at an increased pump rate? By some chance a coherent wave had been built up (note the use of the word 'chance'). This coherent wave now 'instructs' the atoms: it forces the electrons in the atoms to oscillate 'in phase' with the oscillating wave (fig. 5). These in-structed atoms now make their transitions in the 'right way' and reinforce the present wave in a coherent way. Thus, the wave already present is enhanced without introducing any error or, in other words, the instruction given to the atom goes back to the field. Thus, the build-up of the field occurs in a 'self-instructed manner'. To maintain this process, a sufficient number of excited atoms must be present. If not, the coherent waves escape very quickly and the newly emitted wave-tracks have no optical phase in common with the old wave-track.

The transition of disorder to order in the laser is connected with the following very interesting feature. In the disordered state, the energy output of the laser increases only slowly with the input of energy supplied by the pumplight. Beyond a certain energy input, when the ordered state occurs, the energy output increases much more rapidly. Thus, the efficiency of a laser in the ordered state is much higher than in the disordered state. The analogy with biological systems is again very striking. Were lasers selected on the basis of their efficiency (which would be quite a natural principle), obviously, those operating in the ordered state would be chosen (compare fig. 5a).

In my talk I am faced now with presenting my subject from two different angles. On the one hand, I want to satisfy my colleagues of theoretical physics that a very strong analogy indeed exists between this new class of phenomena in systems far from thermal equilibrium and those in systems in equilibrium. On the other hand, I want to invite the attention of my colleagues in biology to the fact that a certain set of Eigen's equations [5] are identical with our laser equations; thus, we have here—models of selection, and even of evolution.

Let me begin, however, with a word to the physicists. Figure 6 illustrates

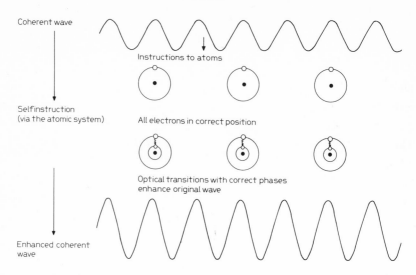

Coherent wave

Instructions to atoms

Selfinstruction
(via the atomic system)

All electrons in correct position

Optical transitions with correct phases
enhance original wave

Enhanced coherent
wave

Fig. 5a. 'Ordered state': self-instruction of the laser light (stimulated emission).

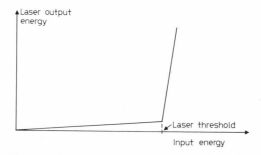

Laser output
energy

Laser threshold

Input energy

Fig. 5b. The laser output energy as a function of the input energy.

the complete one-to-one correspondence between the light distribution of an infinitely extended, one-dimensional laser and the pair-wave distribution in superconductors [6]. In the first part of the figure I have written down the probability-distribution function for the electric field strength E in a laser. This distribution function is given by the exponential function of an integral over the whole space of the laser. The integral contains a term quadratic in the field strength and a term of fourth order. Furthermore, it contains the spatial derivative. d_c is the critical inversion or the critical pumpstrength above which laser action occurs; d is the actual pump or inversion. ω_0 is the frequency of the central emission line. The second part

$$f = N \exp\left(-\frac{B}{Q}\right)$$

$$B = \int \left\{ \tilde{\alpha}|E(x)|^2 + \tilde{\beta}|E(x)|^4 + \tilde{\gamma}|\left(\frac{d}{dx} - i\frac{\tilde{\omega}_0}{C}\right)E(x)|^2 \right\} dx$$

E : electric field strength

$\alpha = \tilde{a}(d_c - d)$; $\tilde{a}, \tilde{\beta}\ \tilde{\gamma} > 0$ are laser constants

C : velocity of light

$\tilde{\omega}_0$: atomic line frequency

N : Normalization factor d: atomic inversion d_c: critical atomic inversion

B : is a generalized thermodynamic potential

Q : is the strength of the fluctuations

Fig. 6a. The distribution function of the laser.

$$f = N \exp\left(-\frac{F}{kT}\right)$$

$$F = \int \left\{ \alpha|\psi(\underset{\sim}{x})|^2 + \beta|\psi(\underset{\sim}{x})|^4 + \frac{1}{2m}|\left(\nabla - \frac{2ei}{c}\underset{\sim}{A}\right)\psi(\underset{\sim}{x})|^2 \right\} d^3x$$

ψ : pair wave function

$a' = \alpha'(T - T_c)$

α', b': superconductor constants

k : Boltzmann's constant

T : absolute temperature T_c: critical temperature

F : the free energy

m, e: mass and charge of electrons respectively

$\underset{\sim}{A}$: vector potential

Fig. 6b. The distribution function of the Ginzburg Landau theory, of superconductivity.

of the figure shows the probability-distribution function of the pair-wave function of superconductivity of the famous Ginzburg-Landau theory. Ψ is the pair wave function; T is the temperature; T_c is the critical temperature; A is the vector potential. A detailed comparison between the two formulas (1) and (2) reveals a complete analogy in the mathematical structure. Thus, we observe the same kind of phase transition and, in particular, similar critical fluctuations in the laser and in the superconductor [7]. We have also found the reason why one may establish such analogies between systems in, and far from—thermal equilibrium. I do not wish to go into further detail, but I hope that the experts have seen that we have an extremely strong analogy in systems in, and far away from —thermal equilibrium.

Addressing myself to our biological colleagues, I should like to present the equations of a laser (of finite extension). An approximation[1] of the laser equations, dealing only with the intensities, is given in figure 7. We decompose the lightfield within the laser into the so-called modes, i.e., into waves each with a definite wavelength. Each wave with wavelenght λ has a certain amplitude E_λ or, a certain intensity $I_\lambda = [E_\lambda]^2$. Instead of the intensity we may use the photon number n_λ which is proportional to I_λ.

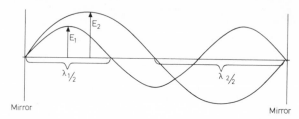

Fig. 7. Exampl of two modes with wavelengths λ_1, λ_2 and amplitudes E_1, E_2.

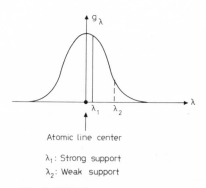

Fig. 8. The gain acts as a filter.

Let us investigate the temporal behavior of each individual mode λ. The emission into each mode is supported by excited atoms so that we have a certain generation-rate of modes described by a gain factor g_λ. On the other hand, each mode may decay because, for instance, it leaves the laser or is scattered by impurity centers in the laser. Thus, we have a loss rate, κ_λ.

---

1   This approximation, which in certain cases is well justified for the laser, allows us to establish a direct relationship with Eigen's theory. On the other hand, we must leave a number of questions aside, which are of importance for the coherence of a single mode.

The gain factor g_λ is of particular interest. First of all the atoms support the different modes in a different manner, depending on the distance of the mode frequency (or wavelength) from the atomic line frequency. This is explained in figure 8. If the frequency of modes is close to that of the atom it supports the mode very strongly, whereas for large distances the mode is supported only weakly. The atoms, therefore, act differently on the modes or, in other words, act as a 'filter'. When I used the word 'filter' in this context in my address at the first Versailles conference, I received objections from biologists claiming that a filter is necessarily something man-made and thus, cannot be invoked in order to explain self-organization. But, as our example shows, the atoms are objects of nature and nevertheless act as a filter. When we take all the different gain and loss factors together, we obtain an equation of the following form

$$\frac{dn_\lambda}{dt} = g_\lambda\, n_\lambda - \kappa_\lambda\, n_\lambda \tag{1}$$

Because modes with largest gain are enhanced most, we find a phenomenon that is called segregation (fig. 9). As we had shown some years ago, the laser equations are also capable of describing selection (which is of great technical importance for the laser) [8]. As we will see in a moment, they apply equally well to selection in biological systems. The physical reason is this: the gain factor g_λ is proportional to the number of excited atoms. When the excited atoms participate in the laser process, they are depleted and thus the gain decreases. Depletion of atoms is porportional to the photons already present. Thus, one has to write more accurately, instead of g_λ, an expression

$$g_\lambda = G_\lambda - G_\lambda \sum_\lambda C_{\lambda'} u_{\lambda'}$$

which takes into account the depletion or, as laser physicists say—the saturation by the term $-G_\lambda \sum_\lambda C_{\lambda'}\, n_{\lambda'}$. Thus, the laser equations now read

$$\frac{dn_\lambda}{dt} = (G_\lambda - G_\lambda \sum_{\lambda'} C_{\lambda'}\, n_{\lambda'})\, n_\lambda - \kappa_\lambda\, n_\lambda \tag{2}$$

It has been shown in the laser case [8], and can be shown quite generally, that these new equations may describe selection. Only the fittest mode survives, while the other ones die out.

The laser equation may be interpreted in an entirely different manner: instead of photon numbers, one may speak of concentrations of certain

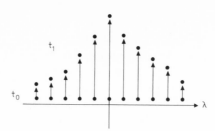

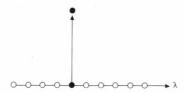

Fig. 9. Segregation: all modes (species) are enhanced, but differently.

Fig. 10. Selection: only one mode (species) survives, all the others die out.

molecules. The equations (2) then describe autocatalytic reactions, and were used recently by EIGEN [5] in completely equivalent form to describe selection of biological molecules. Note that in both cases the n_λ's play the role of order parameters.

To describe laser action in a more accurate manner one has to introduce the so-called fluctuating forces. Indeed, the equations allow for the trivial solution $n_\lambda = 0$, so that it would appear that a laser never starts working. Laser action, however, is triggered by spontaneous emission which, being a quantum-mechanical process, occurs completely at random (note again the role of 'chance' in physics) and which, mathematically, can be taken care of by introducing fluctuating ('stochastic') forces $F_\lambda(t)$ in the laser equation. These fluctuating forces indicate that photons may be generated at random. But then the stimulated process sets in and with it the selection process. The complete laser equations read:

$$\frac{dn_\lambda}{dt} = (G_\lambda - G_\lambda \Sigma \ C_{\lambda'} n_{\lambda'}) \ n_\lambda - \kappa_\lambda \ n_\lambda + F_\lambda(t) \tag{3}$$

The equations express mathematically the interplay between chance and necessity. The stochastic forces produce at random all kind of 'species' (modes or molecules), but the *systematic* forces (S) select but one.

So far, the laser equations have all the necessary features a system needs to describe selection. Can we also think of physical systems which are analogues or, in a certain sense, analogue computers[2] of evolution? This would mean that the once selected mode is capable of transformations which make it still 'more fit'. This can indeed be achieved in a number of ways. I have just indicated one which is well known in nonlinear optics, e.g., the parametric oscillator. Here, the first selected mode coming out of a laser enters a nonlinear crystal and forms the starting point for a new mode which now may have still better qualities, e.g., for survival—i.e., a higher gain factor than the previous mode. This new situation can be treated by a hierarchy of laser-like equations. By means of the laser equations one may also study quite a different mechanism which may cause evolution as well, namely, the impact of the change of external parameters—e.g., the losses κ_λ — on a novel selection of modes (species) which may occur adiabatically or abrupt. Here, at least in simple cases, we have mathematical tools to calculate the time the new mode (species) requires to develop. This might help in the future to estimate the time-constants inherent in models of biological evolution. Such estimates may be crucial for concrete models of evolution.

In conclusion, in my personal opinion, the example of the laser I demonstrated here as a possible analogue of biological processes is only one of many possible examples. I am convinced that future work will reveal a number of very close analogies between physical and biological systems and that concepts such as broken symmetry and order parameter might play an important role in the understanding and adequate treatment of biological processes.

References

1   There is an extensive literature on phase transitions, see e.g. BROUT, R.: Phase transitions (Benjamin, New York 1965).
2   HAKEN, H.: in MAROIS, M. Theoretical physics and biology (North Holland Publ. Comp., Amsterdam 1969).
3   For a detailed review see HAKEN, H.: Laser theory; in Encyclopedia of physics, vol. 25/2c (Springer, Berlin/Heidelberg/New York 1970).
4   HAKEN, H.: in MADELUNG, O. Festkörperprobleme, Adv. in Solid State Physics, vol. 10, p. 351 (Pergamon, Vieweg 1970).
5   EIGEN, M.: Preprint (1971).

2   The author wishes to thank Prof. D. GLASER, Berkeley, California, for drawing his attention to this aspect.

6 GRAHAM, R. and HAKEN, H.: Z. Physik *237:* 31–46 (1970). For a single mode see also
   [4] and DEGIORGIO and SCULLY, M.: Phys. Rev. *A2:* 1170 (1970).
7 SCHMIDT, H.: Z. Physik *232:* 443 (1970).
8 HAKEN, H. and SAUERMANN, H.: Z. Physik *173:* 261 (1963). The complete mode selection
   occurs for running waves, treated in eq. (2).

Author's address: Dr. H. HAKEN, Institut für Theoretische Physik, Azenbergstrasse 12,
D-7000 Stuttgart (FRG)

Discussions

G. NICOLIS: I would like to make a comment about the analogy drawn by Prof. HAKEN
between the form of the laser equations and the kinetic equations for chemical reactions with
autocatalytic processes. Certainly this analogy is a very interesting one and it is even very
striking. However I think one should also point out that it is mainly analogy between the
forms of these equations. Indeed, the interpretation of the various terms is rather different.
In the laser, Ni equations are essentially microscopic occupations. In the chemical kinetic
case, the concentration variables which appear in the equations are highly-averaged macroscopic quantities. This difference in the nature of the laser and the chemical variables is also
responsible for important differences in the behaviour of the two systems. The state of the
laser, as Prof. HAKEN pointed out, is of course quite far from the Maxwellian distribution.
On the other hand, the state of a chemical kinetic system, even in the presence of autocatalytic
reactions, could be on the molecular scale quite close to a local Maxwellian form. As a result,
instabilities and evolutionary processes in chemical systems cannot be interpreted in general
as second order phase transitions far from equilibrium (see also communication by
I. PRIGOGINE and O. NICOLIS in this meeting).

H. HAKEN: In order to explain the relation between the principle of detailed balance
and the strong formal analogy between systems in- and far-from thermal equilibrium, let
me offer some mathematical details. In a first step, we transform the quantum mechanical
problem to a classical one using a quasi-probability distribution function, F, very similar to
the famous Wigner distribution function. Starting from a density matrix formulation we then
find a Fokker-Planck equation for a classical function F(u), where u stands for all variables.
In the Fokker-Planck equation we have terms up to second order in the derivatives. We
then introduce the joint probability W(u, t; u', t') and postulate its invariance against time
reversal which is our postulate of detailed balance. If this principle is fulfilled, we may write
$F = N \exp(-\Phi)$, where N is a normalization factor and the 'generalized thermodynamical
potential' Φ is determined by mere quadratures. The existence, of Φ forms the basis of the
analogy between lasers and superconductors discussed in our talk.

Detailed balance means that we are close to some sort of equilibrium and this is perhaps
a point where some biologists will tell us that this is not the case and in reality we are far
away from it. But I personally don't believe it because (and this is a comment on evolution)
the whole probability distribution depends also on external parameters. In fact, what has
occurred in evolution is a change of external parameters, so one possible explanation is that
we are going more or less slowly from one equilibrium state to another. If we have one state

realized and then the external parameters change, the system is pushed out of its old state by fluctuations and then driven to a new state by systematic forces. This is a possibility but I'm sure that one can object to it.

M. EIGEN: I am sorry that you removed the equation $F = N \exp(-O)$.

H. HAKEN: This certains also one point that I wanted to discuss with you because you objected some years ago what I called a filter and I think that's an important point. So I write it down $F = N \exp(-O)$ this is what I think Eigen calls self instructed sequence.

M. EIGEN: I think that Dr. NICOLIS has already mentioned that this is a formal analogy but there are, of course, certains differences also. This system does something like selection but it is not 'inherently' selective. Now the difference to the properties you need for evolution: at the same time the system selects for a certain mode it should produce steadily new modes. Only a fraction Q of all the instructed process should lead to the same copy, an other fraction 1–Q should lead to error copies.

H. HAKEN: Yes, but this we have here in the fluctuating forces, because they produce new molecules at random and these stochastic forces may depend on the Nj's again.

M. EIGEN: There is a difference, because you can do it in two ways. A random generator could steadily produce new copies, or the new copies could steadily evolve from a given copy. In other words you keep the information as long as you don't have anything better, and only if out of this information you gain a little advantage, you deep that and let the other ones die out.

H. FRÖHLICH: I think this is essential a principle of nonlinearity here, which changes the properties, with excitation. So that is capable of very great generalization.

H. HAKEN: Well, perhaps I can continue this question. The first one was Eigen's. I believe that the structure of these equations is very much the same, so that all you need is the suppression of all unwanted modes and just the 'going-up' with one mode. Then, you may go to the next step of your hierarchy and then insert the other mode. This is one thing. Perhaps the other thing is with respect to the modes. I think that this is an explicit example where one finds stabilized modes as proposed by FRÖHLICH for biological systems, because here, if you pump energy into the system, you really find a stable mode. Here we considered photonumbers, but one may also derive similar equations for oscillations which retain their phase. So I think the laser provides really an example—a realistic example—where a number of these concepts apply.

J. NINIO: I would like to recall a formulation given by CRICK in his book: 'Of Molecules and Men'. He was asking, is there in biology principle not yet found in physics? He answered that if there is such a principle it is the principle of natural selection. And one day perhaps, the people working in the field of chemical engineering will find something like a principle of emergence of replicative structures. Now, on one side there is the formal theory of natural selection and we know what we need for natural selection to operate, although we don't know it very precisely. On the other hand, people working on thermodynamics are looking to something like 'order from disorder' which would be a physical analogue of evolution. But as has been pointed out, somewhere in-between there is something missing which is replication. The question I would like to mention is: is replication given from the beginning? I mean that when you study the systems which are evolving, you can perhaps describe them in a molecular way and put in replication from the beginning. But the way CRICK asked the question is, starting from disorder and having general thermodynamical considerations, will this kind of 'order from disorder' produce something like replication? It seems to me that the more we study DNA, RNA and so on, the less obvious it is that a molecule is replicating by itself. Thus, we should try to look for the principles of likelihood for the formation of what we call complementary replicative structures.

G. NICOLIS: I would like to ask a technical question. What is the type of assumption you have to make on the spectral properties of the Langevin forces in order to get the Fokker-Planck equation? I presume you adopt the usual Markovian assumption as in equilibrium, in spite of the fact that the system has a highly non-Maxwellian distribution.

H. HAKEN: That's right. Exactly.

G. NICOLIS: Did you test the consistency of this assumption?

H. HAKEN: Yes, in all the basic cases we checked that it was consistent. So what we need essentially is that the actual relaxation times of the laser are long compared to the times in which the fluctuations are involved.

H. FRÖHLICH: Excuse me, I don't think we should discuss technical theoretical questions here.

S. LIFSON: First, a question to Professor WEISSKOPF. Do you mean the Bose-Einstein statistics ot the *non* Fermi-Dirac, i.e. including the Boltzmann statistics? Another comment with respect to the similitude of the equations and their implications to reality: it is a general rule that if the same equation represents different physical phenomena, there is a similitude in behaviour of these phenomena provided only that the boundary conditions are also the same. Finally, a comment to those who are sometimes puzzled by the fact that biological processes produce order out of disorder, in apparent contradiction to thermodynamics or statistical mechanics. The law that entropy never decreases is only true for closed systems. The biological systems we are talking about are open systems. If you close these systems by including the radiation field of the sun, as well as the immediate environment which supplies material and energy for metabolism, food digestion etc. the entropy of such systems is always increasing, so there is no reason to be puzzled.

V. F. WEISSKOPF: To your fourth question I really meant Bose-statistics, because I think that the Bose-statistics of photons is the essential point in a laser, that one mode of the radiation field attracts all the light quanta and therefore builds up: the more you have the more you get. Of course, it has some formal similarity with reproduction and this is probably why, it has a sort of vague similarity with population developments. The Bose-statistics requires that more and more particles or units get into one given state, but of course there is always a limit coming from the availability of energy.

H. HAKEN: Call it available energy.

V. F. WEISSKOPF: In a very simple way you can say that a laser evolves quickly to the best possible vibration in the mirror system of the ruby. And if you change the system it evolves right away an other vibration. If you change the mirror the fluctuations that always are present develop and envolve into the suitable vibrational form. But there is no characteristic evolutionary memory which improves each mutation. It jumps immediately into the best solution, which is a feature which is missing in evolution.

H. FRÖHLICH: Well, what is missing is the nonlinear feature, you see. And I think that could be worked out. Now I think we'll have to hurry up and make short remarks.

V. F. WEISSKOPF: One important part in the biological system is the number of possible states which may be populated, which is in very, very large excess to the number of states which actually are populated. What is the situation with respect to laser?

H. HAKEN: You have perhaps 10,000 or perhaps 100,000 modes in a ruby laser in one line width and you have one or a few modes coming out.

B. KNIGHT: A reciprocating steam engine satisfies this equation up here in fact, where N is the action variable of the flywheel, and where P is the pressure-head in the boiler, and L is the drag on the system and the nonlinear term to the right is the choking limitation of the throttle. We may go a step further and treat the steam engine according to quantum theory: A quantized flywheel is formally equivalent to a system satisfying Bose statistics; the engine's chemical fuel is an ensemble of atomic systems with an inversion in its population of energy states. To say it another way: a laser is essentially a steam engine whose cogs are electromagnetic. The steam engine has been analogized to living systems for a very long time. It is a sensible analogy, but we can understand it without going into these embellishments.

H. HAKEN: May I just answer this remark? Any engine is in that sense a laser because it converts many degrees of freedom into just one. Any motor car is one in which you have first in the engine many degrees of freedom but then you just go in one direction. What is important here is that you have these fluctuating forces. They are supposed, first of all to start out of complete randomness. Everything then starts automatically and I believe the same thing must be true for biology.

L. ROSENFELD: I think one should not be hypnotised by the analogy of the laser. For instance, WEISSKOPF's stress on the Bose-statistics, which is certainly essential for the laser, is a bit misleading, because you can have similar families of equations in classical statistics: think of chemical reactions, for instance, such as those discussed by PRIGOGINE, about which we shall hear later.

H. FRÖHLICH: We shall hear more about this later but now I suggest that we fulfil our promise, namely to explain to biologists what kind of experiments could be made or can be made in physics and so we hear CARERI.

Proc. 3rd int. Conf. From Theoretical Physics to Biology, Versailles 1971, pp. 50–53
(Karger, Basel 1973)

Experimental Possibilities to Detect
Electromagnetic Modes of Living Objects

G. CARERI

Institute of Physics, University of Rome

The purpose of this contribution is to show the limitation of the existing instrumentation in the detection of the electromagnetic properties of living systems in the wavelength region where the photon energy is of the same order of magnitude as the ambient thermal energy. We have been interested in the experimental aspects of this problem and have made some measurements using the best apparatus available at present, in order to see whether strong and sharp peaks occur in the spectrum of living systems, as has been suggested by Fröhlich [1, 2].

We found that bacteria, suspended in culture medium, are very convenient because it is easy to be sure that the cells are alive and, by suitable experimental arrangements, long runs can be made on the same collection of cells. There are, of course, some difficulties. The culture medium is largely water and absorbs strongly in the far infrared region. Furthermore, in the Raman experiments, the cell suspension was found to give strong scattering. Nevertheless, in spite of the instrumental difficulties, we preferred to use bacterial suspensions since these constitute a well-defined biological system.

In one group of experiments, by CERDONIO and SAMPOLI [3], the absorption in the far infrared region, from 10 to 70 cm^{-1}, of suspensions of *E. coli* and of yeast was measured using a Michaelson interferometer linked to a Fourier-transform computer. We had already extensively used this type of equipment in studying the far infrared absorption of proteins [4], and it is hard to improve on except for the detecting device. In our case, this was an excellent Golay cell with a limiting sensitivity of 5×10^{-11} W. Typically, a run can be made on a bacterial suspension in one hour with an accuracy of 3% and a resolution of 5 cm^{-1}. The suspensions were placed in the 20 μ gap between two quartz discs. Microscopic examination showed that the cells were intact. Suspensions of high density, i.e., near 1% in volume, were prepared by A. TONOLO, using conventional methods.

Our results so far show that the spectra of the cell suspensions are identical with those of the culture medium. This means that if changes occur in the optical modes of the cells, they must be slight and evenly distributed over the whole frequency range. More exactly, one can exclude the presence of a sharp peak where the cell absorption is larger than the water absorption by a factor of 3 or more. It must be noted that this negative result does not conclusively exclude FRÖHLICH's theory since this does not entail that an intense band should be located in the 10 to 70 cm^{-1} range, nor that it should arise from a transverse electromagnetic oscillation of the kind detected in infrared absorption.

A second group of experiments, using Raman laser scattering of a 1% suspension of E. coli, was carried out by MAZZACURATI, ROMANI and SIGNORELLI [5]. They used a 6 W argon laser and one of the best double monocromators available (working with a resulution of 1 cm^{-1}, the stray light rejection is 10^{-7} at 25 cm^{-1}). This technique has already been used by us to study the Raman spectra of protein solutions [6]. The bacterial suspension was contained in a glass cylinder fitted with optical flat ends of diameter 1 in. We were thus able to focus the laser beam in the centre of the suspension so that spontaneous convection prevented local overheating. Under these conditions, the cells are metabolically active since the turbidity of the suspension increases during a day.

Unfortunately, large density fluctuations in the suspension were found to produce a very intense Tyndall scattering with a large noise. Consequently, a true Raman effect can be detected only at wavelength displacements greater than about 25 cm^{-1}. A fundamental analysis of the optical problems shows that this lower limit can hardly be bettered even using a computer technique to reduce the signal to noise ratio.

So far we have found no difference, in the range 25 − 200 cm^{-1}, between the Raman spectra of the bacterial suspensions and of the corresponding culture medium. This result means that there is no single sharp mode, either transverse or longitudinal, characterizing the cell in this region, where the integrated intensity of the whole cell mass is concentrated.

The above experiments are reported as examples of the experimental limitation of the present apparatus in investigating the electromagnetic properties of living systems. Using this apparatus we can say that the e.m. spectrum of the bacterial cell is not much different from that of water, at least in the frequency range above 5×10^{11} Hz. However, this is not the end of the story, because the range below 5×10^{11} Hz may be the significant one, and, moreover, the new properties may not be characteristic of the whole bacterial cell but only of a minor part of it.

References

1 FRÖHLICH, H.: Phys. Ltrs *26 A:* 402 (1968).
2 FRÖHLICH, H.: in MAROIS, M. Theoretical physics and biology, p. 13 (North Holland Publ. Comp., Amsterdam 1969).
3 CERDONIO, M. and SAMPOLI, M.: Atti Acc. Nay. Lincei (in press).
4 BUONTEMPO, V., CARERI, G. and FASELLA, P.: Phys. Ltrs *31 A:* 543 (1970).
5 MAZZACURATI, V., ROMANI, G. and SIGNORELLI, G.: Atti Acc. Naz. Lincei (in press).
6 CARERI, G.; MAZZACURATI, V. and SIGNORELLI, G.: Phys. Ltrs *31 A:* 425 (1970).

Author's address: G. CARERI, Istituto di Fisica 'Guglielmo Marconi', Università di Roma, Piazzale delle Science 5, *Roma* (Italy)

Discussions

O. MAALØE: There are many details that one could discuss. The main point I would like to ask is that you said 1% bacteria by weight?

CARERI: By volume I would say ... it was really very dense.

O. MAALØE: But this means that your *coli* cells have been in totally anaerobic conditions and that their activity has been quite low compared to what it would have been in a more dilute solution. It is entirely possible that the signal you try to pick up, and which depends on the energy flux in the system would be stronger in a relatively thin suspension of bacteria, in which the individual cell would be metabolically much more active.

G. CARERI: Then one had better work with lower density

O. MAALØE: That is already where a *coli* culture begins not to grow very much more.

H. FRÖHLICH: But this apparently is the difficulty that occurs all the time.

O. MAALØE: Could I ask one question? It is not clear to me and there must be experts in the room who could tell us how to estimate the actual strength of the signal to be expected if we postulate that it is a certain part of the total mass of the cell which generates it.

G. CARERI: Well there is a difference between the two experiments. In the infrared one could get three orders of magnitude by several tricks and a much better detector.

O. MAALØE: My question was, What intensity of signal would you expect if 10% of the mass of the bacterium is participating?

D. GLASER: Perhaps I could reformulate the question? What is the physical mechanism by which you expect the membrane to absorb infrared radiation of this wavelength and which would then presumably allow you to predict the amount that you should get.

G. CARERI: No, let us not take this up now. I'm not interested, at this moment, in that.

D. GLASER: The question is: do you have any idea of a mechanism which would permit you to estimate the magnitude of the effect? A related question I have is: do you know the index of refraction of the cell wall at this wavelength as compared with the index of water. Does one get large Rayleigh scattering in the infrared?

G. CARERI: I have no idea about the possible physical mechanism in a membrane. The index of refraction of the cell in this wavelength region must not be much different from that of pure water because we see no great change in the absorption and, therefore, it must not be an important source of scattering.

D. GLASER: Perhaps the biologist's response to your question, 'is it worth building special apparatus for going further?' has to be: what mechanism are you trying to discover?

O. MAALØE: My question is very simple. Would we, from theory, expect that a signal would have been picked up in this experiment?

D.D. ELEY: Professor FRÖHLICH in his paper [FRÖHLICH, H.: Int. J. Quantum Chem. 2: 641, (1968)] mentioned conductivity electrons as one possible source of charge displacement for his postulated longitudinal coherent electric vibrations. If I may briefly anticipate my own paper later, we have now been able to estimate the concentrations of these electrons in hydrated serum albumin, which I hope may help in taking this question further.

Proc. 3rd int. Conf. From Theoretical Physics to Biology, Versailles 1971, pp. 54–56
(Karger, Basel 1973)

Discussions after the Lecture of W. Fairbank[1]

O. MAALØE: I might present one question of my own. You said that, "if you could make a change in a short time", etc., – did you state how short the short time would have to be?

FAIRBANK: It would be possible to make a measurement in a microsecond. If one could take a living system, say 1 mm in dia. and 100 μ thick and make some change in it and see if there was any change at all in the radiation that was coming out, it seems to me that that would be the most sensitive way to tell whether something was or, was not depending on some living process.

O. MAALØE: There are ways in which you might achieve that. Whether you can do it in practice and under the conditions imposed by the measurements, I don't know. To take a very crude example, if you have bacteria growing with a carbon source that requires complete oxidation, and if you flush such a system with nitrogen all metabolic activities stop very completely. No energy flow can be sustained in the system. By switching from nitrogen to atmospheric air, the system can be turned on rather quickly, but I'm not prepared to say exactly how long it takes – it might be several seconds.

KASHA: I am an experimental physical chemist and I work with electronic properties of molecules, molecular aggregates, and crystals. The discussion I have been listening to sounds unreal, like defining angels' heartbeats.

It would seem to me that to learn how to detect collective modes in molecular systems, the analogous phenomena should be compared for the known cases, such as vibrational modes in molecular solids, for which there is an extensive literature. The frequency range CARERI investigated is exactly the range in which molecular crystal lattice modes are found. The frequency range of 10 to perhaps 200 wavenumbers is one characteristic; the other is a very low oscillator strength.

The very low oscillator strength certainly excludes such modes from creating much of an intermolecular force, since the van der Waals dipole-dipole coupling would be insignificant. This

1 Manuscript not submitted.

was the criticism of the earlier work of JEHLE, H. [YOS, J. M.; BADE, W. L. and JEHLE, H.: Molecular structure and biological specificity: Pub. No. 2, pp. 28–60, American Institute of Biological Sciences, 1957; JEHLE, H.: Proc. nat. Acad. Sci. *43:* 847–855, 1957] who also sought protein surface specificity in antigen-antibody interactions through surface vibrational mapping.

Admittedly, the "membrane modes" described by FRÖHLICH involve charged surfaces of molecular ensembles in collective vibrational modes. But since $1/r^6$ forces are involved, it seems of primary importance to establish the *strength* of any such intermolecular interaction by estimation of the oscillator strength. I would consider the specific frequency of such collective modes as of secondary importance.

H. FRÖHLICH: I think there is a misunderstanding. But I'm afraid it's too late now to talk about this, we can do it afterwards.

A. KATCHALSKY: I wonder why, in order to study the properties and behavior of cellular membranes, you consider whole cells. A whole cell is not only complicated from a biochemical point of view, but comprises a wealth of intracellular membranes, such as those of the mitochondria and of the intracellular reticulum. Since methods are available for the preparation of rather pure protoplasmatic membranes such as those of the red blood cell ghosts, the membrane of the perfused nerve axon, it seems to be advisable to study these preparations for comparison with your theoretical predictions.

KOSHLAND: From what I understood of Dr. FRÖHLICH's theory this morning you can use one enzyme molecule to detect the predicted behavior in another. The lag period of one enzyme activating another should depend on the concentration of enzyme. If you diluted an enzyme, eventually your lag periods would lengthen so you could see them. Is that a possible experiment?

FASELLA: Well, this type of experiment has actually been suggested in a letter published by Dr. FRÖHLICH in "Nature". We have tried to carry it out by studying, in several enzyme systems, the pre-steady state reaction progress. We have investigated several enzyme systems at various concentrations ranging from 10^{-9} to the 10^{-4} molar enzyme and we haven't found anything which could not be accounted for in terms of classical enzyme kinetics. It is possible, however, that we just haven't found the right conditions or the right systems; it is also possible that such phenomena, as those forseen by FRÖHLICH, do occur but only during the very early phases of the interaction between the substrates and the enzyme, i.e. when only a very low percent of the enzyme molecules have met the substrates. In this case they would not have been detected in our experiments.

O. MAALØE: Are there further remarks on this point? If I were to sum up this discussion from a point of view of biology I would say that now we have heard both the prosecution and the defense at some length and that perhaps no verdict should be pronounced. It would certainly be interesting to raise these problems again in a meeting of this kind, provided some clear observations on biological material had been made to suggest that the kind of phenomena we've been discussing play even a minor part in biological activity. This is the view of at least one biologist in this group.

LONGUET-HIGGINS: I think there is no reason whatever to doubt that the mechanism of enzyme reaction will yield to generalization of existing chemical principles. The extremely extensive and thoroughly interesting work that has been done on enzyme reactions, and about which we've hardly heard any reference today, I think confirms one's views that the question does essentially make sense in terms of the ordinary theory of chemical reactions. The function of the enzyme seems to be to lower the free energy of the transition state in the chemical reaction, and one can understand the specificity very well in those terms. I don't think any enzyme chemist . . . and I am prepared to be corrected – would doubt that it's simply a matter of being rather complicated. I don't think that anyone who works in the field suspects that there are basically new principles, or forces which need to be invoked in order to account for these phenomena. I say this simply because it would certainly be news to me if all biologists think we need some new forces and some new concepts in that particular area.

Proc. 3rd int. Conf. From Theoretical Physics to Biology, Versailles 1971, pp. 57–88
(Karger, Basel 1973)

Thermodynamics of Bio-Networks[1]

A. KATCHALSKY[2]

1. Introduction

1.1. The application of classical thermodynamics to biological systems
is restricted to equilibrium processes and to the assignment of energetic
and entropic quantifiers to the chemical components of cells [2]. As a theory
of equilibria, it played only a subordinate role in the development of the
biological outlook on the dynamics of living beings. During the last decade
the newly developed thermodynamics of irreversible processes [3, 3a]
exerted a strong influence on the quantitative treatment of certain physiological processes. Its primary impact was on the correlation of coupled
permeation flows across biomembranes, but it helped in clarifying the
phenomenological nature of active transport and of metabolic cycles [4].

The usefulness of nonequilibrium thermodynamics in biophysics is,
however, limited by the linearity of its phenomenological relations between
thermodynamic flows and forces. Most biological processes are nonlinear
and sufficiently far from equilibrium as to prevent linearization around a
state of equilibrium. Moreover, all living entities are extremely complex
and heterogeneous and require a suitable device to 'hook up' the non-
steady thermodynamic flows and forces. Such a device is provided by the
network representation which allows the treatment of non-steady, nonlinear
flows and forces operating in reticulated systems of a high complexity.

---

1   This paper is an abbreviated and simplified version of the more extensive work of
G. OSTER, A. PERELSON and A. KATCHALSKY [1].
2   Prof. Aharon Katzir-Katchalsky fell victim to the attack on innocent passengers by
terrorists on May 30, 1972, at Lod Airport

1.2. Although a modern network theory stems from the analysis of electrical circuits, it has developed into a general tool for the abstract description of the organization of physical systems. It is no more a graphical representation of technical devices, but has become a conceptual method for the visualization of mathematical structure and topological connectivity [5].

Indeed, twenty years ago KRON [6] has shown that practically all field equations of physics, including those of SCHROEDINGER, of MAXWELL and of NAVIER-STOKES, could be represented by network analogues, and more recently, ROTH [7] and BRANIN [8] have shown that the operational structures of linear graph theory and vector calculus are identical.

The aim of the present paper is to show how the methods of network analysis can be adapted to the thermodynamics of flow, and in particular to systems comprising chemical processes. Chemical reactions are naturally discrete, their topology being contained in the stoichiometric matrix, and are, therefore, eminently suitable for network representation. It has, however, been found by OSTER, AUSLANDER, PERELSON [1, 9] and myself that the conventional way of network representation, the well-known method of linear graphs, is inconvenient for the description of thermodynamic systems. Fortunately, there exists an alternative notation, developed by PAYNTER and his colleagues [10, 10a], denoted as the 'bond graph' representation, which admirably serves our purpose. In this representation, the flow structure of complex systems closely resembles the bond notation of organic chemistry, and provides an intuitive grasp of the system-organization as a whole.

1.3. To introduce this new branch of thermodynamics we shall not proceed deductively from postulates and principles, but will acquire the method through the study of some elementary examples. At a later stage we shall consider general principles of symmetry and laws of evolution, which govern the behavior of thermodynamic networks.

2. Topological Graphs

2.1. Our first example will be a mechanical device composed of two masses, two springs and a dashpot (fig. 1a). To monitor continuously the state of the system, hypothetical instruments which measure the relative velocities, are applied to each element. The points of the attachment a, b, c, d, e, f, 1, 2 and g, are the 'nodes' for insertion of the instrument terminals.

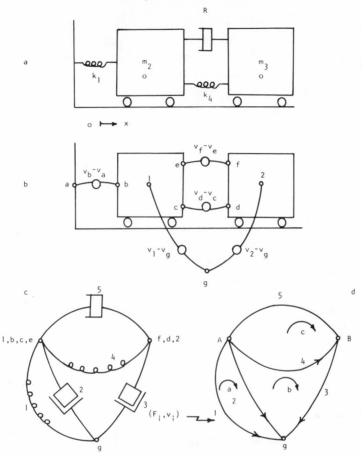

Fig. 1. Topological graph for a mechanical system. The KCL requirement for the system gives

$F_1 + F_2 + F_4 + F_5 = 0$ for node A $\qquad$ $F_4 + F_5 - F_3 = 0$ for node B

which is equivalent to the equations

$k_1 x_A + m_2 \ddot{x}_A + k_4 (x_B - x_A) + b(\dot{x}_B - \dot{x}_A) = 0$

$k_4 (x_B - x_A) + b(\dot{x}_B - \dot{x}_A) - m_3 \ddot{x}_B = 0$

It will be observed that in addition to the relative element velocities, (v_f-v_e and v_d-v_c), velocities are measured versus an external reference node: the ground velocity v_g (or v_a) in figure 1b. Since the nodes 1, b, c and e have the same velocity, they may be combined to a single node (A in fig. 1d) and similarly f, d and 2, as well as a and g, may be united to single nodes (B and g respectively, in fig. 1d). The branches connecting the nodes in figure 1c still carry symbols representing their mechanical properties. Proceeding, however, with the process of abstraction, these symbols may be deleted, and a linear graph remains which demonstrates the topological connectivity between the elements (fig. 1d).

2.2. The same procedure may now be applied to a non-mechanical system such as a membrane, through which a non-electrolyte flow takes place. Our first step will be the consideration of steady flow, depicted in figure 2.

The elements of the membrane-system comprise two reservoirs, 1 and 2, and the membrane itself, m. It is assumed that both reservoirs contain the same permeant at different levels of the chemical potential (μ_1 and μ_2), and the characterization of each reservoir is made by a hypothetical measuring instrument which determines μ versus a reference potential $\bar{\mu}$. The flow of permeant 'through' the membrane (J_m) is a function of the potential difference ($\mu_1 - \mu_2$) 'across' the membrane.

The flow into the membrane is supplied by reservoir 1, acting as a capacitor which is discharged through a resistor. If the 'charge' of this reservoir is identified with the number of moles of permeant n_1 then the *reversible* discharge flow of capacitor 1 is

$$-\frac{dn_1}{dt} = -J_1 \tag{1}$$

which is evidently equal to the membrane flow, or

$$-J_1 = J_m. \tag{2}$$

The output of the membrane charges reservoir 2 in a reversible manner, and the accumulation flow of the second capacitor is

$$\frac{dn_2}{dt} = J_2, \tag{3}$$

where

$$J_2 = J_m. \tag{4}$$

To complete the characterization of the reservoirs it is advisable to assign to each container a chemical capacity

$$C_i = \frac{dn_i}{d\mu_i}. \tag{5}$$

Similar to the differential capacity of electrical devices

$$C = \frac{dq}{dV}.$$

Since an ideal chemical capacitor is reversible, its capacity is according to the laws of thermodynamics positive definite, or

$$C_i \geqslant 0. \tag{6}$$

The definition of C_i leads to a conversion of eqs. (1) and (3) into constitutive relations for the capacitive elements, namely

$$\frac{dn_i}{dt} = \frac{dn_i}{d\mu_i}\frac{d\mu_i}{dt} = C_i\frac{d\mu_i}{dt} = J_i \tag{7}$$

In the same vein the membrane itself behaves like an ideal resistor, which does not lose or accumulate matter, but transmits it in a dissipative process. The constitutive equation attributed to a steady-state membrane relates the thermodynamic force $\Delta\mu = \mu_1 - \mu_2$ to the conjugate flow J_m through a resistive function R_m, so that

$$\frac{\partial\Delta\mu}{\partial J_m} = R_m. \tag{8}$$

In linear cases R_m is independent of J_m, so that eq. (8) reduces to the form used in nonequilibrium thermodynamics

$$\Delta\mu = R_m \cdot J_m; \tag{9}$$

generally, however,

$$\Delta\mu = R(J_m). \tag{10}$$

2.3. It is now possible to proceed with the construction of the topological graph for membrane permeability: the graph is constructed from the

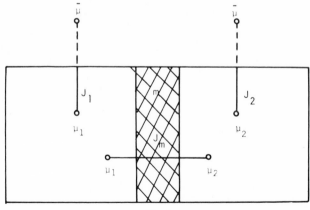

Fig. 2.

branches, representing the elements, each branch being characterized by two terminal potentials and a single flow (c.f. fig. 2). Since the terminals in each reservoir are equal, they may be combined to single nodes, and since the reference potential for the permeant is the same in both reservoirs, the μ values can also be united to one node—hence figure 2 is directly converted into the linear graph of figure 3.

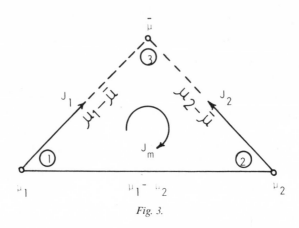

Fig. 3.

Note that we have assigned a single potential value to nodes 1, 2 and 3. This is equivalent to the assumption that the network nodes can be characterized by single valued thermodynamic parameters. This assumption is equivalent to the postulate of local equilibrium underlying the formal structure of the thermodynamics of irreversible processes.

It is useful to assign directions to the branches of a linear graph, as is shown, for example, in figure 3. If a direction is attributed also to the network loop (clockwise in figure 3), then every force following the direction of the loop is regarded as positive, while the opposite direction is considered negative. As expected, the forces around a loop sum up to zero, i.e., $(\mu_1 - \bar{\mu})$ $- (\mu_2 - \bar{\mu}) - (\mu_1 - \mu_2) = 0$, which is a trivial case of Kirchhoff's voltage law (KVL). From a more fundamental point of view, it is shown that the postulate of local equilibrium, or uniqueness of potential at nodes, is equivalent in network topology to KVL.

A further convention is to regard a flow entering a node as negative, while exiting flows are taken to be positive. Thus, eqs. (2) and (4) may be regarded as sums of flows over nodes 1 and 2, and may be written as $J_1 + J_m = 0$ and $-J_m + J_2 = 0$. These are, however, elementary examples of

Kirchhoff's current law (KCL), which is a topological constraint on the structure of thermodynamic networks. Kirchhoff's current law is another expression for laws of conservation, and its incorporation into thermodynamic networks imposes the requirement that the flows chosen by us be conservative. laws of conservation, and its incorporation into thermodynamic networks imposes the requirement that the flows chosen by us be conservative.

2.4. The approach of the previous paragraph can easily be applied also to non-stationary cases, such as a membrane which accumulates or loses permeant until a steady state is reached. In this case the membrane is no longer a pure resistor, and its description would require the consideration of both capacitative and resistive elements. To be sure, both the membrane capacitor and resistor are located in the same place; nevertheless, they can be distinguished conceptually and analyzed as different topological elements. It is further noteworthy that these elements may vary from point to point, so that the membrane should be treated as a series of resistors and capacitors—their number being dictated by the precision of our calculations. For the sake of simplicity we shall consider here a single volume element—composed of an input and an output resistor and a membrane capacitor—inserted between two external reservoirs.

The linear graph for figure 4, which obeys both KVL and KCL, is given in figure 5.

It is worth mentioning that Kirchhoff's current law at the intramembrane node (μ_m) corresponds to $-J_m^1 + J_m + J_m^2 = 0$ or $J_m = J_m^1 - J_m^2$, which represents the 'filling flow' of the membrane. When a steady-state is reached, J_m 0, (or $J_m^1 = J_m^2$ and $J_1 = J_2$), and figure 5 reduces to figure 3.

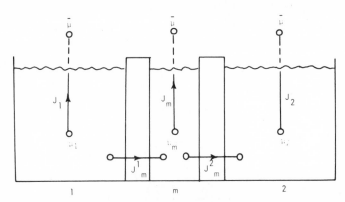

Fig. 4. Assignment of linear graph elements to transport across a membrane.

3. Matrix Representation, Tellegen's Theorem and the Evolution Principle of PRIGOGINE and GLANSDORF [11]

3.1. A major advantage of linear graphs is the ease of their translation into matrices, which provide useful algorithms for computer calculation. Matrix representation is widely used in modern systems approach and need not be considered here; the conditions imposed on the matrices assigned to our networks are, however, that they obey the topological constraints and correspond to physical reality.

The simplest matrix is that of nodes to branches, which represents adequately the connectivity of the network. Every term of the matrix a_{ij} is given the value $+1$ if the j'th branch is directed *out* of the i'th node; -1 if the j'th branch is directed into the i'th node, and 0 if the j'th branch is not incident into the i'th node. To demonstrate the construction of a node to branch matrix A, let us redraw figure 5 in the simplified form of figure 6, which comprises four nodes and five branches. A direct inspection of figure 6 leads to the following matrix:

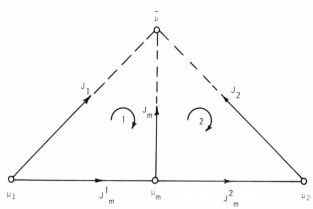

Fig. 5. Topological graph representation of nonelectrolyte transport across a simple membrane.

Branches

		1	2	3	4	5
	1	1	1	0	0	0
Nodes 2		0	-1	1	1	0
	3	0	0	0	-1	1
	4	-1	0	-1	0	-1

$A =$ ... (11)

It can be readily shown how matrix A may represent KCL in the network of figure 6: let us assign a flow J_i to each i'th branch and construct a general flow vector $\vec{J}$, the components of which are all the flows in the network, i.e.

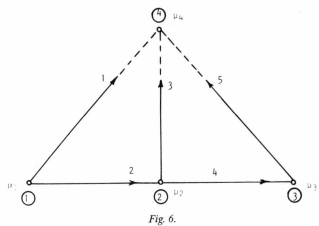

Fig. 6.

$$\vec{J} = \begin{matrix} J_1 \\ J_2 \\ J_3 \\ J_4 \\ J_5 \end{matrix} \qquad (12)$$

Upon multiplying A with $\vec{J}$ we find that

$$A \cdot \vec{J} = \begin{matrix} J_1 + J_2 \\ -J_2 + J_3 + J_4 \\ -J_4 + J_5 \\ -J_1 - J_3 - J_5 \end{matrix} \qquad (13)$$

By Kirchhoff's current law applied to each node, each term in the matrix product of eq. (13) is, however, zero. Hence the general expression for KCL in matricial form is

$$A \cdot \vec{J} = 0 \qquad (14)$$

3.2. Another interesting insight into network properties is gained by considering the transpose of matrix A, namely

$$A^{\mathrm{T}} = \begin{matrix} 1 & 0 & 0 & -1 \\ 1 & -1 & 0 & 0 \\ 0 & 1 & 0 & -1 \\ 0 & 1 & -1 & 0 \\ 0 & 0 & 1 & -1 \end{matrix}$$

If we construct another vector, $\vec{\mu}$, the components of which are the chemical potentials assigned to each node,

$$\vec{\mu} = \begin{matrix} \mu_1 \\ \mu_2 \\ \mu_3 \\ \mu_4 \end{matrix}$$

and multiply A^{T} by $\vec{\mu}$, the following result is obtained:

$$A^{\mathrm{T}} \cdot \vec{\mu} = \begin{matrix} \mu_1 - \mu_4 \\ \mu_1 - \mu_2 \\ \mu_2 - \mu_4 \\ \mu_2 - \mu_3 \\ \mu_3 - \mu_4 \end{matrix} = \vec{X} \tag{15}$$

The components of the new vector, $\vec{X}$, are the forces operating across each of the branches of figure 6. Thus, $\mu_1 - \mu_4 = X_1$ is the force across branch 1, $\mu_1 - \mu_2 = X_2$ is that across branch 2, $\mu_2 - \mu_4 = X_3$ across branch 3, etc. Thus, the vector

$$\vec{X} = \begin{matrix} X_1 \\ X_2 \\ X_3 \\ X_4 \\ X_5 \end{matrix}$$

is a the vector of all forces in the network.

Now that we have characterized the system by a general vector of flow, $\vec{J}$, and a general vector of forces, $\vec{X}$, it is of interest to derive some relations between these vectors and find out their consequences for network behavior. An important conclusion about the $\vec{J}$–$\vec{X}$ relation is obtained by considering the scalar product of their matrix representation

$$
J^t X = \quad (J_1 J_2 \ldots J_5) \quad \begin{matrix} X_1 \\ X_2 \\ 1 \\ 1 \\ 1 \\ X_5 \end{matrix} \quad = \Sigma J_i X_i \qquad (16)
$$

Inserting $\vec{X}$ from eq (15), we find that $\vec{J}^T \vec{X} = \vec{J}^T \cdot A^T \cdot \vec{\mu}$, but according to the rules of matrix multiplication, $\vec{J}^T \cdot A^T = (A \vec{J})^T$, and since according to eq. (14) $A \cdot \vec{J} = 0$, also $(AJ)^T = 0$, and hence

$$
J^T \cdot X = \Sigma J_i X \gtreqless 0 \qquad (17)
$$

Equation (17) is a fundamental consequence of network theory, and is known as Tellegen's Theorem. As has been shown here, it holds also for thermodynamic networks which obey the topological constraints of KCL and KVL. Although it may be derived in certain cases from energy conservation, the range of its validity is wider, as has been recently shown in an extensive study by PENFIELD [12].

Tellegen's theorem may be interpreted in the following way: a thermodynamic flow system is described fully in a 2n dimensional space, based on n-dimensions of flow and n-conjugate dimensions of force. Equation (17) shows, however, that this general space consists of two *orthogonal* subspaces, the subspace of flow to which the vector of flow is confined, and the subspace of forces—orthogonal to that of the flows. One of the consequences of this consideration is the network equivalent of the evolutionary principle of PRIGOGINE and GLANSDORFF [11], derived recently by OSTER and DESOER [13].

3.3. Let us apply Tellegen's theorem to an evolving network in which the forces are allowed to change with time, so that after an interval dt the forces become

$$
X_i + \frac{dX_i}{dt} dt.
$$

Since the flows and the forces lie in fixed orthogonal subspaces, whatever the variation of X, it is required that

$$
dt \, \Sigma \, J_i \frac{dX_i}{dt} = 0 \text{ or,}
$$

since dt is arbitrary,

$$\Sigma J_i \frac{dX_i}{dt} = 0. \tag{18}$$

The sum in eq. (18) comprises both the reversible and irreversible terms, and may be written as:

$$\overset{irrev}{\underset{i}{\Sigma}} J_i \frac{dX_i}{dt} + \overset{rev}{\underset{i}{\Sigma}} J_i \frac{dX_i}{dt} = 0 \tag{19}$$

or

$$\overset{irrev}{\underset{i}{\Sigma}} J_i \frac{dX_i}{dt} = - \overset{rev}{\underset{i}{\Sigma}} J_i \frac{dX_i}{dt}$$

Now the reversible elements obey in general constitutive relations of the type represented by eq. (7),

$$\text{or } C_i \frac{dX_i}{dt} = J_i$$

where $C_i \geqslant 0$. Hence,

$$J_i \frac{dX_i}{dt} = \frac{J_i^2}{C_i} \geqslant 0,$$

and the sum on the right hand side of eq. (19) is negative definite, i.e.

$$\overset{irrev}{\underset{i}{\Sigma}} J_i \frac{dX_i}{dt} \leqslant 0 \tag{20}$$

Equation (20) is the evolutionary criterium of PRIGOGINE and GLANSDORFF [11] derived from a general consideration of the dissipative function

$$\phi = \overset{irrev}{\underset{i}{\Sigma}} J_i X_i. \tag{21}$$

The change of ϕ with time may be written as

$$\frac{d\phi}{dt} = \overset{irrev}{\underset{i}{\Sigma}} J_i \frac{dX_i}{dt} + \overset{irrev}{\underset{i}{\Sigma}} X_i \frac{dJ_i}{dt} \tag{22}$$

In the range of linear dependence of flows on forces, and with constant phenomenological coefficients,

$$\Sigma J_i \frac{dX_i}{dt} = \Sigma X_i \frac{dJ_i}{dt},$$

so that

$$1 \frac{d\phi}{dt} = 2 \, \Sigma J_i \frac{dX_i}{dt},$$

and with eq. (20)

$$\frac{d\phi}{dt} < 0 \tag{23}$$

which is the celebrated principle of minimum dissipation proposed by PRIGOGINE [14].

Equation (23) holds, however, only in linear cases which are rather rare in biology. On the other hand, eq. (20) has a much wider range of validity, and predicts correctly the evolution of all biological networks which can be characterized by thermodynamic descriptors. Equation (20) holds for both linear and nonlinear constitutive relations, and may be used in the realm of quasi-equilibrium as well as for far-from-equilibrium processes.

4. Bond Graphs and Network Reciprocity

4.1. Despite the mathematical elegance of linear graph theory, topological graphs become cumbersome and unwieldy when applied to more complex biophysical systems. Thus, the flow of a single permeant through a double membrane requires already a three-dimensional topological graph, and graphical representation becomes impossible for multiple flows across a complex membrane system. As has been pointed out in the introduction, a more suitable method of representation is provided by the bond graphs of PAYNTER et al. [10, 10a].

The basic element of the bond graph is the 'ideal energy bond' which transmits power instantaneously, without energy loss. All time-dependent processes and all dissipative transformations are localized conceptually in capacitative and resistive elements, respectively, attached to the energy bond. A very useful innovation introduced by the bond graph method are two ideal junctions: a 'parallel' or zero-junction (0-jct), denoted as

$$1 \frac{\text{ideal}}{\text{bonds}} \quad 0 \underset{n}{\overset{2}{\underset{4}{\overset{3}{<}}}} \quad \text{zero junction;}$$

and a 'series' or one-junction (1-jct), denoted as

$$
1 \xrightarrow{\quad \text{ideal} \quad} \overset{2}{\underset{\text{n}}{\left| \overset{3}{\diagup} \atop \diagdown_4 \right.}}
$$

The 0-junction is defined by the condition that all forces or efforts, on all the bonds hooked onto the junction, are equal—or if the efforts are measured as chemical potentials

$$
\mu_1 = \mu_2 = \mu_3 = \ldots = \mu_n. \tag{24}
$$

Since no power accumulates on any of the ideal junctions, the definition of a 0-junction implies that the sum of all the flows over a 0-junction is zero, or

$$
\sum_1^n J_i = 0. \tag{25}
$$

In other words, a 0-junction obeys KCL and is identical with a node of the topological graph.

The 1-junction is defined by the condition that all flows entering or exiting from the junction are equal, or that

$$
J_1 = J_2 = \ldots = J_n. \tag{26}
$$

Equation (26), in addition to the requirement that no power accumulates at a 1-junction, leads to the conclusion that

$$
\sum_1^n \mu_i = 0, \tag{27}
$$

which is equivalent to the statement that a 1-junction obeys Kirchhoff's voltage law, and is indeed a novel feature of the bond graphs.

4.2. With these observations we can readily transform any topological representation, such as figure 5, into a bond graph. It will be recalled that figure 5 depicts in reality a system of three capacitors (two outer reservoirs and one intramembrane capacitance) and two resistors (the input and output resistances of the membrane element) as shown in figure 7a. According to the recipe of the previous paragraphs, the nodes of figure 5 have to be substituted by 0-junctions. More interesting, however, is the representation of the capacitors by C elements and the resistors by R elements hooked onto the bond graph by 1-junctions. If we attach now the chemical potentials and flows to each bond according to the conditions of eqs. (24), (25), (26) and (27), we obtain the bulky, but information-rich picture 7c. Upon 'reading' figure 7c we find, for instance, that the intra-membrane capacitor is filled through the 0-junction by a flow J_m which obeys KCL, so that $J_m = J_m^1 - J_m^2$. The filling is governed by the constitutive

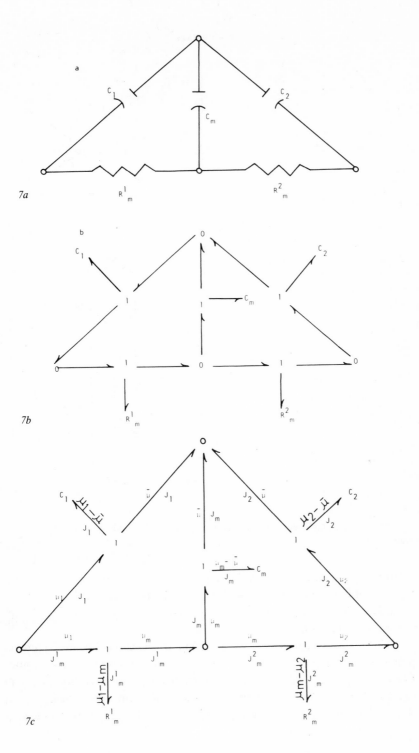

7a

7b

7c

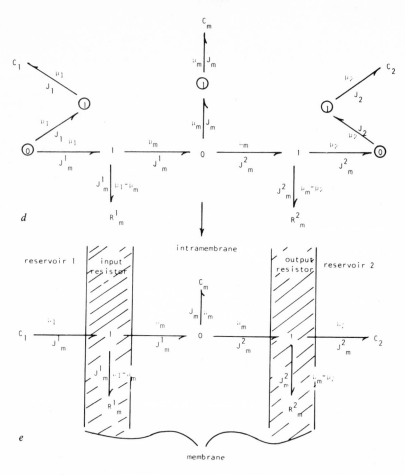

Fig. 7. Bond graph representation of nonelectrolyte transport through a simple membrane.

relation C_m which determines the dependence of the flow on the potential $\mu_m - \bar{\mu}$. Similarly, the flow through the dissipative input resistance (J_m^1) is driven by the force $\mu_1 - \mu_m$ and the flow—force relation is governed by the constitutive function R_m^1.

The graph is considerably simplified if we chose for the reference potential $\bar{\mu}$ a value of zero. In this case, no flow of power can be attributed to the ideal bonds leading to the top 0-junction, and these bonds can be deleted. Opening up the graph, it assumes the form drawn in figure 7d. But it can immediately be recognized that the efforts and flows on each junction which carries only two bonds, are equal for both bonds. Hence, also the junctions marked by ○ can be deleted with no loss of information. Thus, the bond graph condenses finally to the simple and suggestive form of figure 7e. It is noteworthy that in steady flows some of the capacitative flows vanish. Thus, in the case treated in par 2.3 the capacitative flow $J_m = 0$, and the bond graph corresponding to figure 3 assumes the simple form $C_1 - 1 - C_2$.

4.3. The simplification and convenience attained by the use of bond graphs may be seen in the description of the transport of a single permeant through a parallel array of two membranes. The construction of the topological graph for this process should be evident from figure 8.

If all the terminals of equal μ's are united to single nodes, and all the reference potentials $\bar{\mu}$ are taken to be equal, the topological graph becomes the three-dimensional structure depicted in figure 8c.

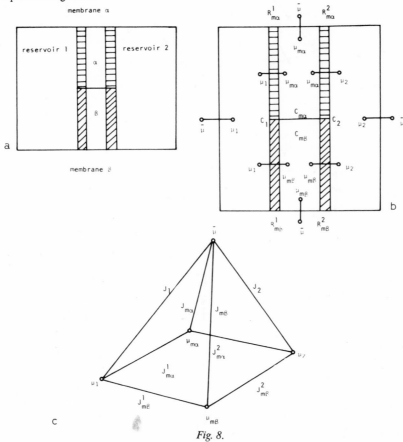

Fig. 8.

The conversion of figure 8c to a bond graph is carried out in the same way as described above: the nodes are converted into 0-junctions, while the C and R elements are hooked on through 1-junctions to the pertinent bonds (fig. 9a). The reference potential is now set as zero ($\mu = 0$), the graph opened and the redundant junctions deleted to give the lucid representation which closely depicts the physical connectivity (fig. 9b).

Finally, the graph reduces to an extremely simple form in a steady-state when the intramembrane capacitors $C_{m\alpha}$ and $C_{m\beta}$ do not fill or empty during the permeation process. In this case, the bonds leading to $C_{m\alpha}$ and $C_{m\beta}$ may be deleted, and the input and output resistances clamped to single resistors for the α and β components of the parallel membrane array. The result is shown in figure 9c.

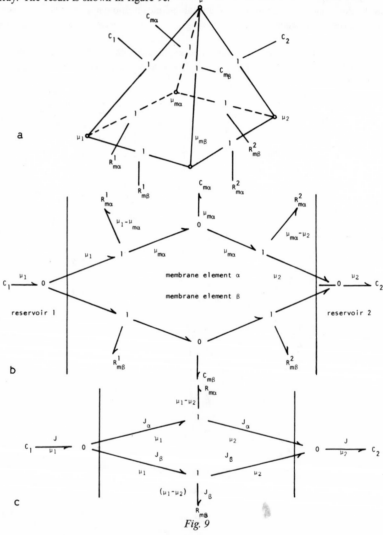

Fig. 9

4.4. The power of the bond graph representation will be demonstrated in the analysis of the symmetry properties of coupled, nonlinear, permeation flows. This analysis will serve as an example for the general statements on the reciprocity of thermodynamic networks.

Consider a single membrane through which a stationary transport of substances A and B takes place. Following the method discussed in the previous paragraph, the flows may be represented by the bond graph shown in figure 10a.

If, however, the flows of A and of B are coupled hydrodynamically, a coupling element (CPL) should be hooked onto the 1-junctions of both flows.

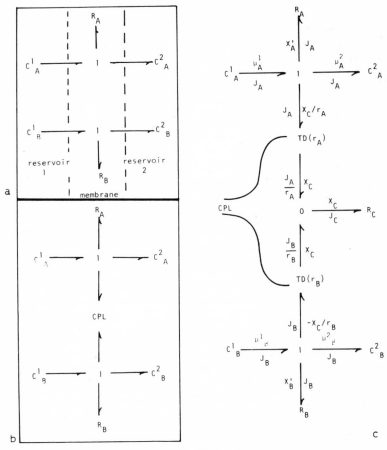

Fig. 10.

The attachment of CPL to the 1-junctions is based on the fact that the flow out of these junctions into CPL is the same as the flow into the 1-junction. Moreover, the hooking up onto 1-junctions implies that coupling exists only as long as flow continues.

It is now our intention to open the 'black box' of CPL and write its constituents in an explicit form which would permit a quantitative evaluation of the coupling parameters. The starting point of our consideration is that coupling is dissipative so that CPL must comprise a resistor R_C. Since the dissipation in CPL depends on the relative flow of A versus B, the resistor R_C should be attached through a zero junction. It is further clear that not all the flows J_A and J_B participate in the coupling dissipation, but that only an 'interacting' fraction is involved in the process. This means that a *transduction*, which converts the power input into flow and force values adapted to the CPL requirements, has to be considered. This gives us an opportunity to introduce another all-important element of network thermodynamics—the linear *transducer* TD depicted in bond graph formation by

$$\frac{X_1}{J_1} \text{ TD } \frac{X_2}{J_2}$$

The transducer converts energy from one form to another but conserves power, so that

$$X_1 J_1 = X_2 J_2 \tag{28}$$

Its operation is characterized by a modulus r, which may be a function of the parameters of state, such as temperature or concentration, but is independent of flows and forces.

The scaling of flows and forces by the transducer gives:

$$X_2 = rX_1 \text{ and } J_2 = \frac{J_1}{r} \tag{29}$$

so that the power remains the same after passing TD.

The detailed bond graph for the coupled flows may now be written in the form of figure 10c.

The read out of figure 10c leads to the following observations: the coupling element CPL is flanked by two transducers which convert the flows of A and of B into

$$\frac{J_A}{r_A} \text{ and } \frac{J_B}{r_B}$$

respectively. These two flows combine at the zero junction of CPL to the coupled flow J_C given by

$$J_C = \frac{J_A}{r_A} + \frac{J_B}{r_B} \tag{30}$$

The relation between the force X_C and the flow J_C, on the coupling dissipative element need not be specified, and may be written in a general way

$$X_C = R_C(J_C) \tag{31}$$

In a similar manner, a nonlinear relation will be assumed for the constitutive relations of the dissipative elements R_A and R_B

i.e., $X'_A = R_A(J_A)$ and $X'_B = R_B(J_B)$ \hfill (32)

Upon summing up the efforts of the A and B flows around the 1-junctions, we now obtain, according to KVL

$$-\mu_A^1 + X'_A + \mu_A^2 + \frac{1}{r_A}X_C = 0$$

which gives upon inserting eqs. (31), (32) and (29)

$$\mu_A - \mu_A^2 = \Delta\mu_A = R_A(J_A) + \frac{1}{r_A}R_C(J_C) \tag{33}$$

and similarly

$$-\mu_B^1 + X'_B + \mu_B^2 + \frac{1}{r_B}X_C = 0$$

$$\Delta\mu_B = R_B(J_B) + \frac{1}{r_B}R_C(J_C) \tag{34}$$

4.5. Equations (33) and (34) are the nonlinear phenomenological relations between the external driving forces of permeation flow ($\Delta\mu_A$ and $\Delta\mu_B$) and the conjugate macroscopic flows (J_A and J_B). It is well known, that the most important property of the linear phenomenological equations is the symmetry of the matrix of the Onsager coefficients [15]. Since in nonlinear cases phenomenological coefficients do not exist, symmetry has to be tested by more general criteria. The mathematical procedure is based on the consideration of the symmetry of the Jacobian of forces versus flows

2   *Note:* In a model calculation discussed in ref. 1 $r_A = c_A$ and $r_B = -c_B$. Since the flows are $J_A = c_A\vec{V}_A$ and $J_B = c_B\vec{V}_B$; $J_C = \vec{V}_A - \vec{V}_B$, i.e., it is the relative velocity of A and B.

$$J = \frac{\partial (\Delta\mu_A, \Delta\mu_B)}{\partial (J_A, J_B)} = J^T, \tag{35}$$

which in our case would mean that if

$$\left(\frac{\partial \Delta\mu_A}{\partial J_B}\right)_{J_A} = \left(\frac{\partial \Delta\mu_B}{\partial J_A}\right)_{J_B} \tag{36}$$

the system is reciprocal. To be sure, the symmetry of Onsager's matrix is a special case of eq. (35).

The validity of eq. (36) can be readily shown for the coupled permeation flows analyzed above. Since $R_A(J_A)$ and r_A are independent of J_B

$$\left(\frac{\partial \Delta\mu_A}{\partial J_B}\right)_{J_A} = \frac{1}{r_A} \frac{\partial R_C}{\partial J_B} = \frac{1}{r_A} \frac{dR_C}{dJ_C} \left(\frac{\partial J_C}{\partial J_B}\right)_{J_A}$$

But according to eq. (29)

$$\frac{\partial J_C}{\partial J_B} = \frac{1}{r_B},$$

so that

$$\left(\frac{\partial \Delta\mu_A}{\partial J_B}\right)_{J_A} = \frac{1}{r_A r_B} \frac{dR_C}{dJ_C} \tag{37}$$

In an equal manner

$$\left(\frac{\partial \Delta\mu_B}{\partial J_A}\right)_{J_B} = \frac{1}{r_B r_A} \frac{dR_C}{dJ_C}$$

so that

$$\left(\frac{\partial \Delta\mu_A}{\partial J_B}\right)_{J_A} = \left(\frac{\partial \Delta\mu_B}{\partial J_A}\right)_{J_B} \text{ q.e.d.}$$

The workers in membrane permeability are familiar with the fact that the symmetry of coupled phenomena holds in a wider range than expected from Onsager's treatment of quasiequilibrium processes. In the light of the present analysis, it is suspected that the reciprocity observed represents in many cases the network structure of the investigated systems.

A general treatment of network reciprocity has been recently presented by BRAYTON [16]. BRAYTON has demonstrated that a network composed of reciprocal elements is reciprocal whatever be the interconnection between the elements within the network. Since the majority of the elements of thermodynamic nature are reciprocal, BRAYTON's work indicates that reciprocity is expected in most systems of biophysical interest, and hence, coupled flows treated above may serve as an example for the general behavior of biological networks.

5. Chemical Processes

5.1. This paper started by stressing the importance of chemical processes in all living processes, and the difficulty of incorporating them into a linear thermodynamics of irreversible processes. Moreover, it was mentioned that the stoichiometry of chemical reactions makes them naturally reticulate and easily adapted to a network formalism. Before proceeding, however, with the analysis of chemical processes, we had to acquaint ourselves with thermodynamic topology and realize that despite geometrical unity a reaction can be separated topologically into different elements, the conceptual connectivity of which is the object of this study. The conventional approach assumes that a chemical process is purely dissipative and can be characterized by a flow of reaction J_r. In the present treatment it is imperative that capacitative attributes be assigned to the components of the reaction and the relation between these capacitors and the dissipative elements will lead to the organizational pattern of the reaction system.

Let us start with the definition of some common descriptors of chemical change: in the general reaction scheme

$$v_A A + v_B B + v_C C + \ldots \underset{k_r}{\overset{k_f}{\rightleftharpoons}} v_D D + v_E E + \ldots$$

the stoichiometric coefficients v_i are negative for the reactants and positive for the reaction products. The rate of change of the i'th component

$$\frac{dn_i}{dt} = J_i \tag{38}$$

is in reality a reversible flow, measuring the filling up or emptying of a conceptual capacitor of the i'th substance. It is clear, however, that if no diffusion takes place in the system,

$$\frac{dn_i}{dt}$$

is related to the irreversible flow of reaction J_r. The rate of reaction is generally measured as the time derivative of the advancement of the process

$$\frac{d\xi}{dt}$$

where

$$\frac{d\xi}{dt} = \frac{1}{v_1}\frac{dn_1}{dt} = \frac{1}{v_2}\frac{dn_2}{dt}$$

or in general

$$\frac{d\xi}{dt} = \frac{1}{v_i}\frac{dn_i}{dt} \tag{39}$$

Equations (38) and (39) provide the transform of the capacitive to the resistive flow, namely

$$J_i = v_i J_r. \tag{40}$$

Thus for example, in the simple reaction

$$2A \underset{k_r}{\overset{k_f}{\rightleftharpoons}} 3B,$$

$$\frac{dn_A}{dt} = J_A = -2(k_f c_A^2 - k_r c_B^3) = -2J_r,$$

while

$$\frac{dn_B}{dt} = J_B = 3(k_f c_A^2 - k_r c_B^3) = 3J_r.$$

The fact that for all components

$$\frac{J_i}{v_i} = J_r$$

has the same value, means that in bond graph notation all their contributions will center around a 1-junction. This statement is amplified by the recognition that the driving force of chemical processes is the affinity

$$A = -\Sigma v_i \mu_i \tag{41}$$

Thus, it will be clear that the bonds leading to the central 1-junction of the reaction should be characterized by effort quantifiers equal to $v_i\mu_i$ (fig. 11a).

Since a 1-junction obeys KVL, $v_1\mu_1 + v_2\mu_2 + \ldots + v_n\mu_n + A = 0$, or $A = -\Sigma v_1\mu_1$, as expected. The resistor function of the reaction

$$R_r = \frac{\partial A}{\partial J_r} \tag{42}$$

is generally nonlinear and approaches a constant value only close to equilibrium.

Figure 11a does not comprise the capacitative elements of the components which provide material for the reaction or take up its products. These elements have evidently the form

$$C_i \frac{\mu_i}{J_i}$$

and are related to the bond of figure 11a through the transducer equation (42). Noting that the transducer modulus in the present case is the stoichiometric coefficient, we may supplement each branch by the connections of figure 11b, and thus obtain the complete bond graph of a chemical process (fig. 11c) and the reaction $2A + B \rightleftharpoons 3C$, for example, can be written as figure 11d.

5.2. It is noteworthy that a relatively simple consideration of the bond graphs 11b and 11c provides a quantitative expression for the relaxation time of a chemical reaction. The constitutive relation for the capacitative element C_i is as before

$$C_i \frac{d\mu_i}{dt} = J_i$$

but inserting the relation (40),

$$\frac{d\mu_i}{dt} = \frac{v_i}{C_i} J_r. \tag{43}$$

Multiplying eq. (43) by $-v_i$, and summing over i,

$$- \sum_i v_i \frac{d\mu_i}{dt} = -(\sum \frac{v_i^2}{C_i}) J_r.$$

The sum

$$- \sum v_i \frac{d\mu_i}{dt}$$

however, is, equal to

$$\frac{dA}{dt}$$

by the very definition of the affinity (eq. 41), so that

$$\frac{dA}{dt} = -(\sum \frac{v_i^2}{C_i}) J_r \tag{44}$$

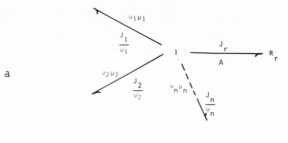

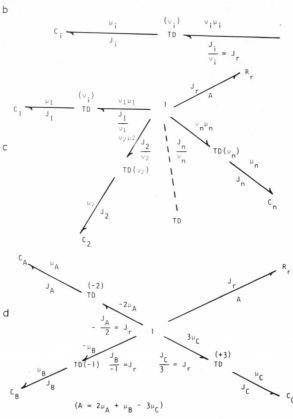

Fig. 11.

Since the affinity changes here only by the progress of the reaction we may state that:

$$\frac{dA}{dt} = \frac{dA}{dJ_r}\frac{dJ_r}{dt} \tag{45}$$

or inserting R_r from eq. (42)

$$R_r \frac{dJ_r}{dt} = -\left(\sum \frac{v_i^2}{C_i}\right) J_r$$

so that

$$\frac{dJ_r}{dt} = -\left(\frac{1}{R_r}\sum \frac{v_i^2}{C_i}\right) J_r. \qquad (46)$$

Equation (46) has, however, a typical relaxation form and may be rewritten as

$$\frac{dJ_r}{dt} = -\frac{J_r}{\tau_r}$$

where the relaxation time τ_r is given by

$$\frac{1}{\tau} = \frac{1}{R_r}\sum \frac{v_i^2}{C_i} \qquad (47)$$

It is rather interesting that also in the chemical case, the network approach casts the relaxation time in the form of an RC product, well known from electrical circuitry.

In simple and ideal cases, eq. (47) reduces to well-known expressions, as may be shown for the case of a monomolecular transformation

$$A \underset{k_r}{\overset{k_f}{\rightleftharpoons}} B.$$

Here,

$$C_i \frac{\partial n_i}{\partial \mu_i} = \frac{n_i}{RT} = \frac{c_i V}{RT},$$

where c_i is the concentration and V the volume of the reaction vessel. Assuming $V = 1 \text{ cm}^2$ and noting that $v_A^2 = (-1)^2 = 1$ and $v_B^2 = (1)^2 = 1$, we obtain close to equilibrium

$$\sum \frac{v_i^2}{C_i} = RT\left(\frac{1}{\bar{c}_A} + \frac{1}{\bar{c}_B}\right),$$

where $\bar{c}_A$ and $\bar{c}_B$ are the concentrations at equilibrium. It can be readily shown that the linear resistance R_r close to equilibrium is

$$R_r = \frac{RT}{k_f \bar{c}_A} = \frac{RT}{k_r \bar{c}_B}.$$

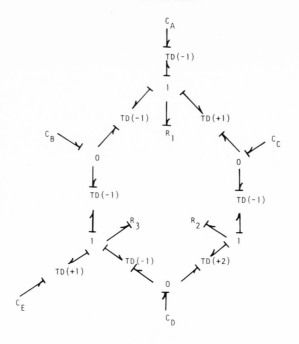

Fig. 12. Bond graph representation of a system of coupled reactions

$A + B \rightleftharpoons C$
$\quad C \quad \rightleftharpoons 2D$
$D + B \rightleftharpoons E$

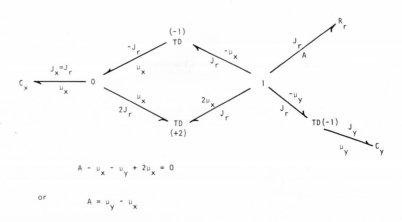

$$A - \mu_x - \mu_y + 2\mu_x = 0$$

or

$$A = \mu_y - \mu_x$$

Fig. 13. Bond graph representation of an autocatalytic reaction $x + y \rightarrow 2x$

Inserting the last couple of expressions into eq. (47), we find that

$$\frac{1}{\tau_r} = k_f + k_r, \text{ or } \tau_r = \frac{1}{k_f + k_r},$$

as expected. The general eq. (47) provides, however, information about any reaction in the nonlinear range and far from equilibrium.

5.3. The bond graph representation can be readily extended to comprise also multiple coupled reactions. Thus, if the i'th substance participates in several chemical reactions, its change in the k'th process will be given by

$$\left(\frac{dn_i}{dt}\right)_k = v_{ik} J_r^k, \tag{48}$$

where v_{ik} is its stoichiometric coefficient in this process. The total chemical transformation of the i'th component is clearly

$$\frac{dn_i}{dt} = \sum_k v_{ik} J_r^k \tag{49}$$

Equation (49) is a sum of several flows and should be represented by an 0-junction on the capacitor C_i of the i'th substance. Such an 0-junction divides the flow delivered to the different reactions but retains the same chemical potential of this substance in all chemical processes.

As an example of coupled reactions we shall show the bond graph of the system

$$A + B \rightleftharpoons C; C \rightleftharpoons 2D; D + B \rightleftharpoons E \tag{50}$$

In concluding this chapter, it is worth noting the mode of representation for autocatalytic reactions. What characterizes these processes is that the product of the reaction flows into the same capacitor from which the reactant is withdrawn. The bond graph comprises, therefore, a cyclic element over an 0-junction, as shown for the case

$$x + y \rightarrow 2x$$

This graph is *kinetically* different from that of the reaction

$$y \rightarrow x \tag{52}$$

which has the same affinity as reaction (51). For reaction (52), the bond graph is as in figure 14.

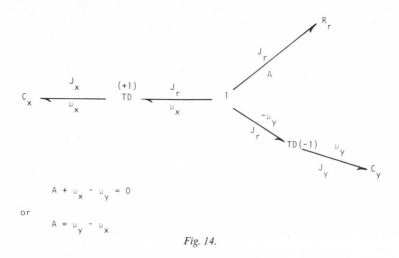

$$A + \mu_x - \mu_y = 0$$

or

$$A = \mu_y - \mu_x$$

Fig. 14.

6. Concluding Remarks

6.1. Since the field of thermodynamics of networks is rather new, it is difficult to estimate the practical advantages derived from the use of topological or bond graph representation. In a few refractory cases, however, which have recently been tested by us, the bond graph method has proved to be of interpretative value and has provided a quantitative description of the observed phenomena. Thus, OSTER and AUSLANDER [9] have studied the case of double membranes composed of two sheets of oppositely charged ion exchanger. Such bilayers, interposed between two solutions of electrolytes, exhibit marked rectification of electrical flow. Although the rectification could be predicted for nonequilibrium thermodynamics, a quantitative treatment requires more detailed information, which was provided by the bond graph representation. And, indeed, the computer calculation based on the algorithm derived from the bond graph gave good agreement with the experimental results reported in the literature.

Another case of more fundamental interest is that of the oscillations described by TEORELL [17]. Since oscillatory phenomena transgress the range of the thermodynamics of irreversible processes, it was of interest to test whether network thermodynamics, which treats both reversible and irreversible elements, can describe quantitatively this well-known experiment. It is worth recalling that TEORELL inserts a glass filter between two solutions of potassium chloride of different concentrations. Silver-silver chloride electrodes are introduced into the reservoirs and an electrical current is passed in such a direction as to cause an electroendoosmotic flow, to bring the dilute solution into the membrane. This process increases the membrane resistance and leads to an increased electroendoosmotic flow. The system, is, however, self-amplifying and ultimately reaches a point of instability which causes an oscillation in the water level of the reservoirs, in the salt concentration of the glass filter, and in the membrane potential. This is a typical case of a network with regulative feedback which can be readily described by a bond graph and numerically evaluated. The results of the calculation are presented in figure 15, which closely resembles the data of TEORELL.

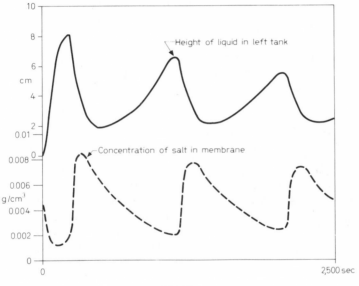

Fig. 15.

6.2. The main importance, however, of the network approach to bio-thermodynamics is of a conceptual nature. Network thermodynamics serves as a bridge between classical thermodynamics and the general dynamic theory of modern physics, and allows the introduction of thermodynamic concepts into the systems approach [18] devoted to the fundamental problems of biological organization.

References

1 OSTER, G.; PERELSON, A., and KATCHALSKY, A.: Network thermodynamics. Nature, Lond. *234:* 393 (1971); Network thermodynamics: the analysis of biological systems. (Lawrence Radiation Lab. Publications); Network thermodynamics: dynamic modellina of biophysical systems. Quart Rev. Biophys. *6:* 1 (1973).

2 MOROWITZ, H. Y.: Energy flow in biology (Academic Press, New York 1968).

3 PRIGOGINE, I.: Thermodynamics of irreversible processes, 3rd. ed. (Wiley, New York 1967).

3a GROOT, S. R. DE and MAZUR, P.: Non-equilibrium thermodynamics (Interscience Publishers, New York 1962).

4 KATCHALSKY, A. and CURRAN, P.: Non-equilibrium thermodynamics in biophysics (Harvard University Press, Cambridge, Mass., 1965).

5 DESOER, C. and KUH, E.: Basic circuit theory (McGraw-Hill, New York 1969).

6 KRON, G.: Physiol. Rev. *64:* 126 (1943); J. appl. Mechn. *11:* 149 (1944); J. Areon, Sci. *12:* 221 (1945); J. Chem. Phys. *14:* 19 (1946).

7 ROTH, J. P.: Proc. nat. Acad. Sci., Wash. *41:* 518 (1955).

8 BRANIN, F.: in Proc. Symp. Generalized Networks, Brooklyn, New York, pp. 453–491 (Polytechnic Press of the Polytechnic Institute, Brooklyn, New York 1966).

9 OSTER, G. and AUSLANDER, D.: J. Franklin Institute, *292:* 1 and *292:* 77 (1971).

10 PAYNTER, H.: Analysis and design of engineering systems (M.I.T. Press, Cambridge, Mass., 1961).

10a KARNOPP, D. and ROSENBERG, R.: Analysis and simulation of multiport systems (M.I.T. Press, Cambridge, Mass., 1968).

11 PRIGOGINE, I. and GLANSDORFF, P.: Physica *30:* 351 (1964).

12 PENFIELD, P.: Tellegen's theorem (M.I.T. Press, Cambridge, Mass., 1970).

13 OSTER, G. and DESOER, C.: J. theoret. Biol., *32:* 219–241 (1971).

14 PRIGOGINE, I.: c.f. [3].

15 ONSAGER, L.: Physiol. Rev. *37:* 405 (1931); *38:* 2265 (1931).

16 BRAYTON, R. K.: IBM Report RC 2606 (No. 12927) (1969).

17 TEORELL, T.: Exp. Cell Res., Suppl. *3:* 339 (1955); Acta Soc. Med. Uppsala *62:* 60 (1957); J. gen. Physiol. *42:* 831 (1959a); J. gen. Physiol. *42:* 847 (1959b); Ark. Kemi *18:* 401 (1961); Acta physiol. scand. *62:* 293 (1964).

17a KATCHALSKY, A. and SPANGLER, R.: Quart. Rev. Biophys. *1:* 127 (1968).

18 LANGE, O.: Wholes and parts (Pergamon Press, New York 1965).

Author's address: A. KATCHALSKY, The Weizmann Institute of Science, *Rehovot* (Israel)

Proc. 3rd int. Conf. From Theoretical Physics to Biology, Versailles 1971, pp. 89–109
(Karger, Basel 1973)

Fluctuations and the Mechanism of Instabilities

I. Prigogine and G. Nicolis

Faculté des Sciences de l'Université Libre de Bruxelles, Belgium, and Center for
Statistical Mechanics and Thermodynamics. University of Texas, Austin, Tex.

1. Introduction

It is customary to associate large systems involving a macroscopic
number of degrees of freedom, with a deterministic causal description.
However, in a number of instances such a description may not be adequate.
The main reason is that the very existence of many degrees of freedom
implies automatically the appearance of *fluctuations,* i.e., of spontaneous
deviations from some average, macroscopic behavior. In most cases, the
appearance of a fluctuation may be treated as a random event obeying
probabilistic laws. On the other hand, once the fluctuation is produced,
the system responds according to definite macroscopic laws.

In general, one expects that fluctuations although measurable, should
remain small compared to the macroscopic averages. An important result
of equilibrium statistical mechanics shows that this cannot be so in the
neighborhood of critical points of phase transitions (liquid-vapor transition,
etc.). In these cases, small thermal fluctuations are amplified, attain a
macroscopic level and drive the system to a new phase. The evolution to
the new phase occurs as an abrupt change *beyond an instability point* of
the reference state. In the critical region around the instability the system
exhibits a markedly coherent behavior which is frequently combined with
an increase in spatial order [1].

Several examples of similar phenomena in systems maintained far from
thermal equilibrium are also known. An ionized gas with a highly non-
Maxwellian velocity distribution becomes unstable with respect to infinitesimal thermal fluctuations and evolves to a regime known as plasma
turbulence [2]. More recently, HARKEN *et al.* [3] have shown that the gene-

ration of laser light may be interpreted quantitatively as a phase transition. In both cases the important feature is that there exists a critical point of transition (threshold) in the neighborhood of which the least microscopic fluctuation is amplified and drives the system to the new state. From this point of view, these phenomena appear to be analogs of second order phase transitions. Clearly, in these processes the study of fluctuations becomes a problem of great importance, as the latter determine the macroscopic behavior of the system as well as the new state beyond instability. In other terms, there is no more a clear-cut distinction between macroscopic averages and fluctuations.

Let us now consider a very different type of situation corresponding to instabilities in fluid dynamics and chemical or biochemical kinetics of open systems [4]. We have here systems subject to constraints such that their *macroscopic state* is beyond a critical distance from equilibrium and which may exhibit instabilities in their steady-state solutions. Hydrodynamic instabilities (e.g., the onset of thermal convection or of turbulence) are well-known examples of these phenomena. More recently, it has been shown [4] that similar transitions may arise in purely dissipative systems, such as chains of nonlinear (e.g., autocatalitic) chemical reactions. Beyond instability, these systems may evolve to states showing spatial, temporal or functional order, which have been called *dissipative structures* [4, 5]. In both cases one has, as a rule, systems which locally are quite close to equilibrium. We may express this property by the condition that the momentum distribution function of the particles is a local Maxwellian. This, however, does not prevent the system from reaching an overall state which is very far from equilibrium owing to the existence of *macroscopic constraints,* e.g., composition gradients on chemical affinities. It is well established [4] that this *local equilibrium* regime is realized, once the system is dense enough for elastic collisions to restore continuously a Maxwellian distribution that would otherwise be perturbed by transport processes or reactive collisions. Clearly, this condition is fulfilled in fluid dynamics where typical instabilities occur also in liquid phase. A similar situation is expected to arise in chemical and biochemical kinetics. The biological environment of the part of a living cell undergoing a typical biochemical reaction cycle is a dense medium, in which elastic collisions between substrates or monomers and macromolecules guarantee that local equilibrium is maintained to a very good approximation.

At first sight, it may seem surprising that systems which, in the molecular scale, are purely dissipative and thus tend to damp all thermal fluctuations, undergo large-scale instabilities which completely change their macroscopic structure. Intuitively, we expect that the origin of these transitions is quite different from that of plasma instabilities or of laser thresholds. This problem will be taken up and analyzed in Sections 2 and 3. Independently of all questions of interpretation we see, however, that again the study of fluctuations becomes very important and the distinction between macroscopic averages and fluctuations breaks down.

The relevance of the preceding considerations in branches other than physics is presently well recognized. Population genetics, ecology and epidemics are some of the best known examples of situations where probabilistic considerations are intrinsic to the problem. For instance, in stochastic epidemic theory one shows that the predictions of deterministic and probabilistic approaches to the same problem may be qualitatively different [6]. Among more fundamental biological problems, situations involving a large number of degrees of freedom—like the kinetics of long polymer molecules including the problem of biopolymerization, or the time-dependent properties of neural networks—have, almost exclusively, been treated stochastically [7, 8].

A still different question which we want to discuss in this paper, refers to the importance of randomness as an element which is inherent in the very foundations of biology [9]. To be more specific: what is the role of randomness compared to that of deterministic laws in the formation of spatial, temporal or functional order? Clearly, this question embraces problems such as prebiotic evolution, the emergence of biological structures or the formation and functioning of regulatory circuits in living cells. In this paper we will approach the problem from the point of view which has been developed during the last few years by GLANSDORFF, PRIGOGINE and coworkers and which may be summarized as follows [4, 5, 10]. A new structure or organization (such as a dissipative structure in a chemical autocatalytic system) is always the result of an instability. It originates in a fluctuation, i.e., in a fundamentally stochastic element. Usually, a fluctuation is followed by a response that brings the system back to the original state and which is a purely deterministic process. It is only at the point of formation of a new structure that *fluctuations are amplified,* reach a macroscopic level, start evolving, and stabilize to a new regime representative of the structure arising beyond instability. We see that in a physical phenomenon involving instabilities and subsequent evolution through fluctuations, both chance and determinism are necessary in the description. This general feature is a direct consequence of the nonlinearity of the equations of evolution which implies the existence of many possible solutions. Thus, in order to discuss quantitatively the process of evolution beyond instabilities it is necessary to develop a theory of fluctuations around nonequilibrium states in nonlinear systems.

In Section 2 an outline of this theory is presented for nonlinear chemical kinetic systems. This choice is motivated by the fact that the behavior of living objects is largely determined by a number of key chemical reactions and transport processes. Now, in biochemical kinetics nonlinearity is a general rule and may appear in a practically infinite number of varieties: autocatalysis, activation or inhibition, and so on. The importance of these processes in prebiotic evolution of macromolecules has also been stressed recently by EIGEN [11]. The main point is that it is likely to have instabilities and successive transitions to more and more spatially or functionally ordered states. Before discussing quantitatively fluctuations in such systems, it turns out that one has to replace the usual stochastic description based on a 'birth and death' type model by a more detailed phase space description. This extension is discussed in detail in Section 2 and in Appendices I and II and illustrated on an example which is also of interest in ecology. Section 3 is devoted to the application of the results in the behavior of fluctuations in evolving systems. A plausible microscopic mechanism for the emergence of instabilities is outlined in which the departure from the steady state is governed by large fluctuations of macroscopic size. On the other hand, small fluctuations are still described by a generalization of EINSTEIN's fluctuation theory. Nonequilibrium chemical instabilities seem, therefore, to bear strong similarities to first order phase transitions. The implications of these ideas in the problem of evolution in biology and in ecology are discussed in Section 4.

2. Fluctuations in Nonequilibrium Systems

The theory of fluctuations has been developed extensively for systems near thermodynamic equilibrium. The classical approach contains a combination of ideas of linear thermodynamics of irreversible processes and of equilibrium considerations [12]. Equivalent to this formulation is the Langevin method [13] which amounts to introducing *a priori* the fluctuations in the equations of evolution through suitably correlated stochastic force terms. A potentially more powerful method consists in the derivation of stochastic master equations based on the assumption that the kinetic equations define a markovian process. As a rule, these equations represent a birth and death process[1] [14]. Finally, a still more general approach is based on statistical mechanics, with a kinetic equation in phase space, such as the Boltzmann equation, as starting point [15].

An important result, which has only recently been proved in full generality [16], establishes that around equilibrium all three methods lead to equivalent results. In particular, the one-time fluctuations around equilibrium are shown to satisfy in the limit of small fluctuations, a distribution first conjectured by EINSTEIN:

$$P(\{\delta X\}) \propto \exp. \left[\frac{(\delta^2 S)_e}{2k}\right]$$

where $\{\delta X\} = \{X - X_e\}$ are the fluctuations of a set of state variables $\{X\}$ around the equilibrium values $\{X_e\}$; k is Boltzmann's constant, and $\frac{1}{2}(\delta^2 S)_e$ is the second order entropy excess (due to fluctuations) evaluated around the equilibrium state. An important consequence of (2.1) is that in an ideal reacting mixture the mean quadratic fluctuations are:

$$\overline{(\delta X)^2} = \overline{X} = X_e \tag{2.2}$$

where the average is taken over $P(\{\delta X\})$. As a result, in a system involving many degrees of freedom the relative fluctuations $\overline{[(\delta X)^2/X_e^2]^{1/2}}$ are very small, of the order of $X_e^{-1/2}$. In other terms, for such systems there is a clear-cut distinction between macroscopic averages and fluctuations.

The behavior of fluctuations around *steady nonequilibrium states* is much less known [17]. It is only in the domain of *linear systems* that one has been able to derive general conclusions comparable to (2.1) and (2.2). Most of the models studied explicitly refer to electric current fluctuations [18] or to composition fluctuations in open, chemically reacting mixtures [19]. In all cases, the equivalence of the Langevin, birth and death and kinetic equation approach, is confirmed. Moreover, it is shown that small fluctuations are described by a generalization of EINSTEIN's formula [19].

$$P(\{\delta X\}) \propto \exp. \left[\frac{(\delta^2 S)_0}{2k}\right] \tag{2.3}$$

---

1   The first application of stochastic theory to nonlinear chemical kinetics seems to be due to DEHBRÜCK (1940). More recently, EIGEN has applied stochastic theory to the problem of selforganization and evolution of biological macromolecules [11].

$$\overline{(\delta X)^2} = \overline{X} = X_0$$

$(\delta^2 S)_0$ is now evaluated around the nonequilibrium state[2]

As we saw in the Introduction, in order to understand the onset of a macroscopic transition such as a chemical instability, it is necessary to study fluctuations in nonlinear systems very far from thermodynamic equilibrium. In this domain, the validity of the Langevin type of approach is no longer guaranteed[3]. One is thus left with the master equations based on birth and death processes and with the more general kinetic equation approach.

Until recently, the study of fluctuations in chemical systems was based almost exclusively on the former method [14]. However, the applications were limited to linear systems or to systems close to equilibrium. We have recently worked out several applications of the birth and death type formalism to nonlinear, noequilibrium systems. The main conclusions of this investigation are as follows [20]:

(i) Systems for which a steady-state solution exists which is asymptotically stable, attain in the limit of a large number of particles, a steady-state regime for the fluctuations. The probability distribution in this regime does not reduce, however, to the generalized Einstein formulae (2.3). In particular, the mean quadratic fluctuations of a chemical constituent are now given by

$$\overline{(\delta X)^2} = \alpha X \tag{2.4}$$

where the numerical factor α depends on the detailed properties of the reaction steps. Thus, there seems to be no universal law governing small fluctuations in nonlinear nonequilibrium systems.

(ii) Systems whose steady-state lacks the property of asymptotic stability (oscillatory or unstable systems) do not admit a steady-state distribution for the fluctuations. This is briefly illustrated in Appendix I for the particularly interesting case of the Volterra-Lotka model which, as is well known, has been applied extensively in ecology [21], neural dynamics [8] and regulation of biosynthetic reactions in cells [22]. From the analysis reported in Appendix I it is seen that the distribution of fluctuations is always time dependent and the mean square correlations $\overline{(\delta X)^2}$, $\overline{(\delta Y)^2}$ are increasing

2   The models treated in ref. 17–19 are all stable with respect to arbitrary perturbations.
3   This statement has also been made by VAN KAMPEN [14].

functions of time, even for infinitesimal fluctuations. Stochastically, therefore, the steady state (cf. Eqs.) [AI.4] is unstable and the system exhibits abnormal 'critical' fluctuations.

The following argument shows that this conclusion cannot be valid in general. Let us recall from our discussion of the different types of instabilities in the Introduction that in chemical kinetics one usually deals with systems described by a *local equilibrium theory* in which the state functions (entropy density, etc.) are described, locally, by the same independent variables as in equilibrium. It is, thus, natural to expect that in this case relations such as (2.3) should still apply, provided they are used locally. Now the sufficient condition for the validity of this theory is that the momentum distribution function deviates only slightly from the local Maxwellian form. This, however, implies the existence of *two largerly separated time scales* [20]: a short relaxation time scale between frequent elastic collisions, which restore continuously an average Maxwellian distribution; and a longer macroscopic scale over which the chemical composition changes as a result of reactive collisions which tend to perturb the Maxwellian distribution. Now the effects related to the relaxation time cannot be accounted for in a description based on the birth and death type master equations, where the internal states (e.g., the values of the momenta) are discarded. We are, thus, led to the conclusion that fluctuations must be discussed on the basis of a *kinetic equation* such as the Boltzmann equation, containing the effect of both reactive and elastic collisions.

The detailed study of this phase space approach to fluctuations leads to the following results [20].

First, *small thermal fluctuations are always damped,* both for systems with or without asymptotic stability. This result, together with an outline of the phase space approach is reported in Appendix II for the particular case of the Volterra-Lotka model. In addition, the distribution of small fluctuations is shown to satisfy the generalized Einstein relations (2.3).

On the other hand, large fluctuations of macroscopic size seem to behave differently. In particular, for systems without asymptotic stability, such as the Volterra-Lotka model, they may evolve in time and drive the system to a new macroscopic regime far from the original reference state.

In conclusion, it appears that in nonlinear, nonequilibrium systems, it is necessary to adopt a stochastic description based on a kinetic equation of the Boltzmann type in order to describe correctly the distribution of thermal fluctuations around steady states. It is only in the limit when the state of the system is close to equilibrium, and also in the case of linear systems (unimolecular reactions) that this more complete description does not contradict the results based on the usual birth and death process formulation. Otherwise, the latter is inadequate. The reason for this discrepancy is seen by a comparison of the two master equations (A.I.7) and (A.II.4) from which it becomes clear that in the birth and death formulation the internal states are treated incorrectly [20].

3. Fluctuations in Evolving Systems. 'Generalized Turbulence'

The results outlined in the preceding Section and in Appendices I and II provide a valuable insight in the problem of the onset of oscillations and instabilities in systems at a local equilibrium regime. Indeed, a system in the neighborhood of an instability behaves in a very similar way to the Volterra-Lotka model in the sense that, macroscopically, small perturbations are not damped or amplified. Now the phase space description shows that *small thermal fluctuations* are damped. At first sight, this result contradicts the conclusions of the macroscopic description outlined in Appendix I. Indeed, because of the infinity of periodic trajectories around the steady-state in the Volterra-Lotka model (cf. Eqs. A.I.1) one would expect that the system may have no criteria to choose the 'right' or 'most probable' regime and that this inability would manifest itself by abnormal fluctuations of some kind. This seems to be ruled out by the conclusions of the phase space description of Section 2 and Appendix II. How then can a system at a state of marginal stability be driven to the new regime beyond instability by thermal fluctuations? We believe that the answer to this question, which is related to the very nature of the onset of structure in dissipative systems, may be as follows.

(i) First, it has to be realized that the lack of asymptotic stability observed in nonlinear dissipative systems (chemical instabilities, etc.) is a purely macroscopic phenomenon which has no molecular counterpart as long as the system (including fluctuations) is maintained in a local equilibrium regime. Thus, it is reasonable to expect that systems undergoing such *macroscopic* instabilities cannot evolve from a given macroscopic reference state by a mechanism of thermal fluctuations of usual size (i.e., very small). This explains why, in the phase space description of Appendix II, small fluctuations are damped. Again, the difference with instabilities in the velocity distribution which was mentioned in the Introduction, should be emphasized.

(ii) A change in the macroscopic state of a system in a local equilibrium regime can, therefore, arise only from a mechanism of *large thermal fluctuations* of macroscopic size. Macroscopic instabilities, thus, seem to bear some analogies with first order phase transitions. The probability of fluctuations of macroscopic size is always very small. As a result, even during the macroscopic evolution to a new regime, the generalized Einstein relation would be practically satisfied around the *true* (time dependent) average. The time required for the formation of this evolving mode would, thus, necessarily be macroscopic, i.e., very long as compared to the relaxation time between molecular collisions. In fact, it should be comparable to the 'hydrodynamic' scale of macroscopic evolution. From a practical viewpoint, therefore, it is most likely that the system will be driven to the instability by a random external disturbance, before the molecular mechanism described in this section becomes effective. This is particularly true in biological systems, which are often subject to a violently varying environment.

(iii) The probability that such a finite fluctuation occurs everywhere in a macroscopic system should be practically vanishing. Rather, the most natural molecular mechanism underlying a macroscopic evolution should be as follows. A small subsystem (containing, say, 10^2 particles) begins to evolve as a result of a finite fluctuation. This creates an 'embryon' in the form of a local inhomogeneity which, in usual circumstances, would tend to be damped by diffusion. However, if the system is near the threshold of a macroscopic transition, diffusion should rather provide a mechanism for propagation over large distances of the disturbance created by a fluctuation.

In order to substantiate these conjectures it would be necessary to solve the complete master equation in phase space for the model treated in Appendix II in a way which takes into account, self-consistently, the simultaneous evolution of the fluctuations *and* of the macroscopic state. This means that one has to study the time-dependent solutions of the master equation for arbitrary fluctuations. This study, which should provide such information as, e.g., the critical size of fluctuations beyond which the system starts to evolve and the time required for the formation of this evolving mode, is presently in progress.

Summarizing the results of Sections 2 and 3, one may say that the separation between macroscopic behavior and fluctuations is related to the distance from thermodynamic equilibrium. In a far from equilibrium regime corresponding to the onset of oscillations or instabilities, this separation is not possible in general and the evolution of the average values depends *explicitly* on the fluctuations. The situation reminds of the familiar phenomenon on fluid dynamics arising beyond the point of instability of the laminar flow, and may be called appropriately *generalized turbulence*. In the phase space description of fluctuations outlined in Sections 2 and in Appendix II, the fluctuations entering explicitly in the macroscopic description are not the infinitesimal thermal fluctuations but, rather, the fluctuations having a size beyond some critical value. In contrast, in systems described by a birth and death type formalism, 'turbulent' behavior is exhibited also with respect to infinitesimal fluctuations in chemical composition. In either case, the time scales of fluctuations and of average values are no longer separated. The evolution acquires an essentially statistical aspect and the usual description on the basis of the macroscopic equations of conservation breaks down.

4. Fluctuations and the Problem of Evolution

In the preceding sections we have shown that fluctuations play an important role whenever there is onset of a new regime at a point of marginal stability. In general, their behavior depends on the internal relaxation processes in the system. Thus, in the Volterra-Lotka model, the analysis of Appendix I and Appendix II gave quite different results: in the birth and death process formulation the internal relaxation processes are discarded. As a result, the system exhibits abnormal critical fluctuations. Ecologically, this situation has a clear interpretation. The individuals of the prey population are consumed indifferently by the predator. The steady-state prey distribution is never stable because there is no internal mechanism which reestablishes equilibrium once the latter is perturbed by the predator.

In the phase space description outlined in Appendix II the situation is different. Ecologically, it corresponds to the more realistic picture of small predator *versus* prey ratio. In this case, the most natural competition consumes preferentially those prey individuals having small values of some 'fitness' parameter which measures the ability to resist to or escape from the predator. It is, thus, natural to expect that in such systems, in addition to the effect of the predators, the internal processes determining the fitness distribution within the prey species should play an important role. Qualitatively, this situation may well be described by an equation of the same type as the kinetic equations (A.II.2) of Appendix II, provided one reinterprets suitably the parameter α determining the internal state. One of the consequences of these processes is to permit an *evolution* of the prey species in which the unfit individuals are eliminated continuously from the population. Another plausible interpretation of equations of the form (A.II.2) is that they describe the influence of environment on the growth of one species X when the individuals are again differentiated in terms of the fitness parameter α.

The result that a model of competing populations where internal relaxation processes are taken into account may give rise to evolution through fluctuations, is also interesting from the point of view of the evolution of biological macromolecules in prebiotic conditions. As recently emphasized by EIGEN, a system of macromolecules which take part into separate autocatalytic cycles and, at the same time, may serve as templates for the synthesis of 'error copies', undergoes essentially a dynamics of competing populations. Again, if one interprets the parameter α as the label of one type of macromolecular copy, one sees that the equations written by EIGEN have a similar structure to the kinetic equations of Appendix II. Stochastically, therefore, one might expect that the macroscopic state will remain stable with respect to small fluctuations but that for certain types of nonlinear kinetics, large fluctuations will increase in time and will drive the systems to a new regime. Further study is necessary in order to appreciate fully the impact of these probabilistic effects in this and other related biological problems.

The analogy drawn in Section 3 between instabilities in dissipative systems and first order phase transitions may be of interest for the thermodynamic interpretation of 'all-or-none' effects in biology, e.g., for the problem of membrane excitability. Recently, BLUMENTHAL et al. [23] have shown that a membrane in the polarized state may be understood as a dissipative structure separated from the excited state by an instability. On the other hand, KOBATAKE [24] suggested that the transition between the two states occurs at the point where the Maxwell rule (familiar from first order phase transitions) is satisfied. Qualitatively, this conjecture is not contradicted by the interpretation outlined in Section 3. However, in order to establish unambiguously the transition point for excitation, it will be necessary to study, following the methods described in this paper, the distribution of fluctuations around the polarized state. The transition threshold will correspond to a point where the probability for occupation of the excited state becomes macroscopic.

Acknowledgements

We wish to express our gratitude to Professors M. EIGEN, P. GLANSDORFF, H. HAKEN, M. KAC for fruitful suggestions and stimulating exchanges of ideas, and to the Welch Foundation, Houston, Texas and le Fonds de la Recherche Fondamentale Collective, Belgium for financial support. Part of the work reported in this paper was done during the stay of one of us (I.P.) at the General Motors Research Laboratories. We would like to express our appreciation to Drs. A. BUTTERWORTH and R. HERMAN of these laboratories for their interest in this work and for fruitful discussions.

Appendix I

The Volterra-Lotka Model

In this Appendix we discuss the properties of fluctuations around the steady-state for the Volterra-Lotka model which, as we shall see, lacks the property of asymptotic stability. Originally, this model was proposed for describing the competition between a number of predator—prey biological species. More recently, it has been analyzed in great detail from a statistical mechanical point of view by KERNER and MONTROLL *et al.* [21] and applied to the study of systems such as neural networks [8] or the control of protein synthesis in cells [22].

We shall be particularly interested in the case of two interacting species. The Volterra-Lotka equations describing this system read [21]:

$$\frac{dX}{dt} = \varepsilon_1 X - k_2 XY$$

$$\frac{dY}{dt} = k_2 XY - \varepsilon_3 Y \qquad (A.I.1)$$

As usual [21], we have assumed that the X-Y coupling term appears with opposite sign in the two equations. Let us set

$$\varepsilon_1 = k_1 A$$

$$\varepsilon_3 = k_3 B \qquad (A.I.2)$$

It is easily seen that Eqs. (A.I.1) are then the conservation of mass equations of the following set of irreversible autocatalytic chemical reactions (in the limit of an ideal mixture):

$$A + X \xrightarrow{\ k_1\ } 2X$$

$$X + Y \xrightarrow{\ k_2\ } 2Y$$

$$Y + B \xrightarrow{\ k_3\ } E + B \qquad (A.I.3)$$

As we shall see soon, this chemical analog of the Volterra-Lotka model will prove very useful for understanding the mechanism of fluctuations around instabilities. System (A.I.3) is assumed to be open to large reservoirs of A, B, E. Inverse reaction rates are neglected. The system thus operates automatically far from thermodynamic equilibrium.

Let us briefly recall the salient features of the evolution equations (A.I.1) [21]:

(i) The system admits a single non-zero steady-state solution

$$X_o = \frac{k_3 B}{k_2}$$

$$Y_o = \frac{k_1 A}{k_2} \qquad\qquad (A.I.4)$$

(ii) Small perturbations around (X_o, Y_o) exhibit undamped oscillations with a universal frequency

$$\omega_o = (k_1 k_3 AB)^{1/2} \qquad\qquad (A.I.5)$$

(iii) For arbitrary perturbations, Eqs. (A.I.1) admit a constant of motion

$$V(X, Y) = X + Y - \frac{k_3 B}{k_2} \ln X - \frac{k_1 A}{k_2} \ln Y \qquad\qquad (A.I.6)$$

which defines for all $X, Y > O$ a set of closed curves. Thus, finite perturbations are also periodic in time with periods depending on the initial conditions. The trajectories corresponding to this motion are all orbitally stable (but *not* asymptotically stable). We see that the Volterra-Lotka system is unable to damp perturbations. From this point of view the model may be considered as prototype of any system undergoing an unstable transition.

In the macroscopic description, Eqs. (A.I.1), fluctuations are neglected. We now adopt a more refined description and assume that Eqs. (A.I.1) define a markovian birth and death process in the space of the total numbers of particles of the constituents A, B, E, X, Y. In chemical kinetics[4] it is usually implied that these variables provide a consistent description, independently of the detailed behavior of the microscopic properties such as momenta. The system is now described in terms of a function P(A, B, E, X, Y, t) which gives the probability for finding given values of the particle numbers at time t. The equation of evolution for this function can be established easily by applying the methods of the theory of markov processes [6, 14, 19].

We have to consider all possible transitions in a time interval $(t, t + \Delta t)$ which lead from the state (A, B, E, X, Y) into other states and the transitions which bring the system from other states into $(\bar{A}, B, E, X, Y)$. In the reduced form where a summation over the initial and final product variables (A, B, E) is performed, the result reads [19]:

$$\frac{dP}{dt} = A(X-1) P(X-1, Y, t) - AXP(X, Y, t) +$$

4   For the purpose of comparison with the results of Appendix II we here formulate the problem of fluctuations for the chemical analog (A.I.3) of the Volterra-Lotka model. The results can be applied straightforwardly to the ecological model.

$$+ (X + 1)(Y - 1) P(X + 1, Y - 1, t) - XYP(X, Y, t)$$
$$+ B(Y + 1) P(X, Y + 1, t) - BYP(X, Y, t) \tag{A.I.7}$$

The system has been assumed to remain uniform in space. A, B are now the average values of the number of particles of A, B. All reaction rates have been set equal to unity. Finally, Eq. (A.I.7) implies the validity of a decoupling procedure, discussed in detail in ref. 19, according to which (A, B, E) which play the role of *external reservoirs* vary in a different scale than the state of the subsystem of intermediate products (X, Y).

The finite difference Equation (A.I.7) is handled most conveniently by introducing the *moment generating function* [6, 14, 19] F (s_x, s_y, t):

$$F(s_x, s_y, t) = \sum_{x,y=0}^{\infty} s_x^x s_y^y P(X, Y, t) \tag{A.I.8}$$

It is easily shown that F is related simply to the average values of various powers of X and Y. In the notation of Section 2:

$$[F(s_x, s_y)]_{s_x = s_y = 1} = 1$$

$$\left[\frac{\partial F(s_x, s_y)}{\partial s_x} \right]_{s_x = s_{yb} = 1} = \overline{X}$$

$$\left[\frac{\partial}{\partial s_x} s_x \frac{\partial F(s_x, s_y)}{\partial s_x} \right]_{s_x = s_y = 1} - \left[\frac{\partial F(s_x, s_y)}{\partial s_x} \right]_{s_x = s_y 1}^2 = \overline{(\delta X^2)} \tag{A.I.9}$$

and similar relations for Y.

In most problems of interest in chemical kinetics, ecology or biology, one deals with systems involving a large collection of entities (of order $N \to \infty$). It is, therefore, useful to separate leading terms in N from corrections of order 0 (1). To this end we introduce the cumulant generating function [6]:

$$F(s_x, s_y, t) = e^N \psi(s_x, s_{y,t}) \tag{A.I.10}$$

This definition and Eq. (A.I.9) imply that $\psi = O(1)$.

Substituting (A.I.8) and (A.I.10) into the original Equation (A.I.7) one obtains:

$$\frac{\partial \psi}{\partial t} = \alpha \xi(\xi + 1) \frac{\partial \psi}{\partial \xi} - \beta \eta \frac{\partial \psi}{\partial \eta} + (\eta + 1)(\eta - \xi) \times$$

$$\times \left[\frac{\partial \psi}{\partial \xi} \frac{\partial \psi}{\partial \eta} + \frac{1}{N} \frac{\partial^2 \psi}{\partial \xi \partial \eta} \right] \tag{A.I.11}$$

We have set:

$$A = \alpha N$$
$$B = \beta N$$

$$\xi = s_x - 1$$

$$\eta = s_y - 1 \tag{A.I.12}$$

clearly $\alpha, \beta = 0(1)$.

In (A.I.11) the second derivative term is multiplied by $1/N$ and, thus, may be neglected as $N \to \infty$. Eq. (A.I.11) reduces then to a nonlinear, first order, partial differential equation which admits solutions of the form:

$$\psi = a_1 \xi + a_2 \eta + b_{11} \frac{\xi^2}{2} + b_{12} \xi \eta + b_{22} \frac{\eta^2}{2} \tag{A.I.13}$$

Notice that by Eqs. (A.I.9) and (A.I.10) one has:

$$X = Na_1, \quad \overline{Y} = Na_2$$

$$\overline{(\delta X)^2} = \overline{X} + Nb_{11}, \quad (\delta Y)^2 = \overline{Y} + N b_{22}$$

$$\overline{\delta X \delta Y} = Nb_{12}$$

Substituting into (A.I.1 1) and identifying equal powers of ξ^2, η^2, $\eta \xi$ one obtains the following nonlinear differential system for the expansion coefficients:

$$\frac{da_1}{dt} = a_1 \alpha - a_1 a_2 - \frac{b_{12}}{N} \tag{A.I.15}$$

$$\frac{da_2}{dt} = - a_2 \beta + a_1 a_2 + \frac{b_{12}}{N}$$

$$\frac{db_{11}}{dt} = a_1 \alpha + \alpha b_{11} - (a_1 b_{12} + a_2 b_{11}) \tag{A.I.16}$$

$$\frac{db_{22}}{dt} = a_1 a_2 - \beta b_{22} + (a_2 b_{12} + a_1 b_{22}) + \frac{b_{12}}{N}$$

$$\frac{db_{12}}{dt} = \alpha b_{12} - \beta b_{12} - a_1 a_2 + (a_1 b_{12} + a_2 b_{11}) -$$

$$- (a_2 b_{12} + a_1 b_{22}) - \frac{b_{12}}{N}$$

In the limit $N \to \infty$ the first two equations reduce to the macroscopic deterministic equations (A.I.1), provided b_{12} is of O (1). If one accepts this conjecture, then Eqs. (A.I.15) admit a steady-state solution

$$a_2 = \alpha, \quad a_1 = \beta \tag{A.I.17}$$

and Eqs. (A.I.16) become:

$$\frac{db_{11}}{dt} = \alpha \beta (1 - b_{12})$$

$$\frac{db_{22}}{dt} = \alpha (\beta + b_{12})$$

$$\frac{db_{12}}{dt} = -\alpha\beta + \alpha b_{11} + \beta b_{22}$$

(A.I.18)

The important point is now that Eqs. (A.I.18) do *not* admit a time-independent solution. In spite of (A.I.17), b_{11}, b_{22} increase in time and, thus, by Eq. (A.I.14) the mean quadratic fluctuations $\overline{\delta X^2}$, $\overline{\delta Y^2}$ deviate from the Poisson regime (cf. Eq. [2.3]) and cannot reach a new steady state. Stochastically, therefore, the steady state (A.I.17) or (A.I.4) is meaningless even in the limit of small fluctuations and the system exhibits abnormal, critical fluctuations. The latter will ultimately alter the order of magnitude of b_{12} which could no more be neglected in (A.I.15). As a result, *the macroscopic averages will be driven by the fluctuations* to a time-dependent regime far from the steady state [20].

Appendix II

Phase Space Description of the Volterra-Lotka Model

The results outlined in Appendix I are direct consequences of the assumption that the Volterra-Lotka equations (A.I.1) define a markovian birth and death process in (X, Y) space. On the other hand, the discussion of Section 2 implies that the chemical analog (A.I.3) of the ecological model requires a phase space description, at least in the range of validity of local thermodynamics of irreversible processes. In this description, it is implied that fluctuations define a markovian birth and death process only in the complete phase space, including internal states.

As we deal with dilute mixtures, our starting point will be a kinetic equation of the usual Boltzmann form. We adopt the notation $\overline{F_\alpha^\gamma}$ for the Boltzmann probability density of component γ corresponding to an internal state α, and assume for simplicity that the spectrum of α is discrete. α may stand for the position r and the momentum p of a particle. The bar over F_α^γ reminds that in the Boltzmann equation, description F_α^γ represents an *average* quantity. Assuming the system is maintained uniform in space, one can easily write the Boltzmann equation corresponding to model (A.I.3) [20]:

$$\frac{d\overline{F_\alpha^x}}{dt} = -\Sigma A_{\alpha jkl} \overline{F_\alpha^x} F + 2\Sigma A_{jk\alpha l} \overline{F_j^x} \overline{F_k^A} -$$

$$-\Sigma B_{aaikl} \overline{F_\alpha^x} \overline{F_y^y} + \left(\frac{d\overline{F_\alpha^x}}{dt}\right)_{el}$$

(A.II.1a)

$$\frac{dF_\beta^y}{dt} = -\Sigma B_{\beta jkl} F_j^x + 2\Sigma B_{jk\beta l} F_j^x F_k^y -$$

$$r - \Sigma C_{\beta jkl} F_\beta^x F_j^B + \left(\frac{dF_\beta^y}{\alpha t}\right)_{el}$$

(A.II.1b)

In the r.h.s. of these equations, $(d\overline{F}_\alpha^x/dt)_{el}$, $(d\overline{F}_\beta^y/dt)_{el}$ describe the effect of elastic collisions and the remaining terms refer to reactive collisions. B_{ijkl}, A_{ijkl} and C_{ijkl} are the transition probabilities per unit time for scattering between two molecules in states (kl) into two molecules in states (ij) for the reactions corresponding to the three steps in (A.I.3). They satisfy a number of conditions imposed by the mechanibs of a scattering process:

$$A_{ij,\,kk} = 0$$

$$A_{ii,\,kl} = A_{ji,\,kl} = A_{ij,\,lk} \geqslant o \qquad \text{for (kl)} \neq \text{(ij)}$$

$$\sum_{kl} A_{ij,\,kl} = o \tag{A.II.2a}$$

If, in addition, one requires microscopic reversibility,

$$A_{ij,\,kl} = A_{kl,\,ij} \tag{A.II.2b}$$

The factor 2 in front of the quadratic term in $F^x\ F^y$ expresses the fact that the population of Y in the β state may be increased in two ways, corresponding to either of the two Y molecules in the second step (A.I.3) being in the energy level j or k.

We now place ourselves in the limit of very frequent elastic collisions. According to the remarks made in Section 2, this implies:

$$\overline{F}_\alpha^x \simeq \qquad\qquad = \text{local Maxwellian}$$

$$\text{and similarly for A, B, E . Y} \tag{A.II.3a}$$

i.e.

$$\left(\frac{d\overline{F}_\alpha^x}{dt}\right)_{el} \simeq o$$

$$\left(\frac{d\overline{F}_\beta^y}{dt}\right)_{el} \simeq o \tag{A.II.3b}$$

We are, thus, permitted to neglect the explicit effect of elastic collisions in the kinetic equations. Of course, the influence of these collisions remains implicitly in the reactive terms through the fact that the molecular speed distributions are now Maxwellian and that a distinction is made between molecules occupying different momentum states. In fact, Eqs. (A.II.1) now express that, as a rule, the system remains stationary on the microscopic scale because of the frequent elastic collisions, but from time to time it is slightly perturbed by reactive collisions between molecules which are sufficiently energetic to overcome the potential barrier for binding. In a sense, reactive collisions are 'exceptional' events associated with the tail of the Maxwellian velocity distribution.

It is interesting to observe that the structure of Eqs. (A.II.1) is quite different from Eqs. (A.I.1) or their generalization to many components [21], in spite of the fact that on averaging (A.II.1) over the internal states one obtains (A.I.1) identically. The most striking difference is that in (A.I.1) dX/dt, dY/dt are proportional to X and Y, respectively. On the other hand, in (A.II.1) dF_α^x/dt is not proportional to $\overline{F}_\alpha^x$ owing to the terms expressing, e.g., that 2 molecules of X in state α may be created from X and Y in different internal states.

Thus, one should not expect Eqs. (A.II.1) to give rise to a constant of motion as in the Volterra or in the Kerner analysis [21]. Bearing also in mind the remarks made in Section 2, one could anticipate that the phase space description of fluctuations for the chemical Volterra-Lotka model (A.I.3) would be quite different from the picture outlined in Appendix I.

Let us set $\overline{F}_\alpha \, (\Delta_r \Delta_p)_\alpha = \bar{f}_\alpha$, $\bar{f}_\alpha$ being the average number of molecules in the phase space volume $(\Delta r \Delta p)_\alpha$ around (r_α, p_α). In the kinetic equation description, the fluctuations of f_α around this average are neglected. We now go to a more refined description in terms of the probability for having, at time t, given occupation numbers for the various internal states. An equation of evolution for this function may be derived by assuming that Eqs. (A.II.1) define a markovian process in the complete phase space. The procedure, in a different context, has been discussed in detail by KAC and by SIEGERT [15]. One obtains, straight forwardly, a reduced master equation of evolution for the distribution of f^x, f^y summed over the reservoir variables:

$$
\begin{aligned}
\frac{dP(\{f^x\}, \{f^y\}, t)}{dt} \\[2mm]
= \sum_{ijkl} A_{ijkl} \, \bar{f}_i^A [(f_j^x + 1) \, P \, (f_j^x + 1, f_k^x - 1, f_l^x - 1, \{f^y\}, t) - \\[1mm]
- f_j^x \, P \, (f_j^x, f_k^x, f_l^x, \{f^y\}, t)] \\[2mm]
+ \sum_{ijkl} B_{ijkl} [(f_i^x + 1) \, (f_j^y + 1) \, P \, (f_i^x + 1, f_j^y + 1, f_k^x - 1, f_l^y - 1, \{f^y\}, t) - \\[1mm]
- f_i^x \, f_j^y \, P \, (f_i^x, f_j^y, f_k^x, f_l^y, \{f^y\}, t)] \\[2mm]
+ \sum_{ijkl} C_{ijkl} \, \bar{f}_i^P [(f_j^y + 1) \, P \, (f_j^y + 1, \{f^y\}, t) - f_j^y P \, (f_j^y, \{f^y\}, t)]
\end{aligned}
\tag{A.II.4}
$$

$\{f^y\}$ denotes the occupation numbers of the states which are not implied in the reaction steps. It is easily verified that when Eq. (A.II.4) is multiplied by f_α^x, f_β^y and averaged over all f's, it yields Eqs. (A.II.1), provided a factorization assumption is made on P. We shall come back to this point later in this Appendix. Finally, it should be pointed out that the reduced equation (A.II.4) is obtained from a more general master equation involving fluctuations of the reservoir composition by summing over the reservoir variables and assuming that the latter vary over a different time scale compared to the variables of the system corresponding to constituents (X, Y). The validity of this *decoupling assumption* is discussed in detail in ref. [19].

The difficulty in solving Eq. (A.II.4) arises from the infinite number of coupled terms contained in the two sums over internal states in the r.h.s. For this reason, we shall presently study this equation in the limit of small fluctuations. We express this limit by setting [14]

$$
f_k^x = \overline{f_k^x} + \delta f_k^x = \overline{f_k^x} + \varepsilon \chi_k
$$

$$
\varepsilon \, \alpha \, | \chi_k / \, f_k^x \, | << 1
\tag{A.II.5}
$$

and similarly for f_j^y. The deviations χ_k are now due to thermal fluctuations.

Expanding consistently both P and the coefficients in Eq. (A.II.4) in powers of ε and keeping the dominant terms in ε, one obtains two equations of the Fokker-Planck type, provided [20]:

(i) a factorization assumption is made on the initial state $P(\{f_\alpha\}, t = 0)$:

$$
P (\{f_\alpha\}, 0) = \prod_i P_1 (f_i, 0)
\tag{A.II.6}
$$

(ii) the averages $\overline{F^x}$, $\overline{F^y}$ satisfy at the steady state the Boltzmann equation (A.II.1). This is consistent with the limit of small fluctuations considered in this Appendix.

The final equations read:

$$\frac{\partial P_1(\chi_\alpha, t)}{\partial t} = \frac{\partial}{\partial \chi_\alpha} \chi_\alpha \, P_1\,(\chi_\alpha, t)\left[2\frac{1}{\overline{f_\alpha^x}} \sum_{ijl} A_{ij\alpha l}\, \overline{f_i^x}\, \overline{f_j^A}\right.$$

$$\left. - \frac{\partial \ln \overline{f_\alpha^x}}{\partial t}\right] +$$

$$+ \frac{\partial^2 P_1(\chi_\alpha, t)}{\partial \chi_\alpha^2}\left[2 \sum_{ijl} A_{ij\alpha l}\, \overline{f_i^x}\, \overline{f_j^A} - \tfrac{1}{2}\frac{\partial \overline{f_\alpha^x}}{\partial t}\right] \qquad\qquad\text{(A.II.7a)}$$

$$\frac{\partial P_1(y_\beta, t)}{\partial t} = \frac{\partial}{\partial y_\beta} y_\beta \, P_1\,(y_\beta, t)\left[2\frac{1}{\overline{f_\beta^y}} \sum_{ijl} B_{ij\beta l}\, \overline{f_i^x}\, \overline{f_j^y} - \right.$$

$$\left. - \frac{\partial \ln \overline{f_\beta^y}}{\partial t}\right] +$$

$$+ \frac{\partial^2 P_1(y_\beta, t)}{\partial y_\beta^2}\left[2 \sum_{ijl} B_{ij\beta l}\, \overline{f_i^x}\, \overline{f_j^y} - \tfrac{1}{2}\frac{\partial \overline{f_\beta^y}}{\partial t}\right] \qquad\qquad\text{(A.II.7b)}$$

It sould be pointed out that, strictly speaking, Eqs. (A.II.7) are *coupled* through the average values $\overline{f^x}$, $\overline{f^y}$ which have to satisfy the self-consistency conditions

$$\overline{f_\alpha^x} = \sum_{\{f^x\},\,\{f^y\}} f_\alpha^x\, P\,(\{f^x\}, \{f^y\}, t)$$

$$\overline{f_\beta^y} = \sum_{\{f^x\},\,\{f^y\}} f_\beta^y\, P\,(\{f^x\}, \{f^y\}, t) \qquad\qquad\text{(A.II.8)}$$

This coupling, however, is quite different from that implied by Eq. (A.I.7) which does not not admit factorizable solutions. In addition, to a first approximation (i.e., as long as one remains close to the reference state) one is allowed to identify the $\overline{f}$s to the macroscopic averages appearing in the kinetic equations (A.II.1). It follows that the coefficients of the first derivatives in Eqs. (A.II.7) are always positive and, therefore, these equations *admit a steady-state solution which is stable with respect to small thermal fluctuations*. To calculate the probability distribution in this state, it is sufficient to set $\partial P_1/\partial t = 0$. Eqs. (A.II.7) reduce to the form:

$$\frac{\partial}{\partial \chi_\alpha} \chi_\alpha \, P_1(\chi_\alpha) + \overline{f_\alpha^x}\frac{\partial^2 P_1(\chi_\alpha)}{\partial \chi_\alpha^2} = 0 \qquad\qquad\text{(A.II.9a)}$$

$$\frac{\partial}{\partial y_\beta} y_\beta \, P_1(y_\beta) + \overline{f_\beta^y}\frac{\partial^2 P_1(y_\beta)}{\partial y_\beta^2} = 0 \qquad\qquad\text{(A.II.9b)}$$

Notice that the transition probabilities A_{ijkl}, B_{ijkl} cancel in these equations. The solution of (A.II.9) is identical to $(2 \cdot 3)$, with

$$\overline{(\delta f_\alpha^x)^2} = \overline{f_\alpha^x}\ ,\ \ \overline{(\delta f_\beta^y)^2} = \overline{f_\beta^y} \qquad\qquad\text{(A.II.10)}$$

We see that one recovers the generalized Einstein relations as expected from the qualitative arguments advanced in Section 2. Thus, the analysis based on the birth and death type formulation of Appendix I is incompatible with the phase space description outlined in this Appendix. Comparing Eqs. (A.I.7) and (A.II.4) we see that the inadequacy of the former is due to the fact that, for instance, in the first step of the reaction scheme (A.I.3) it is implied that two molecules of X are produced in the same state as the molecule of X which combines with A. Now Eqs. (A.II.2) imply that in a macroscopic system in the thermodynamic limit, the probability of this event is negligible compared to the probability of the process described by the first term of Eq. (A.II.4). Hence, Eq. (A.I.7) cannot describe correctly the fluctuations in a thermodynamic system. Alternatively, a closed master equation in the number of particles space cannot describe a markovian birth and death process.

References

1 BROUT, R.: Phase transitions (Benjamin, New York 1964).
2 BALESCU, R.: Statistical mechanics of charged particles (Interscience Publ., New York 1963).
3 HAKEN, H.: Laser light as example of phase transitions far from thermal equilibrium; Ann. Met. American Physical Society, New York 1971.
4 GLANSDORFF, P. and PRIGOGINE, I.: Thermodynamics of structure, stability and fluctuations (Wiley/Interscience Publ., New York 1971).
5 PRIGOGINE, I.: in 1st and 2nd Int. Symp. Theoretical Physics and Biology (North Holland Publ. Co., Amsterdam 1969, 1972).
6 BARTLETT, M. S.: Stochastic processes (Cambridge Univ. Press. Cambridge 1955); BARUCHA-REID, A. T.: Elements of the theory of Markov processes and their applications (McGraw-Hill, New York 1960).
7 See e.g. GIBBS, J.: Adv. chem. Phys. *15:* 185 (1969).
8 See e.g. COWAN, J. D.: Paper presented at IUPAP Int. Conf. Stat. Mechanics, University of Chicago Press (1971).
9 For a recent discussion of the role of randomness, we refer to MONOD, J.: Le hasard et la nécessité (Seuil, Paris 1970).
10 PRIGOGINE, I. and NICOLIS, G.: Quart. Rev. Biophys. *4,* 107 (1971).
11 EIGEN, M.: Naturwissenschaften *58,* 465 (1971).
12 ONSAGER, L. and MACHLUP, S.: Phys. Rev. *91:* 1505 (1953); LANDAU, L. D. and LIFSHITZ, E. M.: Statistical physics (Pergamon Press, London 1959).
13 WAX, N. (Ed.): Selected papers on noise and stochastic processes (Dover, New York 1954).
14 A very clear presentation of this approach as applied to chemical kinetics is given by McQUARRIE, D.: in Suppl. Rev. Ser. Appl. Probab. (Methuen, London 1967). Different applications have been developed in VAN KAMPEN, N. G.: Adv. chem. Phys. *15:* 65 (1969).
15 KAC, M.: Probability and related topics in physical sciences (Interscience Publ., New York 1959); SIEGERT, A. J. F.: Phys. Rev. *76:* 1708 (1949).
16 See e.g. CALLEN, H. B.: in DONNELLY, R.; HERMAN, R., and PRIGOGINE, I. Non equilibrium thermodynamics, variational techniques and stability (Univ. of Chicago Press, Chicago, Ill. 1965).

·17 PRIGOGINE, I. and MAYER, G.: Bull. Cl. Sci., Acad. Roy. Belg. *41:* 22 (1955); LAX, M.:
     Rev. mod. Phys. *32:* 25 (1960); SCHLÖGL, F.: J. phys. Soc., Japan, Suppl. *26:* 215 (1969).
 18 See e.g. GANTSEVICH, S. V.; GUREVICH, V., and KATILIUS, R.: Sov. Phys., JETP *32:* 291
     (1971).
 19 NICOLIS, G. and BABLOYANTZ, A.: J. chem. Phys. *51:* 2632 (1969); HAWKINS, R., and
     RICE, S. A.: J. theor. Biol. *30:* 579 (1971).
 20 NICOLIS, G. and PRIGOGINE, I.: Proc. nat. Acad. Sci., (US) *68,* 2102 (1971); NICOLIS, G.:
     J. stat. Phys. *6,* 195 (1972).
 21 KERNER, E. H.: Bull. Mathem. Biophys. *19:* 121 (1957); ibid. *21:* 217 (1959); MONT-
     ROLL, E. *et al.:* Rev. mod. Phys. *43:* 231 (1971).
 22 GOODWIN, B.: Temporal organization of cells (Academic Press, New York 1963).
 23 BLUMENTHAL, R.; CHANGEUX, J. P. and LEFEVER, R.: J. Membr. Biol. *2:* 351 (1970).
 24 KOBATAKE, Y.: Physica *48:* 301 (1970).

Author's address: I. PRIGOGINE, Faculté des Sciences, Chimie-Physique II, Université
Libre de Bruxelles, 50, avenue F. D. Roosevelt, *Bruxelles* (Belgium)

The First Steps of Evolution
and the Nature of Life

Proc. 3rd int. Conf. From Theoretical Physics to Biology, Versailles 1971, p. 112
(Karger, Basel 1973)

Introduction

A. Szent-Györgyi

My first words, as a chairman, can only be words of thanks to the Institut de la Vie for having brought us together in this princely fashion.

It is befitting that the Institut de la Vie should devote one day to life itself, life as we know it, and how it originated.

Discussing the nature of the living state, the difficulty is that we cannot study life itself but have to study some part of its machinery. The smaller the part we study the more it behaves as a pure physico-chemical system, but also the less alive it is. So that what I hope we will do here is not only to discuss various details we have studied but also to try to correlate our findings with the whole, the emphasis being laid on this correlation. What we have published will be less important than what is residual—in the back of our minds. The success of our discussions will depend on the freedom with which you will disclose your thoughts.

Proc. 3rd int. Conf. From Theoretical Physics to Biology, Versailles 1971, pp. 113–124
(Karger, Basel 1973)

Water, Motion, Muscle and Evolution

A. Szent-Györgyi

Marine Biological Laboratory, Woods Hole, Mass.

Being a biochemist my first question about the living machinery has to be: what is it made of? Life originated in the ocean and so could build its machinery only of the three things it found there: dissolved molecules, ions and endless amounts of water. Accordingly, it is believed that life has built its machinery from these molecules, using ions as triggers and for balancing osmotic and electric differences. Water was the medium. I am unable to accept this view because water is the most extraordinary substance, all the properties of which are anomalous and I cannot believe that life should not have made use of them. I think that it is the anomalous properties of water which made life possible. Water is part and parcel of the machinery, and not merely its medium.

The anomalous behavior of water is revealed already by its two classical constants, melting point and boiling point. According to the size of the molecule the melting point should be $-100°$ C; it is $0°$ C. The boiling should be $0°$ C; it is $100°$ C. The shape of its molecules is also very unusual.

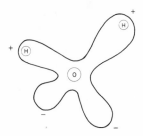

Fig. 1. Shape of water molecule [redrawn from R. A. Horne: Marine chemistry, Wiley-Interscience New York, N. Y. 1969].

It is represented in figure 1. On the one side, the two H-s stick out, while on the other, the two non-bonded electron pairs of oxygen protrude, lending to the molecule a strong dipole moment and enabling it to make 4 H bonds with about 6 cal of binding energy each. Accordingly, water tends to build solid structures if it finds a solid support. If the water structures surrounding two particles fuse, they will hold the two structures together. With L. DEMÉNY, we wondered about the strength of this bonding, and measured the forces by which water can bind two glass plates together. We found them surprisingly high. They depended on the thickness of the water layer. Below a thickness of 2000 Å we found it to be more than 1/2 kg per cm². Over 10,000 Å we found it very low, approaching zero, so that we can say that below a distance of 1000 Å water can act as a solid, and can make part and parcel of the living machinery (Fig. 2). I think that what holds our body together is, to a great extent, water – 'Water-bonds'. What obscured the issue till now is that there is a great difference in tendency of various surfaces to build water structures. Life could not exist if water would behave always only as a solvent or only as a glue.

That water plays a central role in the energetics of the biosphere is common knowledge. There are two ways only in which living systems can store energy, and then use it. The one is by separating the elements of water

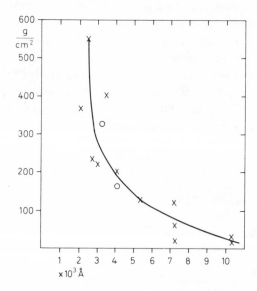

Fig. 2. Relation of thickness of water layer plotted against force by which it holds two glass plates together.

(photosynthesis), and then putting them together again (biological oxidation). The other way consists of taking out a water molecule of an ADP from between two phosphate molecules. What drives life directly is the energy released by putting the water back again.

This is well known. What I want to discuss is, how far water makes actual part of the living machinery which transduces chemical energy into work, be it mechanic, electric or osmotic. These transformations are the central features of life. Transduction into mechanical work, that is, motion seems the simplest of them. Its organ, in the animal, is muscle, and what I would like to discuss is how far water is involved in its function.

Being a biochemist my first question about muscle has to be again: what is it made of? My laboratory found that the contractile matter was chiefly made of two proteins, actin (1) and myosin which can form a complex: 'actomyosin' (2). The most striking quality of this complex is its extreme hydration. 3 g of actomyosin can hold 97 g of water, and hold it very strongly. (You can also say that 97 g of water holds 3 g actomyosin strongly.) If ATP is added to the hydrated system, it is split and, simultaneously, actomyosin turns into a completely hydrophobic matter and shrinks enormously. It not only shrinks, its mycels actually shorten. If we make a thread of actomyosin and arrange the mycels parallel to the axis, on addition of ATP the thread shortens and becomes wider without decreasing volume, does what muscle does. This hydration and dehydration is the central event of muscle contraction.

The next question was: what hydrates and dehydrates? Is it actin or myosin? It was easy to show that it was myosin which hydrates and dehydrates under the catalytic influence of actin. Later the myosin molecule was shown to consist of two parts which were called 'heavy' 'H' and 'light' 'L' meromyosin, corresponding to their sedimentation constants [2]. The 'light' had no special relation to actin and water (A. G. SZENT-GYÖRGYI). The dynamic part is the 'heavy' [4]. It has been shown lately [RICE, 5; HUXLEY, 6; SLAYTER and LOWEY, 7], that the H is a double helix of two molecular threads with two globules at the end. It is the globule which interacts with actin [MÜLLER, 8], then splits ATP, and it is the stalk which hydrates.

The function of muscle is to contract, shorten, that is, bring its two end points closer together. I should like to start with discussing how, on the molecular level, two points, A and B, can be brought closer together (fig. 3). The simplest way would be to take a molecular filament and bind its two ends to A and B. Such filaments have a strong tendency to increase

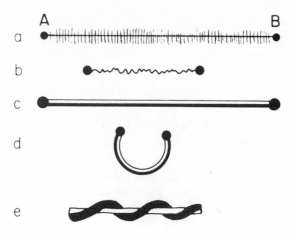

Fig. 3. Explanation see text.

their entropy by folding. So I would have to stretch my thread, tie it to A and B and then let it go. The thread would shorten and bring A and B together. The problem is how to keep the thread straight while I tie it to A and B. The only way I know is hydration, putting on a stiff water jacket, represented on my figure by shading. On dehydration the folding and contraction should occur (fig. 3c). The situation will be different if I take two threads out of which only the one can hydrate and dehydrate. Together, they would form a so-called 'bimetallic strip'. If the one would dehydrate, the system would bend. Muscle has two linear colloids, myosin which hydrates and actin which does not. If the dehydration would proceed along the double thread as a wave, at the point of hydration curving would be produced and a wave motion would be generated. If the two threads would be twisted, then a giratory screw motion would be generated (fig. 3e).

Coming back to muscle, according to our present knowledge, the situation is this: myosin is present in the form of so-called 'thick filaments' and actin as 'thin filaments' [HUXLEY, 9]. The former consist of L-myosin (fig. 4). The H is attached with a hinge [HUXLEY, 10]. In the resting muscle the globules are repelled by actin, or more exactly, by the troponine of EBASHI [11], a globular protein attached to the actin thread. In excitation Ca is released, is bound by the troponine and neutralizes its repulsion, whereupon the H-globules touch actin, become enzymatically active, split ATP (fig. 4b). This makes the stalk lose its hydrate shell and fold up, pulling the actin thread relative to myosin, which makes muscle shorten.

The thin stalk of the H-meromyosin constitutes only about $^1/_5$ of the total weight of actomyosin, but all the same, all the enormous amount of water is bound and released by it. The chemical energy is released in the H-globule by the splitting of ATP, but the energy transduction itself takes place in the stalk of H. What actually moves muscle is the entropy change of the stalk of H and that of the water. What pays for it is the energy of ATP, the splitting of which starts up some sort of a wave going through the stalk and releasing the water on its way. The nature of the wave is unknown. We are faced here with a hitherto unknown phenomenon which may be one of the most fundamental processes of life. It is difficult to believe that such a violent change in hydration could take place without changes in conformation, the molecule turning its hydrophylous groups alternately outside or inside. ASTBURY, BEIGHTON and WEIBULL [12] found indications of a transition from an α helical to a β cross configuration. The wave may be a wave of a configurational change.

My next problem was: how can nature develop such a very complex system, as a muscle fiber? Evidently, it had to start from some simpler beginnings. I think I can answer this question. If you look at a flagellate, moving bacterium, say, a vibrio which moves by wagging its tail, you notice that the tail ends with two globules. It is difficult not to recognize in this system the ancestor of the heavy meromyosin. The motion is generated by the waves passing along the tail, so you have here a direct visual evidence for the existence of waves passing along a protein fiber. ASTBURY, BEIGHTON

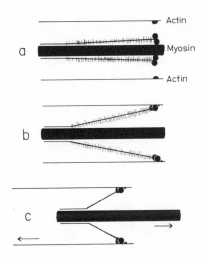

Fig. 4. Contraction cycle in muscle. Explanation see text.

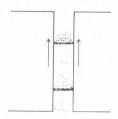

Fig. 5. Schematic representation of pore in membrane with waves travelling on wall. Explanation see text.

and WEIBULL [12] who studied flagelli of bacteria, called them 'monomolecular muscles'. Their x-ray spectrum resembled that of myosin but also showed periodicities of actin. Thus, I think that the primitive instrument of motion of monocellular organisms is the ancestor of muscle. Nature took the essential part, the tail, and put around it long filaments which integrate the motion of the single H-s to muscle contraction. These waves may be one of the most basic phenomena of life. A wave passing along the wave of a pore in a membrane (fig. 5), and producing structured water, may push dissolved molecules along, if these are less soluble in structured than in random water. So it may underlie transports of molecules. Such waves may be responsible for "protoplasmic streaming".

Before ending, I should like to touch upon a problem which reaches over into the problems which are to be considered later in this symposium. We know that nature evolves by random mutation. How can random mutation lead to the development of such a highly ordered and meaningful structure? I do not think that the length of time available, 3 billion years, is the answer. There must be some principle operating which can distinguish between useful and senseless mutation and throw out the latter. DARWIN, already, was looking for it.

One of the most striking differences between living and non-living machines is that the non-living are worn out by use, while the living ones deteriorate if they are not used. If you use your car too much it becomes worn out, while your legs atrophy.

There is a simple, hundred-year old physiological experiment which is analogous to this relation. This is the Bowditch [13] staircase. The experiment is this: we make the muscles of a frog heart dependent for excitation on electric impulses. When we stop these for a while, say half a minute or so, we expect that the first beat after the pause will be stronger, the heart having rested. The contrary will happen. The first beat will be weaker, and the beats will get gradually stronger as the heart works. HAJDU and I [14, 15] were interested in the mechanism of this phenomenon, and found that in the resting heart, potassium diffused into the fibers, its distribution became more random: entropy increased. The K was pumped back by the resumed activity. So, function decreases entropy, generates order and structure, while inactivity increases entropy till the whole system goes into a shapeless soup. Function generates structure and structure generates function. They stabilize one another. Even half a minute of inactivity or even a few seconds are enough to start disintegration. So it is believable that *non sense* mutations which do not lead to function, have to disappear, while mutations which improve function will be stabilized. This means that there is some very deep laying thermodynamic principle which stabilizes favorable mutations and throws out non-sensical ones, making random changes generating improved function and structure.

My feeling is that we will have to introduce a new concept into biology, that of *'biological stability'* which, contrary to physical stability, has its maximum at the minimum of entropy and the maximum of free energy.

References

1 STRAUB, F. B.: Papers from the Institute of Medical Chemistry, Univ. Szeged, vols. 2 and 3 (1942).
2 SZENT-GYÖRGYI, A.: The chemistry of muscular contraction (Academic Press, New York 1947).
3 MIHALYI, E. and SZENT-GYÖRGYI, A. G.: J. Biol. Chem. *201:* 189, 211 (1953).
4 SZENT-GYÖRGYI, A. G.: Arch. Biochem. Biophys. *42:* 305 (1953).
5 RICE, R. V.; BRADY, A. S.; DEPUE, R. H., and KELLY, R. E.: Biochem. Z. *345:* 37 (1966).
6 HUXLEY, H. E.: T. Mol. Biol. *7:* 281 (1963).
7 SLAYTER, H. S. and LOWEY, S.: Proc. nat. Acad. Sci., Wash. *58:* 1611 (1967).
8 MÜLLER, H.: J. biol. Chem. *240:* 3816 (1965).
9 HUXLEY, A. F.: Progr. biophys. Chem. *7:* 255 (1957).
10 HUXLEY, H. E.: Science *164:* 1356 (1969).
11 EBASHI, S. and ENDO, M.: Progr. Biophys. molec. Biol. *18:* 125 (1968).
12 ASTBURY, W.T.; BEIGHTON, and WEIBULL: Soc. exp. Biol. Symp. No. 9, p. 282 (1955).
13 BOWDITCH, H. P.: Ludwigs Arb. *6:* 759 (1871).
14 HAJDU, S. and SZENT-GYÖRGYI, A.: Amer. J. Physiol. *168:* 171 (1952).
15 HAJDU, S.: Amer. J. Physiol. *174:* 371 (1953).

Author's address: Dr. ALBERT SZENT-GYÖRGYI, Laboratory of the Institute for Muscle Research at the Marine Biological Laboratory, *Woods Hole, MA 02543* (USA)

Discussions

G. Ling: I would like to comment on Dr. Szent-Györgyi's fine presentation by adding two slides which give further support to the general concept that reversible organization of water in the cell plays a major role in cell physiology. More specifically, we attempt to answer the questions: Is water in the living cells truly different from water in a normal aqueous solution? If it is, does this different physical state change during muscle contraction? In this particular instance, we chose to study the freezing pattern as a criterion. Figure A1, taken from Hallett, shows how normal water freezes in two dimensions when it is supercooled to $-3°$ C and seeded with a single crystal on the left edge. Figure A2, taken from Menze and Luyet, shows how ice forms in an actomyosin gel, which has been dried to about 11% and is thus roughly at the same strength as found in the living muscle cells. The pattern of ice formation is distorted but not fundamentally different from ice formation in normal water. However, as shown in figure BA ice formation is quite different in a living muscle cell [Miller and Ling: Physical Chem. Phys. 2: 495, 1970]. Here it only grows in one dimension. It never grows sideways, and never branches. By touching the exposed supercooled cytoplasm with seeding crystals, we can initiate many parallel running spikes. The direction of ice crystal growth is delineated by the cytoplasmic proteins which also run in essence in the longitudinal direction. When the muscle fiber is twisted, the ice filaments have become similarly twisted, yet at all times never touching each other (fig. BB). In figure BC, you have at one end of the muscle fiber these spikes growing in a regular straight manner until the ice spike approaches the right side of the muscle fiber. At this end the muscle has been chemically excited with caffein (5 mM). Ice formed in this region is in a totally different fashion. Figures BD and BE suggest that the unusual physical state of water is an essential ingredient in cell functions.

I would like to make a further comment on the idea that deep layers of water can exist in a physical state different from normal water. There are other evidences in addition to Prof. Györgyi's demonstration of strong cohesion between glass plates filled with water. One specific piece of work I would like to cite is that published by Hori from the Hokkaido University Low Temperature Station. Hori found that at the distance of separation of $0.1~\mu$, water between flat glass plates becomes very different from normal water: firstly it will not freeze at $-90°$ while in contact with ice crystals; secondly, it has no measurable vapor pressure even at $300°C$ [Ling, 1970, loc. cit.].

Finally, I would like to make still another comment. A theory has already been proposed on how this water can be polarized by the formation of successive polarized multilayers [Ling, 1965, loc. cit.; 1970, loc. cit.]. A theory has also been suggested as to how the propagation of signals along the length of protein-water filaments such as that discussed by Prof. Szent-Györgyi, could be achieved by a falling-domino-like short-range interaction. This theory has been called the 'association-induction hypothesis' [Ling, 1962, loc. cit.].

C. Levinthal: I would like to ask a question, partly of Dr. Szent-Györgyi and partly of Dr. Palma. It is my understanding that when Dr. Palma speaks about proton resonances in an agar suspension, he is measuring the proton in the water. One can estimate that the distances between the agar molecules in a 2% solution is about 250 Å. Apparently what he is finding is, that at these distances, the water is behaving as though it is frozen, as you are suggesting. I wonder if experiments of that kind can be done using some solute other than agar in which one might suspect there is less direct solute-solute interaction. The cooperativity Dr. PALMA was discussing appears to be a normal sol-gel transformation. But the question is whether one can use experiments of this kind to measure the distance over which water is essentially frozen or to determine whether the positioning of the solute molecules is determined by the direct interaction with each other or by the ordering of the water. Experimentally one might be able to distinguish the latter possibility by determining whether types of solute could be ordered in the same way.

A. SZENT-GYÖRGYI: The question about proton resonance can better be answered by Dr. PALMA. I hope that in the not too distant future the water structures will become visible. Dr. FERNANDEZ MORAN works at the temperature of liquid helium. At this temperature the water structures must be solid and may be made visible in the electron microscope.

VOICE: If I understood it right, when you take the water away, the fiber shrinks, it pulls up. What exactly is the process of this folding and why did you call it entropy folding?

A. SZENT-GYÖRGYI: Muscle shows many analogies to rubber. My earlier calculations suggested that what is driving muscle is the increase in entropy. Eventually the splitting of ATP has to foot the energy bill.

VOICE: The mechanism of the folding, is it the reversible chemical reaction between parts of the fiber?

A. SZENT-GYÖRGYI: I do not know the mechanism of folding. I only know that folding of protein molcecules is one of the most fundamental processes in biology. All globular proteins are formed as fibers and fold up later, acquiring their specific properties. What is driving and directing this folding we do not know. Possibly, water structures play an important role.

U. PALMA: May I comment on LEVINTHAL's comment. Of course, that is a typical sol-gel transition; however, the point I tried to make yesterday (also in my reply to ONSAGER) is that, in order to understand this type of transition, it is not sufficient to introduce cooperativity on a phenomenological basis; one also needs a microscopic model. Now, just because the distance between two agar molecules is the one you have calculated, it would be difficult to explain the very conspicuous cooperative character of the transition in terms of a direct molecular interaction only. There must be a channel, so to say, to spread the structural information, to account for a host of effects, such as helix formation and stability, cation selectivity in gel-promotion, extraordinarily large effects on the dielectric constant, let alone the NMR data I have discussed.

C. LEVINTHAL: It seems to me that the agar molecules being very long and thin could easily be making a cross-linked network which is rigid without invoking any ordering of the solvent to explain the properties of the gel.

U. PALMA: If this were true, then there would only be thin layers of bound water around the agar molecules, and the rest of the water would be completely free. Apart from the effects I have just mentioned (to which I would add the thixotropic and 'healing' behaviour), this situation would cause a linear dependence of T_2 upon the agar concentration, which was not experimentally observed.

C. LEVINTHAL: The reason I raised my question after Dr. SZENT-GYÖRGYI's talk was just the fact that he also has evidence that the ordering of water extends over more than a monomolecular layer and that is the way you are interpreting your experiments. If I understand correctly, this is really the conclusion of both yours and Dr. SZENT-GYÖRGYI's experiments whether or not your explanation for the sol-gel transformation is correct.

U. PALMA: I think so, too, thank you. I would also like to add that the way we think of this order is not in terms of identifiable layers of water which orderly add one on top of the order, but rather in terms of an overall cooperative phenomenon. May I also ask Professor SZENT-GYÖRGYI how he measured the 97% loss of water upon addition of ATP.

A. SZENT-GYÖRGYI: Simply by measuring it putting the fibers into the desiccator and measuring the loss of weight. In the contracted fiber there is still about 50% water, the quantity one would expect to be present as an inclusion. This is free water. The bound water is lost.

U. PALMA: So, your specimen could indefinitely stay in the dessicator without suffering any loss of water, unless ATP is added, and only then the 97% of water goes.

A. SZENT-GYÖRGYI: You can calculate the quantity of the water lost on addition of ATP by measuring the volume change. On adding ATP there is a great loss of volume. Only a small volume is left, half of which is occupied by the protein.

A. KATCHALSKY: Some of the aspects of the picture drawn by Prof. SZENT-GYÖRGYI are closely related to dissipative structures, on which I will have more to say during my talk. It is the close relationship between structure and function which makes dissipative structures an interesting model for the properties of living systems. Here I would like, however, to mention another organizational aspect underlying the structure of the muscle. A long-standing physical question is how the actin and myosin of muscle retain their beautiful quasi-crystalline array, with a precise spacing and without local collapse. Moreover, it is rather odd why the molecules, which do not maintain contact, do not dissolve readily in aqueous media. There must operate an equilibration principle which keeps the muscular actomyosin at the right distances, without collapsing and dispersing. Recently G. F. ELLIOTT in England tried to figure out the equilibrium pattern of the muscle by the assumption that the distances are determined by an exact match of electrostatic repulsion and Van der Waals attraction [J. theoret. Biol. 21: 71, 1968]. For electrostatic repulsion he assumed polyelectrolyte forces of the type calculated a decade ago by LIFSON and myself[1], while the Van der Waals attraction was evaluated according to the procedure of VERWEY and OVERBEEK.[2] It is rather gratifying that the calculated equilibrium distance is close to the experimental values. It seems, therefore, reasonable to postulate that the interplay of these two types of forces plays an important role in the organization of many biological structures.

1 LIFSON, S. and KATCHALSKY, A.: J. Polymer Sci. 13: 43 (1953).
2 VERWEY, E. J. W. and OVERBEEK, J. Th. G.: Theory of the stability of hydrophobic colloids (Elsevier, Amsterdam 1948).

D.D. ELEY: I wonder how far Stefan's law for the viscous flow of a body between two plates can contribute to Professor SZENT-GYÖRGYI's very interesting results.[3] The force should be proportional to the viscosity, fourth power of the plate radius, and inversely as the second power of the separation distance[3]. With regard to Prof. MENDELSOHN's question, although the formation of a random coil from stretched out chains, has a positive ΔS in rubber, it seems to me the rather precise folding of a protein chain might have a ΔS more nearly zero. However, if this precise folding is accompanied by the loss of a monolayer of bound water this might give a large positive ΔS in itself. It may not be necessary to postulate long range structural changes in water, and I am a little critical of the evidence that has recently been brought forward for such structures, especially those associated with the name, 'polywater'.

G. LING: I have recently had the experience of reviewing the field of hydration of macromolecules in general [LING: Hydration of macromolecules; in HORNES Structure and transport processes in water and aqueous solutions, John Wiley, New York, in press]. In the process of reviewing and reading the papers published, a curious picture emerged. There are two diametrically opposed theories proposed. Each one of them, as time went by, has gathered more and more experimental supports. The first theory should be credited to LINUS PAULING, who argued that proteins hydrate only at their polar side chains and that the backbone NHCO groups do not form centers of hydration. The other theory can perhaps be traced to DOROTHY JORDAN LLOYD in which she and her coworkers had produced evidence that the backbone as well as its polar side chains are the sites of hydration. As I was reading these papers supporting one or the other view, I noticed that all those people who came to support LINUS PAULING's theory were more or less professional protein chemists working in an academic institution on pure, largely globular proteins. On the other hand, those supporting the alternative view were all more or less industrial chemists involved with studies of wool and other protein fibers. So, the conclusion suggested itself: all globular proteins hydrate to the extent of no more than a single layer of water on the outside surface of a roughly spherical molecule with the backbone NHCO groups folded inside. Fibrous proteins, on the other hand, have definite new centers of hydration, i.e., the backbone amide groups. Now, the backbone NHCO groups can, of course, offer a one-dimensional or even two dimensional array of positive NH and negative CO sites. It is the possession of these alternating positive-negative centers that is shared between proteins and polished glass surfaces. Apparently, in order to influence water in deep layers, you need these checkerboard positive and negative sites.

U. PALMA: Let me quckly comment on diffusion properties. Take for a moment the 'flickering clusters' model for liquid water: we can think of this flickering as excitations propagating throughout the liquid. Similarly, any other model of liquid water will allow us to think in terms of excitations. If we have, as you said, molecular surfaces acting as templates to imprint a structure on water, this is equivalent to saying that we change the boundary conditions, thus changing the excitations spectrum of water. This may also result in a concerted orientation or arrangement of macromolecules, giving rise to a water excitations spectrum having fairly defined symmetry properties. Now, it must be seen, where does the bottleneck for diffusion lie? It seems more than reasonable to assume that it involves a local activation energy that is not very much affected by the overall excitations spectrum. On the other hand,

3 STEFAN, M. J.: S. Ber. Akad. Wiss. Wien, math.-nat. Kl Abt II 69: 713 (1874).

this spectrum may profoundly affect e.g., the T_2 relaxation time in NMR. So, as you see, completely different situations can correspond with substantially unchanged diffusion properties.

L. ONSAGER: Evidence brought for hydration is ambiguous, as is some of the evidence against. The diffusion of small molecules in protein crystals could go just as fast if some other water is bound and the small molecules move in the part that is not. A better test would be the cell self diffusion of water; either oxygen 18 or tritium will do it for labels. Another test of binding, and the strength of binding and—one of the more significant ones, is infrared spectra. Here incidentally, we must bear in mind that water binds rather strongly to itself. Some notions that water binds strongly to monovalent ions must be tempered by an observation made over 30 years ago by GANZ, that actually among the monovalent ions, the one that practically leaves the frequencies unchanged is sodium. Sodium chloride does very little to the infrared spectrum of water. Replace sodium by lithium and you raise the frequencies; replace sodium by potassium, rubidium or cesium and frequencies are decreased. In other words water sticks probably less to those ions than it sticks to itself. Similarly, fluoride tightens the binding and bromide and iodide loosen it. Another criterium, more subtle and somewhat paradoxical is the dielectric relaxation time. Here, in dealing with ions you have to bear in mind that the motion of the ions assists in the relaxation of the solvent. So you have the curious thing of the decrease of the dielectric constant, combined with the shortening of their relaxation time—for most small ions. The ones that seem to lengthen the relaxation time by some secondary effect are big things like tetraethylammonium and no one would suggest that tetraethylammonium ties down the water. It does hamper the motion of the water, in the same way that the vapor does—I mean the *empty* space—because the water molecules want to coordinate with each other; if on the one side, the water 'experiences' this, that, in itself is a severe constraint on the motion of the water molecules, that on the one side there is a 'nothingness'. If on one side there is something, e.g., oil—that is 'in-between'. But in either case, it causes the water molecules to tie up with each other and actually imposes some constraint associated with 'hydrophobic bonding'. In general, however, it tends to lengthen relaxation times. That is not a matter of bonding; that is a matter of other constraints. Then, there are similar situations in some other gas hydrates where the water molecules from little tetrahydra tied up with each other and that is a somewhat different structure than in the ices and so forth. Aside from all the details, bear in mind that there are various subtleties involved in the whole picture.

Proc. 3rd int. Conf. From Theoretical Physics to Biology, Versailles 1971, pp. 125–132
(Karger, Basel 1973)

On the Prebiotic Synthesis of Peptides

A. KATCHALSKY [1]

The subject of this paper is a possible mechanism for the biosynthesis of peptides. I am going to assume that everyone is informed about the basic findings of the last two decades, with regard to chemical evolution and the first step of prebiotic synthesis. It will, therefore, be superfluous to review here the possible mechanisms for synthesis of the low molecular weight compounds which might have taken place in the primeval ocean. We shall consider, however, in some detail a possible mechanism for the chemical synthesis of the biopolymers, and particularly polypeptides; I hope that this will provide food for thought on the evolution of biopolymers towards information-rich macromolecules and the origins of cellular organization.

The problem of the prebiotic synthesis of polypeptides was considered already several years ago by Fox [1]. He succeeded in polymerizing amino acids under dehydration conditions and obtained his famous "proteinoides", which resemble, from a certain point of view, protein molecules. CALVIN [2] and his coworkers, and in particular STEINMANN [3], felt, however, that it is difficult to assume that life started in an anhydrous medium, and carried out an interesting study on polymerization in aqueous media. The project is rather difficult on *a priori* grounds, since all biopolymers—proteins, nucleic acids and polysaccharides—are based on the removal of water from monomers and it seems to be a rather hopeless task to dehydrate in an aqueous medium. To make the process possible, the monomers have to be transformed into energy-rich forms which would polymerize with liberation of energy and overcome the dangers of

1 Professor Aharon Katzir-Katchalsky fell victim to that attack on innocent passengers by terrorists on May 30 1972 at Lod Airport

hydrolysis. As a starting point for this work, CALVIN and his coworkers made use of the observation of KHORANA that DCC (dicyclocarbodiimide) can remove water from monomers in an aqueous medium, leading to the formation of such interesting polymers as the polynucleotides—using mononucleotides as starting material. Close chemical relatives of DCC are dicyandiamide and its transformation forms, which are readily synthesized spontaneously under prebiotic conditions. It was, therefore, natural to try and prepare polypeptides in aqueous media using dicyandiamide and related compounds as the dehydrating agent. It was found, however, that although peptides could be obtained in this manner, their molecular weights were very low and they required acid media for their synthesis. Since the primeval ocean was rather alkaline it does not seem that this process solves the problem of the origin of peptides.

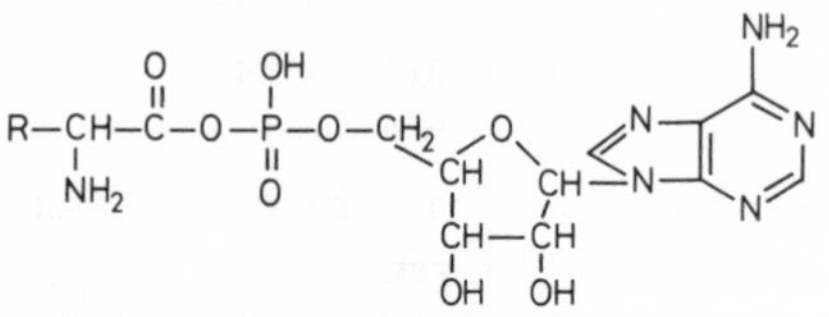

Amino acid—adenylate

Fig. 1. Amino acid adenylate.

If we assume that the product of the first step of protein synthesis in the cell resembles the active forms of amino acids in the primeval poly-merization, it would be natural to look to adenylates of amino acids as the active forms for the polymerization to polypeptides. In the cell, amino acid adenylates are formed by enzymatic interaction of amino acids with ATP (fig. 1). It was a logical idea, therefore, to mix amino acids and ATP at a neutral pH and try and prepare the adenylates by a non-enzymatic process. Since both amino acids and ATP were shown to form spontaneously under prebiotic conditions, such a reaction was expected, in particular since ATP is an energy-rich compound which could activate the amino acid. No interaction was found, however, between amino acids and ATP, even after prolonged incubation at pH 7 and a temperature of 37°C. My coworker, Dr. MELLA PAECHT-HOROWITZ, and I, came to realize that the reaction of the triphosphate groups with the carboxylic end of the amino acid cannot take place as long as the carboxyl is in the ionic form, COO⁻, and conditions had to be found to transform it to the COOH form, despite the neutral pH.

This could be carried out only on heterogeneous surfaces and a suitable group of catalysts was found in the zeolites, in particular Decalco F, which undergoes a membrane hydrolysis in water and carries on its surface free silicic acid groups.

$$> \text{SiO}^-\text{Na}^+ \qquad \xrightarrow[\text{hydrolysis}]{\text{H}_2\text{O}} \qquad > \text{SiOH} + \text{NaOH}$$

Decalco F   Decalco F

in sodium form   in free acid form

The free silicic acid reacts with the amino acid to give

$$\text{SiOH} + \overset{+}{\text{NH}}_3 - \overset{\overset{\text{R}}{|}}{\text{CH}} - \text{COO}^-$$
$$\downarrow$$
$$\text{SiO}^- + \overset{+}{\text{NH}}_3 - \overset{\overset{\text{R}}{|}}{\text{CH}} - \text{COOH}$$

and the acid cation is now available for interaction with ATP, producing readily the adenylate which desorbs from the zeolite surface and appears in the free solution.

Amino acid adenylates indeed polymerize spontaneously and give polypeptides in aqueous media, the polypeptide yield being maximal at a pH slightly higher than 8. The polycondensation reaction, however, has

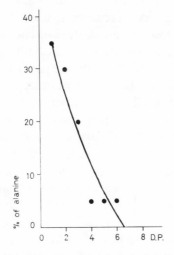

Fig. 2. The degree of polymerization of peptides obtained in homogeneous polycondensation of alanine-adenylate. D.P. = degree of polymerization.

a strong competitor, namely, hydrolysis of the adenylates to amino acid and adenylic acid. Hence, the main product of the reaction is the monomer, and the polypeptides obtained are generally of a low molecular weight as shown in figure 2. For several years, PAECHT-HOROWITZ and I tried in vain to increase the yields and molecular weights, but the results were discouraging. Recently, we came to the conclusion that heterogeneous catalysis is required also for a polycondensation which would lead to suitable starting materials for the evolution of information-rich biopolymers. Reviewing the literature was of little help in suggesting the right catalyst. The only pertinent indication was found in a rather old reference source, namely, the book of Genesis . . . As is well known, the Bible claims that life started from clay, so that we are led to consider various clays as possible catalysts. Indeed, already in 1948, BERNAL [4] came to the conclusion that if life started in the ocean, the polycondensation process of active forms should have taken place on clays. In his famous lecture on "The Physical Basis of Life" he wrote: "The primary difficulty, however, of imagining processes going thus far is the extreme dilution of the system if it is supposed to take place in the free ocean. The concentration of products is an absolute necessity for any further evolution. One method of concentration would, of course, take place in lagoons and pools which are found to have fringed all early coastlines, produced by the same physical factors of wind and wave that produce them today. It has occurred to me, however, that a much more favorable condition for concentration, and one which must certainly have taken place on a very large scale, is that of absorption in fine clay deposits, marine and fresh water. Our recent knowledge of the structures of clays has shown what an enormous role they still play in living processes."

"It is therefore certain that the primary photochemical products would be so absorbed, and during the movement of the clay might easily be held blocked from further possibly destructive transformations. In this way relatively large concentrations of molecules could be formed." The trouble was, however, that the number of clays is enormous and it was rather lucky that a group of swelling clays—the montmorillonites—were found to act as marvelous polycondensation catalysts for adenylates of amino acids [5]. The montmorillonites disperse in dilute solution to single platelets composed of three layers : two bounding layers of silica and an inner layer of aluminium (fig. 3). The high specific area of such platelets permits a very high adsorption of amino acid adenylates. The adsorbed monomers are protected from hydrolysis and the polycondensation product is practically free of monomeric amino acid. An advantageous aspect of the montmorillonite

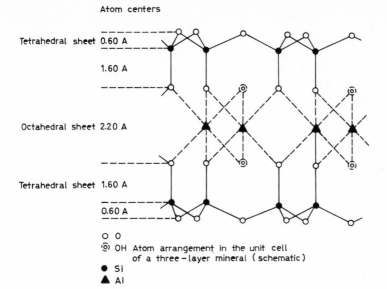

Fig. 3. Ideal structure of montmorillonite.

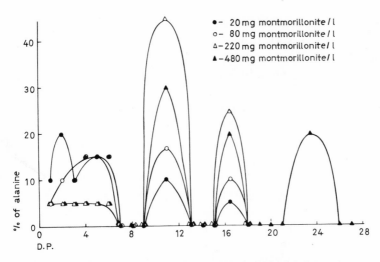

Fig. 4. The degree of polymerization of polypeptides obtained in heterogeneous polycondensation of alanine-adenylate with montmorillonite as catalyst. D.P. = Degree of polymerization.

adsorption is the fact that, while the adenylate adsorbs strongly, neither the adenylic acid nor the polypeptide produced in the polycondensation remain on the surface of the montmorillonite particles and desorb readily into the aqueous medium. The remarkable consequence of the process is, however, that in a short time very high polypeptides of up to 50 amino acids are attained.

A closer study of the product of polycondensation demonstrated that there were other aspects of fundamental interest in the polycondensation on montmorillonites. A conventional analysis of a polymerization process is based on the study of the distribution curve of the molecular weights of the different polymeric chains obtained in the process. Generally, the distribution of molecular weights is Poissonian or Gaussian. In the present case, however, to our surprise we found that the distribution of molecular weights was spectral, arranged in bands of repetition of a basic number: one band of molecular weights was around the hexamers, another band round the dodecamers, a third around the octadecamers, and so on, up .o 48 amino acids, as shown in figure 4. The sharpness of the bands is of interest in itself with regard to the unique distribution of molecular weights in the proteins. There is, however, another great interest in the evolution of the spectral pattern of macromolecular chains. A possible interpretation which is immediately suggested is that the sterical arrangement of groups in the montmorillonite surface favors the initial formation of hexapeptides, which then combine to doublets, triplets, etc., to give peptides of 12, 18 and 24 amino acids. There are, however, some doubts about this mechanism since in some cases the repetitive unit is 7 or 8 amino acids and the spectral lines are not sufficiently sharp. There is, however, a possibility that another mechanism, of a less specific nature, is operative, namely a mechanism in which a superposition of the chemical rate of diffusion determines the distribution. Such a mechanism should lead to a dissipative structure, and I shall discuss this now. Before considering the general aspects, it is worth noting what are the processes involved in the present case: the polycondensation of amino acid adenylates requires an interaction of a carboxyphosphoanhydride group of one monomer with the amino group of another. Such a reaction cannot take place with a non-ionized ammonium group; it requires a free amino group. Hence, it is plausible to assume that on the surface of the montmorillonite particles a sufficient number of free amino groups is formed. In an acid medium, however, the process is inhibited, since protons compete with the heterogeneous catalyst. Now, in the polymerization process itself, adenylic acid is liberated, which tends to inhibit the polymerization. We may, therefore, envisage the polycondensation as composed of bursts of rapid condensation, in which a band of molecular chains is formed. During this burst adenylic acid is formed, which stops the process until the adenylic acid diffuses out, leaving the catalyst surface ready for the next burst and the formation of the following band of molecular weights. If this hypothesis will be found to describe quantitatively the behavior in this system, the prebiotic synthesis of polypeptides will represent a new field of dissipative structures[2].

References

1 Fox, S. W.; Harada, K., and Kendrick, J.: Science 129: 1221 (1959).
2 Calvin, M.: Chemical evolution (Clarendon Press, Oxford 1969).
3 Steinmann, G.; Kenyon, D. H., and Calvin, M.: Biochim. biophys. Acta 124: 339 (1966).
4 Bernal, J. D.: The physical basis of life (Routledge and Kegan Paul, London 1951).
5 Paecht-Horowitz, M.; Berger, J., and Katchalsky, A.: Nature, Lond. 228: 636 (1970).

Author's address: A. Katchalsky, The Weizmann Institute of Science, Rehovot (Israel)

2 The name "dissipative structures", coined by Prigogine, fits a large family of phenomena considered since the middle of the 19th century. Dissipative structures are patterns in space and time which emerge, beyond a range of instability, through the interaction of several coupled flows. It seems that the first who considered spatial organization through coupling of flows was the famous British philosopher Herbert Spencer, who wrote in his First Principles that the basis of life is a complexation related to the transition from homogeneous to inhomogeneous systems. To demonstrate his idea, he suggested a rather simple experiment, in which a drop of shellac in light petroleum is allowed to evaporate on a glass surface. After a while the surface of the shellac is covered with a film which develops into a honeycomb structure. Some fifty years later Benard in France discovered the first pattern based on interaction of flows. Benard heated evenly and homogeneously the bottom of a container in which a heavy liquid (spermacetti) was held. At a critical point a cellular structure developed in the liquid, which filled out the container with a honeycomb pattern. Each cell of the honeycomb structure comprises a circulation of convectional flow, coupled with a heat flow. It is this coupling which breaks the symmetry of the homogeneous liquid and transforms it into an inhomogeneous structured system. Upon continuing the heating, another transition point is found at which the pattern disappears. It is thus recognized that these patterns of flow have a limited range of existence, with an upper and lower bound. The maintenance of these dynamic structures requires energy input, and in contradistinction to equilibrium structures (such as the furniture in the room), must dissipate the energy in order to survive. Since life of organisms is also bounded and since it is based on dissipation, it is intriguing to speculate about the possibility that numerous biological patterns are true dissipative structures.

The theory of the Benard phenomenon was given by Lord Rayleigh (1916), who for the first time evaluated the transition point from one pattern to another. This theory was further advanced by CHANDRASEKHAR in his famous 1945 book on hydrodynamic and hydromagnetic stability. Recent work of the hydrodynamicists extended the experimental range of Benard patterns and there is, at present, good evidence for the transition from one structure to another by a controlled input of energy.

From a biological point of view, the flow patterns based on the coupling of chemical reaction with diffusional flow are more interesting. In a classic paper, published in 1952, A. M. TURING demonstrated that such patterning is possible, and, moreover, he indicated that it may play a profound role in the morphogenetic development of living systems. PRIGOGINE and his school have adduced numerous examples which show the possibility of chemico-diffusional organization of matter both in space and in time, and the fine experiments of ZHABOTINSKY show that dissipative patterns can be obtained by the coupling of a simple reaction with diffusion. There is a growing recognition that such patterns are rather common in biological systems and Benard-like structures were found in bacteria and protozoa and are believed to underlie tissue formation in slime molds and higher organisms. It is on this background that we assume that already in the very early process of prebiotic synthesis of biopolymers the coupling of chemical and diffusional flows on the surface of the early heterogeneous catalysts led to a sharpening of the molecular weights and to a simple selection process of favored macromolecules.

Proc. 3rd int. Conf. From Theoretical Biology, Versailles 1971, pp. 133–144
(Karger, Basel 1973)

The Rapid Evolution of Complex Systems from Simple Beginnings[1]

S. W. Fox

Response to a number of earlier comments will occur spontaneously in this presentation. I want to emphasize, however, that experiments indicate that proteinoids would not have had to evolve to information content; they had this quality from the outset [1]. This information is expressed in selective interactions with enzyme substrates [2], selective interactions with polynucleotides [3], and in selective assembly into systems.

Amino acids represent an early stage in organic evolution; methods for their geological production have been much studied [4]. Our interest here centers on what can happen after the amino acids are polymerized. At this point, we need to consider a lingering difficulty in concept.

CAROTHERS, who gave us the nylons through thermal polymerization of amino acids (if we use the term loosely), proposed that α-amino acids could not be polymerized thermally [5]. When we examine this concept, distinction between structural types of amino acid becomes crucial.

In essence, CAROTHERS polymerized ω-amino acids

$$H_3^+N\,(CH_2)_n\,COO^-\quad n = 4 - 6$$

The amino acids from protein are, however, α-amino acids:

$$H_3^+N\,CHRCOO^-$$

1   The research described has been made possible by funds from the National Aeronautics and Space Administration, Grant No. NGR 10-007-008. Contribution No. 194 of the Institute of Molecular Evolution.

CAROTHERS' inference is correct for most (the "neutral") amino acids. What has been learned beyond CAROTHERS' inference is that amino acids mixtures containing basic amino acids,

e.g. $H_3^+N (CH_2)_4 CHNH_3^+ COO^-$ lysine

or acidic amino acids

$^-OOCCH_2 CHNH_3^+ COO^-$ aspartic acid

$^-OOC (CH_2)_2 CHNH_3^+ COO^-$ glutamic acid

can be copolymerized by heat, to include all of the proteinous amino acids. When such mixtures are heated, the polymers which are produced, more closely resemble the proteins than the polymers CAROTHERS and his students produced in the effort to simulate the protein of silk.

Our purpose in using a mixture of as many as eighteen kinds of amino acid is based on recognition of the potentiality for isomerism which is unique to protein or protein-like molecules [6]. The vast numbers possible during evolution provide a potential matrix for arrays of differing specificities, such as would be needed for enzymes permitting the evolution of metabolism. Moreover, geologically relevant experiments show that amino acids are typically produced in sets, or families [7].

Since polymerization is rapid, the kind of complexity represented by varieties of heteropolyamino acids is the kind, I have been told, that has been of interest to physicists. The relevance to fundamental physical thought is increased by the rapidity and ease with which the polymers assemble into complex microsystems. This kind of study is constructionistic in contrast to much of biology, which proceeds by disassembling complex contemporary systems. I believe that only by assembling small synthetic components to macromolecules and thence to a kind of cell [8] can we hope to appreciate how easily complexity arises from simple beginnings by simple processes.

When the conditions are geologically relevant, as those in our studies are [4], the reactions constitute a model of pre-Darwinian evolution.

Figure 1 demonstrates the kind of result known to CAROTHERS (on the left). On the right is the clean proteinoid (protein-like polymer) obtained by including sufficient proportions of aspartic acid and glutamic acid, heating to 170° for 6 h, and purifying the polymer from a light pigment.

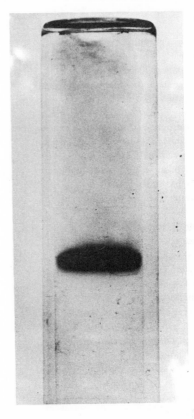

Fig. 1. On left, tube containing mixture of amino acids heated to above the boiling point of water; product is dark and tarry. On right, granular polymer prepared by heating a mixture of amino acids containing sufficient proportions of aspartic acid and glutamic acid; product is light or white depending upon details of preparation.

Fig. 2. Acrylamide gel electrophoresis of hemoproteinoid 83a at pH 8.6 [10]. Colored by Amidoschwarz 10B. The preparation appears to be homogeneous also by the use of electrophoresis at pH 4.5 and gel filtration.

The amino acids were heated in an initially dry state to overcome the energetic barrier which exists in aqueous solution [9]. The temperatures used are sufficient both to distil water from aqueous solutions and to polymerize the dried residue [4]. The requirement of a hypohydrous environment is, in my opinion, also met by the use of montmorillonite, with adenylates. Reactions facilitated at such surfaces are not occurring in dilute aqueous solution.

Table I. Self-ordering in the thermal condensation of amino acids

Evidence	Date and authors
Nonrandom sequences by disparity between N-terminal and total analyses in thermal polymers	*1958:* FOX and HARADA
%s in reaction mixture ≠ in polymer	*1960:* FOX and HARADA
Two peaks from proteinoids on electrophoresis	*1960:* VESTLING
Limited heterogeneity on ultracentrifugation	*1961:* VEGOTSKY
Constant composition on repurification from water	*1963:* FOX, HARADA, WOODS, WINDSOR
Single band on gel electrophoresis for acidic proteinoidamide	*1966:* FOX and NAKASHIMA
Nonrandom elution pattern from DEAE-cellulose	*1967:* FOX and NAKASHIMA
Symmetrical peaks from DEAE-cellulose	
Almost uniform amino acid compositions in various fractions	
Stoichiometric amino acid compositions	
Uniform ultracentrifugal patterns of various fractions	
Almost uniform peptide maps in all fractions	
Single spots on high voltage electrophoresis of fractions	
Single species of "active site" proteinoids	*1968:* USDIN, MITZ, and KILLOS
Single band for gel electrophoresis of basic hemoproteinoid	*1971:* DOSE and ZAKI

While the objective of employing eighteen types of amino acid was pointed mainly toward later evolutionary development [6], the thermal coupling of two to eighteen types of amino acid has been shown to yield polymers with much internal order. The various kinds of evidence, all of which require this inference, are shown in table I. A striking instance is seen in figure 2, kindly provided by Professor KLAUS DOSE [10].

We visualize that the ordering is due to selective interaction of amino acids, also shown in other systems [11, 12], and especially to thermal rearrangement to increasingly thermodynamically stable sequences. The

Table II. Catalytic activities in thermal proteinoids

Reaction and substrate	Remarks	Authors and year
Hydrolysis		
p-Nitrophenyl acetate	Activity of proteinoid greater than of equivalent free histidine	Fox, Harada, and Rohlfing (1962)
p-Nitrophenyl acetate	Thermal polymers most active	Noguchi and Saito (1962)
p-Nitrophenyl acetate	Inhibition by organic phosphates; reversal	Usdin, Mitz, and Killos (1967)
P-Nitrophenyl acetate	General description	Rohlfing and Fox (1967)
p-Nitrophenyl acetate	Reactive site, and inactivation	Rohlfing and Fox (1967)
ATP	Through Zn salt	Fox and Joseph (1965)
p-Nitrophenyl phosphate	A second phosphate hydrolysis	Oshima (1968)
Decarboxylation		
Glucuronic acid	From glucose, CO_2	Fox and Krampitz (1964)
Pyruvic acid	→ acetic acid + CO_2; Michaelis-Menten kinetics	Krampitz and Hardebeck (1966) Hardebeck, Krampitz, and Wulf (1968)
Oxaloacetic acid	Rapid, requires basic polymers	Rohlfing (1967)
Amination		
α-Ketoglutaric acid	Requires both Cu^{++} and proteinoid	Krampitz, Diehl, and Nakashima (1967)
Deamination		
Glutamic acid	Requires both Cu^{++} and proteinoids	Krampitz, Haas, and Baars-Diehl (1968)
Oxidoreductions		
H_2O_2 (catalase reaction)	Activity of hemin lowered when incorporated into proteinoids	Dose and Zaki (1971)
H_2O_2 and hydrogen donors (guaiacol, hydroquinone, NADH, and others) (peroxidase reaction)	Activity of hemin increased up to 50 times in lysine-rich hemoproteinoids	Dose and Zaki (1971)

result is almost at the other extreme from the *a priori* theoretical calculations based on an assumption of randomness. That assumption is unjustified for this coupling process, as the many results show. The significance to the primordial sequence is that the first ordered proteins did not require prior nucleic acids. This experimentally derived fact resolves a number of chicken-egg dilemmas.

One of these dilemmas is how enzymes came into existence when no enzymes existed to make them. The ordering of amino acids in proteins to produce catalytically active polymers is better understood in association with an explanation of that order.

Internal compositional ordering is observed also in the copolymerization of aminoacyl adenylates, which has been studied [13], in an approach to understanding the origin of the genetic code [3].

Table II presents a summary of the enzyme-like activities which have been identified in various thermal proteinoids by various laboratories. These include in a number of instances pH-activity curves of the usual type, Michaelis-Menten kinetics, heat inactivation in aqueous solution, specificity of interaction between proteinoid and substrate, enhancement of five types of reaction, and a basis for the origin of metabolic pathways [2]. Typically, each kind of proteinoid has its array of (weak) enzyme-like activities, and the variety of proteinoids permits an array of arrays.

Comparison of the tests with those of contemporary enzyme proteins has shown, generally, an absence of activities in the evolved protein, except for its specialized power. This has supported the inference that macromolecular evolution could not have begun with any highly specialized evolved protein; a heteropolyamino acid possessing a number of enzyme-like activities in small degree was what was needed.

The way in which water entered and left the evolutionary development also requires precise understanding. We know that contact of thermal polyamino acids (proteinoids) with water produces, in a maximally simple operation, vast numbers of microsystems (fig. 3) with many of the properties of contemporary cells. These properties are listed in table III.

The microsystems produced are uniform in size, numerous ($10^7 - 10^9$ per gram), dynamic (they shrink in response to hypertonic solution and swell in hypotonic solution), they can be made to stain gram-positive or gram-negative, they have the size, shape, and pattern of association of some *cocci,* and their stability overlaps the range of stability of contemporary cells.

The fact that the units are stable has permitted embedding and micro-

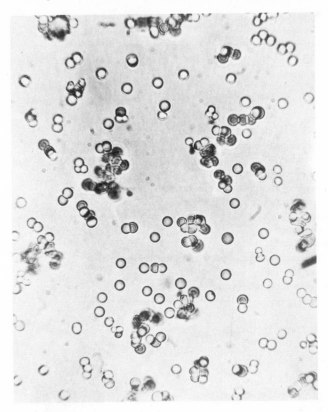

Fig. 3. Proteinoid microspheres produced by heating proteinoid in aqueous solution and cooling. They are typically uniform in the range of 1–2 μm in diameter.

toming them for electron microscopy. An electron micrograph is seen in figure 4.

In figure 5 is presented, through the optical microscope, a kind of replication [1]. It is an explanation of how cells first came to replicate. With this observation, we can construct answers of how a primitive cell [8] could have arisen when no cells existed to produce them, how they contained protoenzymes when no enzymes existed to make them, and how informational macromolecules arose from heated amino acids. With these phenomena organic evolution, as we know it, could have begun.

I would like to turn next to recently reported observations indicating that such microsystems possess an inherent tendency to communicate [14].

These results also illustrate the kind of phenomenon mentioned by Dr.

Table III. Properties of proteinoid microparticles

Stability (to standing, centrifugation, sectioning)
Microscopic size
Variability in shape
Uniformity of size
Numerousness
Stainability
Producibility as gram-positive or gram-negative
Shrinking or swelling in atonic solutions
Structured boundary
Ultrastructure (electron microscope)
Selective passage of molecules through boundary
Assembled from catalytically active polymer
Patterns of association
Budding and fission
Growth by accretion
Ability to propagate through budding and growth by accretion
Ability to form junctions
Ability to communicate information

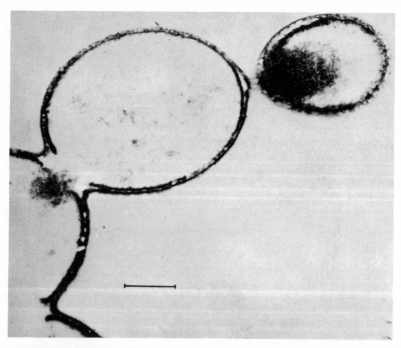

Fig. 4. Electronmicrograph of proteinoid microsphere subjected to minor elevation of
pH. Stained with osmium tetroxide and embedded in methacrylate. Note double layers.

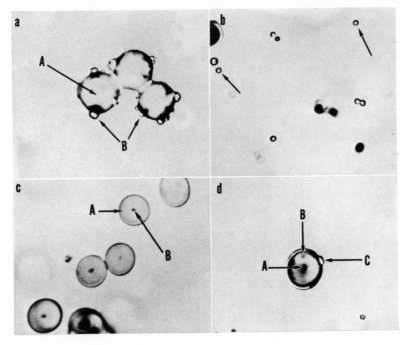

Fig. 5. Replication of proteinoid microspheres.—a) Budded microspheres produced by standing.—b) Buds liberated mechanically.—c) Stained buds around which second generation microspheres have grown by accretion (temperature of solution saturated at 37° has dropped to 25°).—d) Second-generation bud (C) on second-generation microsphere (B) accreted around first-generation bud (A). Microspheres approximately 12–15 μm in diameter.

SZENT-GYÖRGYI; they particularize how biological phenomena might have emerged from the action of structural constraints on random (Brownian) motion.

Figure 6 displays a time-lapse sequence of proteinoid microspheres approaching each other. One pair is joined, another becomes joined, and a third vacillates between junction and separation.

Figure 7 presents the evidence that these junctions are indeed new structures. Figure 8 indicates that the junction permits the transfer from one microsphere to another of smaller particles, *endoparticles*. Since the endoparticles are composed of proteinoid, and proteinoid contains information as earlier defined, this is a model of the transfer of information.

Although Brownian motion may be random, the nonrandom deployment of the portals in the total structure results in ordered behavior.

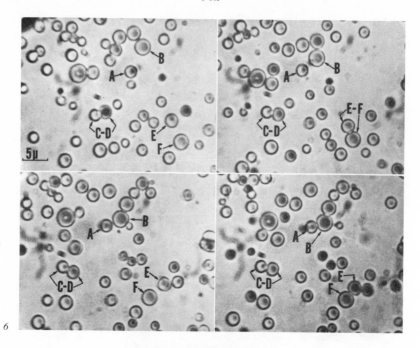

6

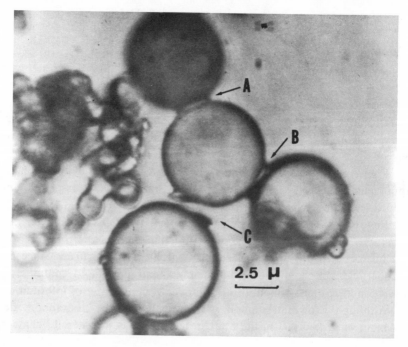

7

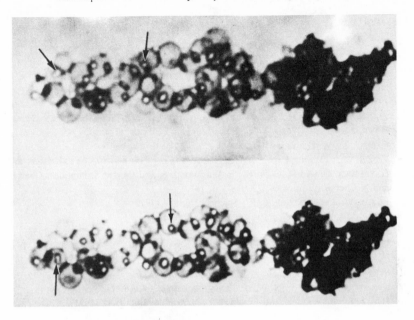

Fig. 8. Two time-lapse frames [14] showing transfer of endoparticles from one proteinoid microsphere to another through hollow junctions.

The possibility that these systems involve Brownian motion renders them appropriate for physical study. In another sense, all aspects of the model described are amenable to physical study. A basic premise that has permitted such phenomenological experiments is that biology is ultimately reducible to physics. The opposing proposition results in no experiments and no new knowledge, and is thus a self-fulfilling premise.

Fig. 6. Proteinoid microspheres forming junctions. 10 sec time-lapse. A and B join and remain joined. C and D are joined throughout. E and F join, separate, join, and separate. This behavior is often seen. Junctions are visible.

Fig. 7. Stained joined microspheres (crystal violet). A is intact, B is cracked, C has separated. Junctions are new structures formed from microsystems composed of preformed polymer.

References

1 FOX, S. W.: Naturwissenschaften *56:* 1 (1969).
2 ROHLFING, D. L. and FOX, S. W.: Adv. Catal. *20:* 373 (1969).
3 FOX, S. W.; LACEY, jr., J. C., and NAKASHIMA, T.: in RIBBONS, D. W. and WOESSNER, J. F. Nucleic acid-protein interactions, pp. 113–127 (North-Holland Publ. Comp., Amsterdam 1971).
4 FOX, S. W.; HARADA, K.; KRAMPITZ, G., and MUELLER, G.: Chem. Eng. News *48 (26):* 80 (1970).
5 CAROTHERS, W. H.: Trans. Faraday Soc. *32:* 39 (1936).
6 FOX, S. W.; WANG, C.-T.; WAEHNELDT, T. V.; NAKASHIMA, T.; KRAMPITZ, G.; HAYAKAWA, T., and HARADA, K.: in WEINSTEIN, B. and LANDE, S. Peptides: Chemistry and biochemistry, pp. 499–527 (Dekker, 1970).
7 FOX, S. W. and WINDSOR, C. R.: Science *170:* 984 (1970).
8 LEHNINGER, A. L.: Biochemistry, p. 782 (Worth, 1970).
9 DIXON, M. and WEBB, E. C.: Enzymes, p. 666 (Academic Press, New York 1958).
10 DOSE, K. and ZAKI, L.: Z. Naturforsch. *26b:* 144 (1971).
11 FOX, S. W.; WINITZ, M., and PETTINGA, C. W.: J. Amer. chem. Soc. *75:* 5539 (1953).
12 STEINMAN, G.: Arch. Biochem. Biophys. *119:* 76 (1967).
13 KRAMPITZ, G. and FOX, S. W.: Proc. nat. Acad. Sci., Wash. *62:* 399 (1969).
14 HSU, L. L.; BROOKE, S., and FOX, S. W.: Curr. mod. Biol. *4:* 12 (1971).

Author's address: S. W. Fox, University of Miami, *Carol Cables, FL 33134* (USA)

Proc. 3rd int. Conf. From Theoretical Physics to Biology, Versailles 1971, pp. 145–146
(Karger, Basel 1973)

Electronic Mobility and Transfer of Energy and Momentum as Time-Dependent Processes

P.-O. LÖWDIN

Quantum Chemistry Group, University of Uppsala, and Quantum Theory Project,
University of Florida

Abstract

A fundamental problem in the study of electronic conductivity of the proteins is the large gap between the filled valence band and the empty conduction band in the polypeptide chain which has been estimated to be about 3 eV. Even in the 'transverse' pathway which is provided by the amid groups linked together by hydrogen bonds[1], the gap is appreciable. It is suggested that the oxidation-reduction enzymes may act as electron bridges for external carriers along these transverse pathways: the external electrons are provided by the molecules in the tissue with a certain momentum at an energy level which fits the lowest empty π—orbitals of the amide groups. Since a single electron is a strong 'radical', the electrons are assumed to travel in the pathway as pairs with antiparallel spins which temporarily form H^- ions at the hydrogen bonds and finally separate at the end point where they enter the chemical reaction $4e^- + O_2 + 2 H_2O \rightarrow 4 OH^-$. The electric neutrality is preserved by having the OH^- ions diffuse in the opposite direction. The behaviour of such a travelling electron pair considered as a double 'wave packet' may be studied by means of time-dependent quantum mechanics.

Author's address: P.-O. LÖWDIN, Department of Quantum Chemistry, Uppsala University, Box 518, Uppsala (Sweden) and Quantum Theory Project, University of Florida, *Gainesville, FL 32601* (USA)

1   P.-O. LÖWDIN, Biopolymers, Symp. *1*, 293 (1964); Ann. New York Acad. Sci. *158*, 86 (1969).

Discussions

K. MENDELSSOHN: The most important statement to my mind that Dr. LÖWDIN has just made is this enormous energy gap. I think that if there is such an energy gap it must be of profound importance for the explanation of the structures of life. Yesterday, I asked MATHIAS to draw up the benzene ring: that we know the strength of the benzene ring is such that the ground-state is extremely stable, one needs a great deal of energy to do something to affect this ground state. So the steady state of life, I think will actually depend on the stability of this ground-state. I should like to re-emphasize to people trying to equate the cooperative phenomena they've seen in anorganic matter to reappear in organic matter in unchanged form: this would be too great a miracle to be likely. But such a basic thing as the stable ground state is an essential feature and probably allows for the peculiar persistence of life.

D. D. ELEY: I should like to ask Professor LÖWDIN whether any quantitative predictions are possible for the conduction path he has considered, e.g. band gap and band width, and what kind of reactions or processes are considered as possibilities for charge injection into the conduction band.

P.-O. LÖWDIN: The gap along the polypeptide chain is the same as before, but the point is, that what has been done is to take external carriers coming from the outside. This is not a question of internal carriers, but external, and conductivity along the amide groups and the hydrogen bonds more or less 'perpendicular' to the chain.

D. D. ELEY: How have you injected the carriers?

P.-O. LÖWDIN: They are coming from the substrate with a certain momentum, and a certain energy, and the only thing you can do, since you don't know the initial conditions, is to try to measure the time it takes for a wave packet with a certain given initial condition, coming in on one side of the protein to go over to the other side, and to study its behaviour in this process.

D. D. ELEY: What is the substrate you used to inject the carriers?

P.-O. LÖWDIN: I thought it was clear that this was an electron donor-acceptor process, where the substrate is an electron donor delivering an electron or electron-pair from its highest occupied molecular orbital to the lowest empty molecular orbital of the protein. For the sake of simplicity we have considered the electrons coming into the protein with a certain energy and momentum; it is still quite a formidable quantum-mechanical problem.

M. KASHA: May I make a comment on ELEY's question? LÖWDIN has an extraordinarily long abstract which gives the details of what he has done and he is really providing a new descriptive mechanism for mobility of electron pairs in polypeptides. So it's a question of a model mechanism rather than specific numerical data.

Proc. 3rd int. Conf. From Theoretical Physics to Biology, Versailles 1971, pp. 147–155
(Karger, Basel 1973)

Electron Transport in Biological Systems

D. D. ELEY

Chemistry Department, Nottingham University

It is now 30 years since SZENT-GYÖRGYI [1] suggested that electron transport between the insoluble oxidation enzymes occurred through an energy-band, as in semiconductors or metals, passing through the fibrous proteins of the supporting membranes. Some 8 years later EVANS and GERGELY [2], using simple LCAO-MO theory, showed that the extended π-electron system of the transverse $C=O\cdot\cdot HN$ hydrogen bonds in the β protein structure would give rise to three narrow energy bands, with a gap of 3.2 eV between the second filled band and the upper third unfilled band. At the same time, experimental work was started which led to the first international conference at Nottingham, at which many of the basic ideas were presented and discussed [3]. Since that time there have been several meetings, and many direct current measurements carried out on biological molecules and macromolecules, their complexes, and even organelles. This brief presentation draws extensively on the author's reviews [4, 5] and the recent monograph of GUTTMAN and LYONS [6], and the author apologises that references to original sources are restricted for reasons of brevity.

The mitochondrion is the organelle containing the insoluble enzymes responsible for oxidation and oxidative phosphorylation [7]. A complex membranous structure, in size a spheroid of perhaps $3.3 \times 1.0\,\mu$ but existing in many morphological states [8], it contains 60–65% protein of which roughly three-quarters is structural and one-quarter electron carrier and coupling enzymes, and 35–40% lipids, mainly phospholipids, with 3.2 double bonds per P atom. Comminution of the mitochondrion allows a particle EP associated with electron transfer to be isolated, which in turn is separable into 4 insoluble complexes [9]:

  I. NADH-CoQ (oxido)reductase;
 II. succinate-CoQ (oxido)reductase;
III. CoQH$_2$-cytochrome C (oxido)reductase;
IV. cytochromed: O$_2$ oxidoreductase.

The lipid soluble mobile CoQ and cytochrome C serve to link the 4 complexes.

Semiconduction in the Molecular Constituents [4, 5]

Most of the background measurements concern dc conductivity on samples, often in the form of compressed discs of polycrystalline substances or films, *in vacuo* or controlled environments, and these usually fit the semiconduction equation

$$\sigma = \sigma_0 \exp(-\Delta\varepsilon/2kT) \; \Omega^{-1} \, cm^{-1}$$

Thus native proteins in the 'dry' state give reproducible values of σ, σ_0 and $\Delta\varepsilon$ *in vacuo,* independent of the electrode metal, and room temperature conductivities are very low, ca. $10^{-17} \, \Omega^{-1} \, cm^{-1}$. The values for cytochrome C$\sigma_{298} = 6.3 \times 10^{-18} \, \Omega^{-1} cm^{-1}$, $\log_{10}\sigma_0 = 4.8$ and $\Delta\varepsilon = 2.60$ eV falls in with some twenty odd proteins and polyamino acids for which $\Delta\varepsilon$ lies in the range 2.57 to 3.13 eV [10]. Denaturation was found to somewhat decrease dry state conductivity and raise $\Delta\varepsilon$, e.g., for haemoglobin from $\Delta\varepsilon = 2.66$ (native) to 2.89 eV [10]. However, cytochrome oxidase is stated to possess a dry state $\Delta\varepsilon$ of only 0.52 eV [11]. The conductivity of proteins may be raised 10^5-fold or more by adsorption of a BET monolayer of water, acting, we believe, as an electron donor, which injects electrons into the hydrogen bond system (N-type conduction) or by complex formation with electron acceptors, such as chloranil which are postulated to inject holes (P-type conduction). Haemoglobin, because of its known structure (M. F. Perutz) has been studied by us in some detail. Water vapour adsorbs on the molecular surface and does not penetrate appreciably within the molecule where the non-polar side-chains are concentrated. From molecular dimensions a close-packed monolayer corresponds to 1160 M water/10^5 g haemoglobin (20.88%), while analysis of the adsorption isotherm by the BET method puts the BET V$_m$ as 319 M water/10^5 g haemoglobin (5.74%). According to Pauling, the V$_m$ value corresponds to water molecules held on polar groups, but the large increase in conductivity suggests some are located on surface peptide groups. The evidence is that at V_m adsorbed water the conductivity is electronic, but around $3\,V_m$ the evidence of hydrogen evolution (MARČIČ) and time-dependent polarisation (LESLIE) points to onset and dominance of proton conduction. This is to be expected since as the adsorbed water molecules become closepacked, protons may traverse the adsorbed layer by a Grotthus mechanism, as visualised by RIEHL [12]. The conduction saturates around 3–$4\,V_m$ which is expected on our model. An alternative view for the increase in electronic conduction attributes this to a classical increase in dielectric constant, due to the adsorbed water, decreasing the intrinsic energy of forming an electron and hole in the hydrogen bond system [13]. Intermediate mechanisms are possible (cf. recent suggestions for DNA [14]). In connection with the action of electron acceptors, it has been reported that the cytochrome C-coenzyme Q complex has a $\Delta\varepsilon$ of only 0.71 eV [15] and a 10^6-fold greater conductivity than cytochrome C itself ($\Delta\varepsilon = 2.60$ eV) [10]. It may be added that the dry-state unsaturated phospholipids have room temperature resistivities $\rho \simeq 10^9\,\Omega$ cm and $\Delta\varepsilon = 1.4$ eV, both values being lowered by adsorbed water [16].

There remain many problems. Modern SCF MO calculations predict $\Delta\varepsilon = 5.2$ eV for the 'dry' state $C{=}O\cdot\cdot H{-}N$ extended π-electron system in proteins [17], much greater than the observed values of around 2.8 eV. However, various factors still remain to be taken into account, such as side-chains, possible residual water molecules, polarisation around the carriers, etc. The observed 'dry' state σ_0 values, if indeed we have intrinsic conduction, yield mobility values of 10^4 cm^2/v.sec if we have free electrons, or 40 cm^3/v.sec for an effective mass of 40 corresponding to a band width of 0.1 eV [10]. To explain these values a model has been advanced of narrow band conduction within the hydrogen bond structure, e.g., within a particular helix, with tunnelling through barriers at each end of the structure between adjacent α helices [10]. In order to test these ideas, mobility measurements are necessary. These have proved impossible by dc Hall or pulse-drift techniques, but have been successfully achieved by microwave techniques (next section).

In connection with the approach to the whole mitochondrial system, observed activation energies for respiration of ca. 0.5 eV are even larger than $\Delta\varepsilon/2$ values for the cytochrome C-CoQ complex, or cytochrome oxidase. There seems every reason to suppose that proteins in association with other components, will provide an adequate conducting pathway. The approach of Dr. PETHIG and myself, started with the assistance of the late

Professor G. H. A. HUBSCHER and his colleagues in the Biochemistry Department at Nottingham, has been to study the microwave Hall mobility
of the charge carriers in the mitochondrion, and the effect of the specific
inhibitors on this mobility. The results so far published, briefly reviewed
below, go some way to support the involvement of semiconduction in
mitochondrial electron transport.

Microwave Hall Mobilities

In the past in our laboratory, ac measurements have been used to derive
what we hoped was a true conductivity value for the crystal in polycrystalline powders. Recently, with Mr. M. H. JONES, we have been reviewing
and extending this work into the 1000 MHz region. In an important paper,
TRUKHAN [18] has shown that to overcome the effect of particle dis-
continuities and derive a true Hall mobility for a microcrystalline system,
such as a protein, it is necessary to work at a frequency greater than a
dispersion value, dependent on particle variables such as size, number and
mobility of charge carriers and dielectric constant, which for typical values
of these parameters is 3×10^8 Hz. The method is applicable to various
shapes of particles, and the main problem is the need to correct for Debye
dipole losses, which otherwise leads to low values of ac resistance and
mobility. Liquid water has a loss maximum at 1.74 cm at 20 °C, which
presumably also applies to water in multilayers, but not to water irrotationally bound in the first adsorbed monolayer. TRUKHAN [19] measured
the Faraday rotation of 10 GHz microwaves, with the specimen as a thin
disc at the electrical antinode of a bimodal cavity operating in the TE_{11}
mode similar to that developed by PORTIS and TEANEY [20]. By observing
the effect of the specimen on the Q of the cavity the ac conductivity of
the specimen is determined. By the application of a magnetic field along
the axis of the cavity it is possible to measure the rotation of the electric
vector in the specimen in terms of the ratio of transmitted to incident power
and from this to calculate the Hall mobility of the free charge carriers.
Details of the method will be found in TRUKHAN's papers, and in our own
papers as they appear. All we need note at present is that the observed
Faraday rotation is proportional to the Hall mobility and the volume of
the conducting particles and it is therefore necessary to know the latter to
calculate the mobility. TRUKHAN [18] reported values for a number of solid
proteins, polyamino acids and DNA, all as hermetically sealed specimens

in polythene and therefore with 'normal' water content. The values observed were all in the neighbourhood of resistivity 10^3 Ωcm and mobility $1\,cm^2/v.sec$.

In our work at 9.2 GHz at Nottingham we calibrated our cavity with germanium single crystals of known conduction parameters and checked out a value for zinc oxide powder of $110 \pm 40\ cm^2/v.sec$ in agreement with that measured by TRUKHAN. In our case the specimen protein discs were not sealed and dry nitrogen gas could be passed over the specimen so as to remove the easily desorbed water. In this way bovine plasma albumin [21], which started with $\rho = 10^2\Omega$cm, after 3 h in the wave-guide reached $10^4\,\Omega$cm ($\mu = 1\ cm^2/v.sec$ N-type) and after 3 days passage of dry nitrogen reached $10^6\,\Omega$cm ($\mu = 26\ cm^2/v.sec$). A logarithmic extrapolation gave a final resistivity of $10^8\,\Omega$cm ($\mu = 40\ cm^2/v.sec$) and the indications are that only the monolayer of strongly and irrotationally bound water remains adsorbed on the BPA. According to ROSEN's solid state dielectric measurements [22] this should correspond to 24.6% water, and according to the solution dielectric work of BUCHANAN et al. [23] BPA binds 18% irrotational water. By analogy with haemoglobin, of similar molecular weight to bovine serum albumin, we expect about 20% water to correspond to one close-packed monolayer of water molecules over the surface of the protein molecule. On this basis we think that TRUKHAN's lower values of around $1\ cm^2/v.sec$ for various proteins (not including BPA) were possibly due to the effect of dipolar losses to which he himself drew attention in his paper and which we have further discussed recently [24]. The point about the present result of $\mu = 40\ cm^2/v.sec$ is that it supports the possibility of narrow band conduction in the protein complex BPA-irrotational H_2O. The difference between the ac resistivity for this system of $10^8\,\Omega$cm and our dc value of $2.5 \times 10^9\,\Omega$cm for BPA-18% adsorbed water may reflect the effects of intercrystalline defects. Finally an N-type mobility of $40\ cm^2/v.sec$ and a resistivity of $10^8\,\Omega$cm corresponds to an *electron* carrier density of $1.6 \times 10^9\,cm^{-3}$. This is referred to in the final section of this paper.

Measurements on the Mitochondrion [25]

According to SNART [15] dc measurements on the whole mitochondrion give a room temperature resistivity of $2 \times 10^{13}\ \Omega$cm and a $\Delta\varepsilon$ of 1.80 eV. We should expect this resistivity to be determined by the most resistive series components, possibly the saturated lipids. On the other hand, the microwave method will cause the charge carriers to oscillate to and fro on a path of length $\mu E/2\pi f$ where E is the voltage gradient at the specimen and f the frequency. This path length may therefore be of order 2–3 Å. However, an indicated mobility of 50 cm²/v.sec would suggest the existence of a well defined energy band, which in itself implies specific interactions of atomic wave functions extending over distances very much greater than 2–3 Å. The microwave method will measure the mobility of the electrons in the most conducting pathway in a complex organelle such as a mitochondrion. The effective ac resistivity of the mitochondria on "drying" in the wave-guide increased from 1.0×10^3 Ωcm to a final steady value of $3.0–5.5 \times 10^3$ Ωcm, a much smaller change than that found for BPA. The limiting Hall mobilities derived by extrapolation were 10 cm²/v.sec for Batch I and 16 cm²/v.sec for Batch II, both N-type, the numerical values being based on the mitochondrial volume. Assuming the electron transfer regions occupy only 20% of the total mitochondrial volume, the corrected Hall mobilities come out to 50–80 cm²/v.sec. We have attempted to analyse the biochemical electron transfer role in the light of an assumed geometry for the electron path, and calculated that the mean resistivity of this path needs to be about 3.8 Ωcm, which is about 10^3 times less than that observed above for the 'dry' state, as it was defined above, although some organic charge transfer complexes do have resistivities as low as this. However, we shall withhold further comments at present.

Perhaps our most significant result was that 1×10^{-4} M KCN (IV) lowered the microwave Hall mobility to one-half or one-third its original value, whereas antimycin A (III) was without effect, and rotenone (I) was effective only at relatively high concentrations. The numbers in brackets refer to the complex on which the inhibitor acts. This suggests we are measuring the mobility of electrons on the biochemical pathway through complex IV, the cytochrome oxidase. The mobility of lipid extract was very much less than that of the mitochondria, as was also that for crystalline cytochrome C. If the cytochrome C lies in the microwave Hall pathway, its conformation must be different in the mitochondrion from the crystal. However, as already mentioned, a part of the cytochrome C is usually ascribed a mobile molecular role in the mitochondrion. If these views are correct, it would seem that we should expect a greater microwave Hall mobility for fraction IV than for the other three fractions and this is presently under examination, in collaboration with R. PETHIG, R. J. MAYER and E. SHEPPARD. Finally, by contrast with the N-type behaviour of mitochondria, spinach chloroplasts show P-type mobility of 0.5–0.80 cm^2/v.sec.

The values observed above all are much greater than the drift mobilities observed in mitochondrial films, of water content <1%, following injection of a pulse of low energy electrons [26]. These values were electrons 5×10^{-2} cm^2/v.sec and positive holes 2×10^{-2} cm^2/v.sec. The difference in part relates to water content, the mobility decreasing at very low values due to a breaking down of hydrogen bond structure [24]. Partly it will also be due to the need for the pulse to traverse regions of high resistivity associated with saturated lipids, thus reducing the average mobility.

Proteins in Solution

We have inferred from semiconduction measurements on globular proteins that adsorbed water molecules inject electrons into the interior of the molecule. However, 1.6×10^9 injected electrons/cm^3 only corresponds to about 1 electron/10^{10} albumin molecules. The conductivity results also point to a comparable number of protons in the saturated adsorbed water layer, in the solid state. On the face of it, one would not expect any appreciable contribution to the charge fluctuation interactions of KIRKWOOD and SHUMAKER [26, 27] from this source. It should perhaps be noted that the source of the conduction protons above may be COOH or NH$_3^+$ groups, but their concentrations still remain to be properly calculated. Furthermore, FRÖHLICH has suggested the possibility of longitudinal electric vibrations arising in protein molecules or similar molecules in solution [28] and pointed out that electrons in non-localised states may take part in such vibrations, provided their concentration is not too high, i.e., $<10^{15}$ cm^{-3} [29]. Our inference that there are injected electrons with a concentration of 10^9 cm^{-3} may fit the case and it will be of great interest to enquire whether they could be responsible for the Raman-like scattering referred to by FRÖHLICH, or for other anomalous optical behaviour. We may estimate a diffusion coefficient for the electrons $D = 1$ cm^{2}sec^{-1} from the Einstein relation $D = \mu kT/e$, where e is the electronic charge. Then using KIRKWOOD's expression $\tau = b^2/D$ where b is the radius of the protein (taking as 30A.), we estimate an order of magnitude for the relaxation time controlling the distribution of injected electrons as 9×10^{-14} sec, which corresponds to a possible dispersion region around a wavelength of $27\,\mu$, i.e. in the infra-red.

Acknowledgements

The author would like to acknowledge many valuable discussions over the years with Dr. M.R. WILLIS. The microwave Hall results owe everything to the collaboration of Dr. R. PETHIG, the valuable design and constructional work to Messrs. W. E. PORTER, R.E. PARSONS and A.G. HANDS, and the valuable advice and mitochondrial preparations provided by Dr. ELIZABETH SHEPPARD and Dr. R.J. MAYER of the Biochemistry Department, Nottingham University Medical School.

References

1 SZENT-GYÖRGYI, A.: Science 93: 609 (1941).
2 EVANS, M.G. and GERGELY, J.: Biochim. biophys. Acta 3: 188 (1949).
3 Disc. Faraday Soc. No. 27: Energy transfer with special reference to biological systems, Nottingham 1959.
4 ELEY, D. D.: in KASHA, M., and PULLMAN, B. Horizons in Biochemistry (Academic Press, New York 1962).
5 ELEY, D.D.: in KATON, J.E. Organic semiconducting polymers (Dekker, New York 1968).
6 GUTTMAN, F. and LYONS, L.E.: Organic semiconductors (Wiley, New York 1967).
7 LEHNINGER, A.L.: The mitochondrion (Benjamin, New York 1965).
8 KORMAN, E. F.; ADDINK, A. D. F.; WAKABASHI, T., and GREEN, D. E.: J. Bioenergetics 1: 9 (1970).
9 See e.g. MAHLER, H.R., and CORDES, E. H.: Biological chemistry; pp. 597 et seq. (Harper and Row, New York 1966).
10 ELEY, D.D. and SPIVEY, D.I.: Trans. Faraday Soc. 56: 1432 (1960).
11 COPE, F. W. and STRAUB, K. D.: Bull. math. Biophys. 31: 761 (1969).
12 RIEHL, N.V.: Z. Fiz. Khim. URSS 29: 1537 (1955).
13 ROSENBERG, B.: J. chem. Phys. 32: 816 (1962).
14 BURNEL, M. E.; ELEY, D. D., and SUBRAMANYAN, V.: 4th Jerusalem Symp., April 1971. The Purines—Theory and Experiment, ed. by E. D. BERGMANN and B. PULLMAN (The Israel Academy of Sciences and Humanities, Jerusalem, 1972).
15 SNART, R. S.: Biochim. biophys. Acta 88: 502 (1964).
16 LESLIE, R. B.; CHAPMAN, D., and HART, C. J.: Biochim. biophys. Acta 135: 797 (1967).
17 SUARD, M.; BERTHIER, G., and PULLMAN, B.: Biochim. biophys. Acta 52: 254 (1961).
18 TRUKHAN, E.M.: Biofizika 11: 412 (1966).
19 TRUKHAN, E.M.: Pribory, tekhn. eksper. 4: 198 (1965).
20 PORTIS, A.M. and TEANEY, D.: J. appl. Physics 29: 1692 (1958).
21 ELEY, D.D. and PETHIG, R.: J. Bioenergetics 1: 109 (1970).
22 ROSEN, D.: Trans. Faraday Soc. 59: 2178 (1963).
23 BUCHANAN, T. J.; HAGGIS, G. H.; HASTED, J. B., and ROBINSON, B. G.: Proc. roy. Soc. Lond. A 213: 379 (1952).
24 ELEY. D.D. and PETHIG, R.: Disc. Faraday Soc.: Conduction in organic solids, Nottingham 1971.
25 ELEY, D.D. and PETHIG, R.: J. Bioenergetics 2: 39 (1971).
26 VANNIKOV, A.V. and BOGUSLAVSKII, L.I.: Biofizika 14: 421 (1969).
26 KIRKWOOD, J.G. and SHUMAKER, J.B.: Proc. nat. Acad. Sci., Wash. 38: 855, 863 (1952).

27 JEHLE, H.; PARKE, W. C., and SALYERS, A.: in PULLMAN, B. Electronic aspects of bio-
    chemistry (Academic Press, New York 1964).
28 FRÖHLICH, H.: in MAROIS, M. Theoretical physics and biology (North Holland Publ.
    Comp., Amsterdam 1969).
29 FRÖHLICH, H.: Int. J. Quantum Chem. 2: 641 (1968).

Author's address: D. D. Eley, Department of Chemistry, University of Nottingham,
Nottingham (England)

Discussions

VOICE: What frequency did you use?

D.D. ELEY: The frequency used in the microwave Hall measurements was 9.2 GHz.

Proc. 3rd int. Conf. From Theoretical Physics to Biology, Versailles 1971, pp. 156–164
(Karger, Basel 1973)

Role of Photons in the Storage
of Energy and Resonance Transfer[1]

M. KASHA

Institute of Molecular Biophysics, Florida State University, Tallahassee, Fl.

In discussing the interaction of light with matter, one normally thinks of single-photon single-molecule interactions. Simple excitation events are appropriate to consider in the absorption of light of moderate intensities by molecules in dilute vapor phase or even in solution. But in nature, especially in biological environments, aggregates of molecules are usually involved, and sometimes high photon intensities may be studied. It is appropriate in studying molecular excitation behavior in natural systems to consider the role of *multiple excitation in composite molecules* [1]. By *multiple excitation* we may describe polyphotonic processes, such as excitation by two photons, including (1) non-linear absorption, (2) successive excitation, and (3) triplet-triplet annihilation. By *composite molecule excitation* we may indicate a pair or cluster of molecules in cooperative interaction with the radiation field, including (4) charge-transfer excitation, (5) molecular exciton phenomena, (6) molecular excimer production, (7) simultaneous transitions, and (8) biprotonic phototautomerism.

In this note, three of these topics will be discussed in interrelation: *molecular exciton* effects in lamellae, the *simultaneous transitions* (and single molecule transitions) in singlet molecular oxygen, and *biprotonic phototautomerism,* or the cooperative transfer of two protons in excited states of hydrogen-bonded dimers.

---

1   Work done under a contract between the Division of Biology and Medicine, U.S. Atomic Energy Commission, and the Florida State University.

Molecular Excitons in Lamellae

The molecular exciton model is well known in interpreting the spectra of molecular crystals [2]. The application of the model to lamellar systems [3] leads to unique but typical results. In the model, the resonance interaction between allowed excited states of neighboring molecules is studied in a dipole-dipole approximation for ordered arrays of pigment molecules. Quantitative expressions are given for the energy of absorption of light in a lamella, in terms of the distance between chromophores, their oscillator strength and orientation, and the energy of absorption in the monomers.

For a 'pin-cushion' array, the *allowed exciton state* corresponds to a strong repulsion (fig. 1a), while for a 'flat-sheet' arrangement of molecular

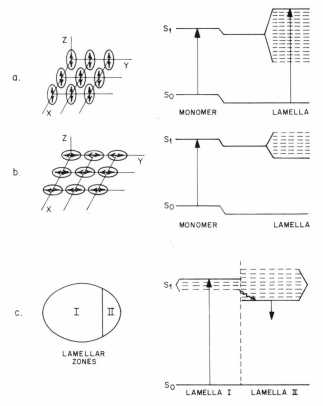

Fig. 1. Effect of lamellar molecular organization on exciton state band structure *(solid, allowed exciton state; dashed, forbidden exciton states).*

transition dipoles, the allowed state is on the whole attractive, so that a lowering of excitation energy ensues (fig. 1b). In other words, the absorption bands of molecules in a lamella are shifted according to the geometry of molecular (transition moment) stacking, the magnitude of the shift in energy being related directly to the intensity of the transition. But if N molecules interact in the lamella, N exciton states result, most of these being forbidden in light absorption but nevertheless are of very great importance in energy storage and migration.

The forbidden exciton states are metastable states and their existence and disposition in the exciton band govern excitation phenomena for a given lamellar organization [3]. In principle, these forbidden states could have infinite lifetime for a perfect lattice. In practice, various relaxation effects, especially geometrical distortions, could shorten the lifetimes to the order of milliseconds. But even excited states of such lifetime must be considered metastable, and susceptible to competitive changes in excitation path.

The generation of a molecular exciton band by 'strong-coupling' interaction [4] implies simultaneously a delocalization of excitation: a whole zone of the lamellar molecular lattice is collectively excited by each photon. This makes such a lamella a powerful antenna for photons.

It is possible [3] to have such a lamellar molecular organization that extensive delocalization is present, with but small displacement of the single molecule absorption frequency (fig. 1c, left). The strong coupling exciton model involves very short energy transfer times, $\ll 10^{-14}$ sec/step, even with such small apparent spectral displacements.

Then the possibility arises [3] of having in one interacting system, two lamellar zones of differing molecular organization. Figure 1c indicates this possibility schematically. Lamellar zone I corresponds to an extensive organized system of strongly absorbing molecular chromophores, but with transition moment organization yielding narrow exciton band width. This zone acts as the photon antenna. Then lamellar zone II represents another extreme of molecular organization, yielding wide exciton band width. Energetically, zone II acts as an excitation trap, and can be very localized. The molecules of lamellar zone II may be so organized that the lowest exciton state in the band is metastable. Thus, zone II not only acts as a primary trap (energetically) with respect to excitation energy of zone I, but also a secondary trap (from its intrinsic metastability) as a pathway for longtime (up to milliseconds or longer) storage of excitation until slower chemical processes may occur utilizing the molecular excited state. In our picture [3] zone II could be the site of the rate determining enzymic initiating reaction for a biological reaction chain.

The model described theoretically represents the molecular basis for empirical classification of System I and System II in chloroplast function in the photosynthetic process.

However, variations of the analysis with different specific details (e.g. helical molecular organization, with molecular heterogeneity as an origin of traps, as well as heterogeneous molecular geometry of organization) are to be expected in numerous other situations of biological interest, e.g. as a detailed molecular model for photo- and radiation damage in DNA.

Singlet Molecular Oxygen

In the past five years there has been a great renaissance of interest in the role of singlet molecular oxygen in biological and chemical systems [5, 6]. These highly metastable excited molecular oxygen species, $^1\Delta$ and $^1\Sigma$ (fig. 2), once thought to be of interest merely in stratospheric astrophysics, have been recently rediscovered as an important and almost ubiquitous chemical species. Singlet molecular oxygen not only can be made in macroscopic amounts by various chemical and physical methods, but also can be detected by specific chemical reactivities and unique physical properties. Even ordinary (triplet) oxygen $^3\Sigma$ (fig. 2) is easily converted to singlet excited oxygen by sensitization processes in radiation systems.

Because of its extreme intrinsic metastability, singlet molecular oxygen can be considered as a transient energy storage species. The times for such storage are not extremely long in aqueous systems, but certainly competitive with many chemical reactions. Estimates of the lifetimes of the two forms of excited singlet molecular oxygen, $^1\Delta$ and $^1\Sigma$, fall in the range 10^{-5} to 10^{-7} sec in aqueous systems, even though the intrinsic lifetimes in gas phase are very much larger, 45 min, and 7 sec, respectively. Direct measurements of actual lifetimes of metastable singlet molecular oxygen in various environments are being studied in our laboratory.

Since the singlet molecular oxygen excited species are so metastable, they become particularly susceptible to *simultaneous transitions*. In a simultaneous transition, the collision of a pair of excited molecules leads to a single photon emission with an energy corresponding to the sum of the two separate molecular excitation energies. On the right side of figure 2 are indicated the three new electronic states which arise from this pheno-

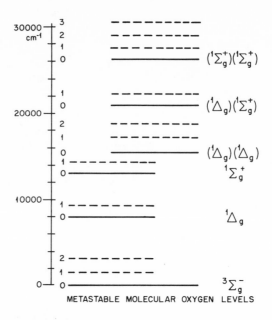

Fig. 2. Single molecule electronic states and simultaneous transition double molecule electronic states in molecular oxygen (pairs), from experimental observations.

menon. It is a rare event in absorption, but a fairly probable event in emission.

Thus it is apparent that two $^1\Sigma$ oxygen molecules, each excited by a 7,620 A.U. photon, can upon simultaneous transition emit a 3,810 A.U. photon (double the energy), or *transfer this excitation energy to an acceptor system* [7].

Such phenomena involving simultaneous transitions bring a new element into molecular excitation phenomena. It is quite possible that simultaneous transition states for quite heterogeneous molecular pairs can occur. For example, new simultaneous transitions have been interpreted [8] involving one molecule of oxygen and one aromatic hydrocarbon molecule.

In the case of molecular oxygen, it is obvious from figure 2 that a whole set of electronic and vibronic states are available for energy transfer processes. The role of these various singlet oxygen single-molecule and molecular-pair states is being actively investigated.

Biprotonic Phototautomerism

Recently an experimental study [9] of the luminescence phenomena in 7-azaindole and its doubly-hydrogen-bonded dimer (fig. 3) has revealed that emission from the dimer at room temperature is from the tautomerized species (fig. 3, bottom right). It has been found that even in cyclically-bonded ethanol solvates of 7-azaindole, emission from the tautomerized base occurs [9]. Experimentally it is found that the normal violet fluorescence of the 7-azaindole becomes a green fluorescence in the H-bonded complex, with a band shift of $-10,000$ cm^{-1}.

The phenomenon described for 7-azaindole dimer has been called biprotonic phototautomerism, and is thought to be a cooperative effect which will be found in many other multiply-hydrogen-bonded systems.

In an extension of the study [10], the temperature dependence of the biprotonic phototautomerism reveals that a Boltzmann "over the barrier"

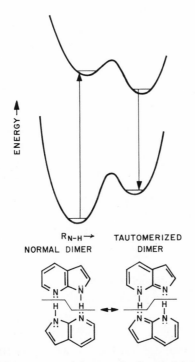

Fig. 3. Double-minimum potential curves for 7-azaindole dimer tautomerization and excitation, schematically drawn according to experimental observations.

mechanism predominates at room temperature in comparison with "proton tunnelling" as a mechanism in this case.

A double-well potential (fig. 3) as has been found to be involved in biprotonic phototautomerism, presents much intrinsic interest. The energetics of each molecular case will vary with electronic structure in each of the four pertinent states. But what is of particular interest is the depth of the potential-well in each case. In particular, photo-excitation may in some cases lead to facile tautomerization, which upon subsequent emission of a photon, could be stabilized (or made metastable) if the ground state tautomer well is deep enough. In this way an inaccessible tautomer might be formed. Such a possibility was conceived qualitatively in the earliest considerations of DNA structure [11], and could serve as a photochemical path for inactivation of DNA.

A hydrogen-bonded molecular chain could lead to a further manifestation of cooperative proton-transfer phenomena in photoexcited systems [12]. Since biological macromolecules depend so fundamentally on hydrogen-bonded networks, the occurrence of proton-transfer cooperative phenomena under suitable excitation or energy transfer conditions can be expected to be prevalent and deserves intensive study.

Interrelationships

One of the main differences between the *in vitro* system of the physical spectroscopist, who must isolate his phenomena to decipher them, and the *in vivo* systems of the biologist, who must grapple with natural phenomena in their full complexity—is that the interrelations of coupled systems must be considered in the latter case.

Of the three multiple excitation events in composite systems selected for consideration in this brief note, it is obvious how these may be inter-related. For example, the chloroplast chlorophyll species will, during excitation, undergo bipartition of excitation among triplet as well as among singlet states. Although the triplet states which are (inadvertently) produced may not participate in phtosynthesis, singlet oxygen may be produced by a sensitized transfer of excitation to triplet oxygen. Singlet oxygen would then constitute an interference to the photosynthetic system, it is thought [13], by virtue of its reactivity. But the carotenoids of the chloroplast can serve [13] as a powerful quencher of singlet oxygen, thus protecting the chlorophyll from singlet oxygen oxidation reactions.

It will thus take a full awareness of the manifold multiple excitation mechanisms in composite systems and their interrelation, and then a discriminative testing of the complex interactions before biological mechanisms involving excited states of molecules can be considered to be satisfactorily understood.

References

1 A comparative study of these phenomena is being prepared by M. ASHRAF EL-BAYOUMI and M. KASHA for *Accounts of Chemical Research*.

2 DAVYDOV, A. S.: Theory of molecular excitons (transl. by M. KASHA and M. OPPENHEIMER, jr.) (McGraw-Hill, New York 1962).

3 HOCHSTRASSER, R. M. and KASHA, M.: Application of the exciton model to monomolecular lamellar systems. Photochem. Photobiol. *3:* 317 (1964).

4 Cf. KASHA, M.: Classification of excitons, in AUGENSTEIN, MASON and ROSENBERG, Physical processes in radiation biology, p. 17 (Academic Press, New York 1964).

5 KASHA, M. and KHAN, A. U.: The physics, chemistry, and biology of singlet molecular oxygen. Ann. N. Y. Acad. Sci. *171:* 5–23 (1970).

6 WILSON, T. and HASTINGS, J. W.: Chemical and biological aspects of singlet excited molecular oxygen; in GIESE, A. C. Photophysiology, vol. V, pp. 49–95 (Academic Press, New York 1970).

7 KHAN, A. U. and KASHA, M.: Physical theory of chemiluminescence in systems evolving molecular oxygen. J. Amer. chem. Soc. *88:* 1574 (1966).

8 RETTSCHNICK, R. P. H. and HOYTINK, G. J.: Collision-induced electronic transitions of oxygen. Chem. phys. Ltrs. *1:* 145 (1967).

9 TAYLOR, C. A.; ASHRAF EL-BAYOUMI, M. and KASHA, M.: Excited-state two-proton tautomerism in hydrogen-bonded N-heterocyclic base pairs. Proc. nat. Acad. Sci. *63:* 253 (1969).

10 INGHAM, K.; ABU-ELGHEIT, M., and EL-BAYOUMI, M. ASHRAF: J. Amer. chem. Soc. *93:* 5023 (1971).

11 WATSON, J. D. and CRICK, F. H. C.: The structure of DNA. Cold Spring Harbor Symp. quant. Biol. *18:* 123 (1953).

12 TERENIN, A.; PUTZEIKO, E., and AKIMOV, I.: Energy transfer in systems of connected organic molecules. Faraday Soc. Disc. No. 27, 83 (1959).

13 FOOTE, C. S.; DENNY, R. W.; WEAVER, L.; CHANG, Y., and PETERS, J.: Quenching of singlet oxygen. Ann. N. Y. Acad. Sci. *171:* 139 (1970).

Author's address: Dr. M. KASHA, Institute of Molecular Biophysics, Florida State University, *Tallahassee, FL 32306* (USA)

Discussions

G. CARERI: What becomes of the oxygen after the emission?

M. KASHA: Both of the oxygen molecules return to their ground electronic states. There is no evidence for any molecular clustering. In the $O_2\acute{u}O_2$ interactions in a simultaneous transition, a collisional electron exchange perturbation is involved.

G. CARERI: Could you comment about the last figure you showed? Is there any correspondence between what happens in carboxylic acid dimers, because there also you can have the double problems in switching, in almost the same way. Is there anything on this subject?

M. KASHA: Yes, such a symmetrical proton switching is well studied, as in the case of carboxylic acid dimers (Weller). The tautomerized dimer is in such a case identical to the normal dimer. However, the necessity for delocalized excitation in such a case is also demonstrated.

L. ONSAGER: There is an exception: formic acid.

M. KASHA: Formic acid dimers would be symmetrical.

L. ONSAGER: Forms change.

M. KASHA: Oh, you mean the forms in equilibrium in liquid solution.

L. ONSAGER: All the fatty acids are associated. Their dielectric properties would suggest that in formic acid you may have extensive chainwise association of the dipoles, while in acetic and higher fatty acids there are just a few short chains and mostly symmetrical dimers.

D.D. ELEY: Is there any chance of excited oxygen arising from the dissociation of oxyhaemoglobin?

M. KASHA: In transannular peroxides and in other compounds in which oxygen is bound covalently, when it is ejected or displaced, excited or singlet molecular oxygen results. This seems to be required by the spin state correlation diagram. But for oxygen bound more loosely by a ligand, the same result would not be expected.

Proc. 3rd int. Conf. From Theoretical Physics to Biology, Versailles 1971, p. 165
(Karger, Basel 1973)

Introduction

M. EIGEN

We shall have three lectures this afternoon, all of which will be concerned with evolution.

First we shall hear about some biochemical facts. In previous meetings we have heard about protein synthesis or polypeptide synthesis, coded by DNA. Mr. LIPMANN has a fine example, wherein an oligopeptide, an antibiotic, is made by a protein complex, without instruction via DNA. We surmise that those processes may have been of great importance in the formation of certain structures before a definite code came about.

The second paper by Dr. MARGOLIASH will tell us about how the evolution of proteins of the "species" occurred. Of course, these steps available to our analysis are relatively late steps of evolution. Therefore, in the final paper I shall try myself, to reflect on the point of view of organization, necessary in early evolution, leading us from the pure chemical-type of evolution into the Darwinian-era.

Proc. 3rd int. Conf. From Theoretical Physics to Biology, Versailles 1971, pp. 166–174
(Karger, Basel 1973)

The Polypeptide Synthesis
of Gramicidin S and
Tyrocidine on a Protein Template[1]

F. LIPMANN

The Rockefeller University, New York

The chief result that has emerged is the evidence that an enzyme protein acts as template for sequentially ordered amino acid polymerization to polypeptide chains of 10–15 amino acids in length. This was found during studies on the biosynthesis of the bacterial antibiotic polypeptides gramicidin S (GS) [1], tyrocidine (Ty) [2, 3], and the not yet completed work on the straight-chain gramicidin A (GA) which contains a sequence of 15 amino acids. The synthesis of these compounds is performed on a poly-enzyme system, the main characteristics of which for GS and Ty are listed in table I. These antibiotics are cyclic decapeptides, GS, however, being made by antiparallel cyclization between two pentapeptides.

GS

$$D\text{-Phe} \rightarrow Pro \rightarrow Val \rightarrow Orn \rightarrow Leu$$
$$\uparrow \qquad\qquad\qquad\qquad\qquad \downarrow$$
$$Leu \leftarrow Orn \leftarrow Val \leftarrow Pro \leftarrow D\text{-Phe}$$

Ty

$$\text{(Trp)} \quad \text{(D-Trp)}$$
$$D\text{-Phe} \rightarrow Pro \rightarrow Phe \rightarrow D\text{-Phe} \rightarrow Asn$$
$$\uparrow \qquad\qquad\qquad\qquad\qquad\qquad \downarrow$$
$$Leu \leftarrow Orn \leftarrow Val \leftarrow Tyr \leftarrow Gln$$
$$\text{(Phe)}$$

In the activation for biosynthesis of GS, Ty, and GA, the amino acids react first with ATP to form aminoacyl adenylates, from which they are transferred to an enzyme-bound –SH group. Polymerization is initiated by

1   The research reported in this paper was supported by a grant from the United States Public Health Service (GM-13972).

Table I. Enzyme activities in the biosynthesis of GS and Ty

Decapeptide synthesized	No.	Mol. wt. of enzymes	Amino acid activated and fixed in sequence indicated	Pantetheine content (M/enzyme)
GS	1	100,000	D-Phe	None
	2	280,000	Pro, Val, Orn, Leu	1
Ty	1	100,000	D-Phe	None
	2	230,000	Pro, Phe, D-Phe,	1
	3	440,000	Asn, Gln, Phe, Val, Orn, Leu	1

the interaction of the smaller enzymes carrying single amino acids (cf. table I), with polyenzymes carrying four (GS) or three and six (Ty)-bound amino acids. It proceeds from N-terminal D-phenylalanine, as indicated in the formulae, by producing thioester-linked polypeptides from the thioester-linked amino acids; it terminates by release through cyclization.

The template function of the enzyme protein was most thoroughly investigated with Ty synthesis [3]. As indicated in table I, it involves the interaction of three enzymes. The first activates and racemizes D-phenylalanine, the second activates proline, L-phenylalanine, and D-phenylalanine. These two enzymes interact to form a tetrapeptide that is transferred to the largest enzyme to continue polymerization of the remaining six amino acids. The place of activation of the second phenylalanine on the intermediate enzyme, which is the third phenylalanine in the over-all sequence, coincides with a racemase. If charged to the corresponding enzymes before initiation, all the amino acids become linked in stoichiometric proportion to the enzyme protein, presumably through a cysteine. We like to consider these amino acids on the large enzymes as peripherally sulfhydryl-linked.

As indicated in table I, there is one molecule of pantetheine in the large enzymes [4, 13], presumably connected by a 4'-phosphate bridge to serine (fig. 1). This appears to collect the amino acids into polypeptides like a swinging arm, analogous to its activity in fatty acid synthesis [6, 7]. In our present discussion we will neglect the role of pantetheine but will return to it at the end.

To study the sequential addition that forms thioester-bound polypeptides, the amino acids were added one-by-one, in the sequence indicated in the Ty formula, to a mixture of the three enzymes, amino acids, and

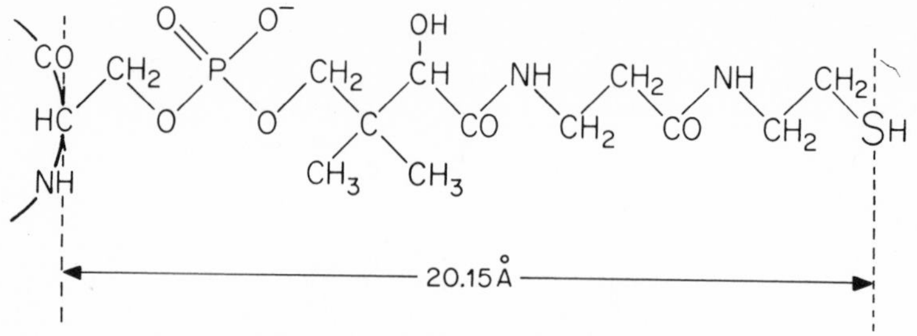

Fig. 1. Enzyme-bound 4'-phosphopantetheine [3].

ATP; ^{14}C-proline carried the marker [3, 8]. Excess amino acids and ATP were removed by Sephadex G-50 filtration, and the enzyme-bound amino acids and peptides were precipitated with trichloroacetic acid (TCA). Eventually, the TCA-precipitable material was collected on Millipore filters and the radioactivity was determined as previously described [3]. The results are given in table II; it shows that for every amino acid added in sequence

Table II. Formation of protein-bound nascent peptide chains by sequential addition of amino acids with ^{14}C-marker in the proline

Peptide chains formed	Bound radioactivity (pM)	Incre-ment
1. ...—[14]C-Pro	6	6
2. *Phe*—[14]C-Pro		
3. *Phe*—[14]C-Pro—*Phe*	20.5	14.5
4. *Phe*—[14]C-Pro—*Phe*—*Phe*		
5. Phe—[14]C-Pro—Phe—Phe—*Asn*	27.6	7.1
6. Phe—[14]C-Pro—Phe—Phe—Asn—*Gln*	39	11.4
7. Phe—[14]C-Pro—Phe—Phe—Asn—*Gln*—Phe[1]		
8. Phe—[14]C-Pro—Phe—Phe—Asn—Gln—Phe—*Val*	43.5	4.5
9. Phe—[14]C-Pro—Phe—Phe—Asn—Gln—Phe—Val—*Orn*	49.3	5.8
10. Phe—[14]C-Pro—Phe—Pre—Asn—Gln—Phe—Val—Orn—*Leu*	54.7	5.4
11. ...—[14]C-Pro...(Asn, Gln, Val, Orn, Leu)	7.0	
12. Phe—[14]C-Pro—Phe—Phe...(Gln, Phe, Val, Orn, Leu)	21	
13. Phe—[14]C-Pro—Phe—Phe—Asn—Gln—Phe—Val—...Leu	43.7	

1   This Phe waiting in position 7 does not cause an increment before Asn, 5, and Gln, 6, are added, indicating that binding out of sequence does not cause elongation. In 11, 12, and 13, omission of an amino acid, indicated by dots, stops further incorporation of ^{14}C-Pro posterior to omission; the amino acids in parenthesis are present on the enzyme in peripheral thioester links but are not polymerized due to the omission.

to the growing chain, the increment listed in the last column is 6 pM $\pm$ 1. The addition of phenylalanine to the proline causes the rather large increment because the di-, tri-, and tetrapeptides formed contain one, two, and three phenylalanines, respectively; on addition of asparagine, the next amino acid, there is a normal increment corresponding to single addition. On addition of glutamine, in view of the now possible inclusion of the already 'waiting' phenylalanine into the sequence, there is a larger increment, indicating addition of two amino acids. From then on, with every further amino acid added there is a single addition increment until the final leucine is reached, when, after formation of the decapeptide, a relatively slow cyclization takes place, indicated by retention of the decapeptide on the enzyme for a reasonably long period.

The template function of the large enzyme clearly appears when a single amino acid in the sequence is omitted (last three lines of table II). Particularly impressive is the result when phenylalanine is omitted: no polymerization occurs, as if proline only were present. On omission of asparagine, polymerization is halted after the phenylalanine in position 4; and when ornithine, the next to last, is omitted, polymerization occurs only to the octapeptide stage (comparable to No. 8 in the upper part of the table).

This nicely illustrates the template nature of the large enzymes, which does not permit progress of polymerization although it is charged with all amino acids except the one omitted from the sequence. The experiment of table II shows, furthermore, that when added in the right sequence, the amino acids will form polypeptides linked to the enzyme by thioester links, as seen from their coprecipitation with the TCA-denatured enzyme protein. The quite constant increment of ^{14}C-proline incorporation per amino acid incorporated indicates that the polypeptides of increasing chain length are bound simultaneously to the enzyme. The presence of one single amino acid and nine peptides on the enzyme when all amino acids have been added is confirmed in figure 2 where they were separated by two-dimensional chromatography [3].

Proof that it really is the thioester-linked amino acids from which polymerization occurs is given by experiments in which the D-phenylalanine thioesters of thiophenol and N-acetyl cysteine were prepared and were shown to react with the 100,000 Dalton fraction which activates and racemizes the phenylalanine [9]. Thus it can be said that the phenylalanine was charged to the enzyme from the chemically prepared thioester instead of by activation with ATP, since the phenylalanyl thioester could replace ATP-dependent phenylalanine charging. We think this experiment is

Ethyl acetate: Pyridine:
Acetic acid: H_2 O →

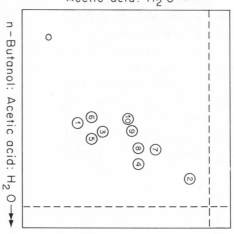

Fig. 2. Two-dimensional thin-layer chromatography of protein-bound peptides after alkaline liberation. The conditions of the experiment were as outlined in table II; the final mixture contained L-^{14}C-phenylalanine, L-proline, L-asparagine, L-glutamine, L-valine, L-ornithine, and L-leucine at 0.1 mM. Chromatography and radioautography were carried out as described [6]. The encircled figures indicate the number of residues in each peptide.

important insofar as it shows that a specified enzyme thiol will accept the amino acid from an organically prepared thioester link. However, the D-phenylalanine appears to be most easily acceptable to the corresponding enzyme. Although such phenylalanine transfer could be performed with both the Ty and GS 100,000 Dalton enzymes, a charging from other aminoacyl thioesters to the corresponding enzymes has met with difficulties.

Recent experiments using a pepsinolytic dissection of the enzyme-polypeptide complex during progress of polymerization has given us some insight into the functioning of the pantetheine [10, 11]; it could be shown that free amino acids are not bound to pantetheine but polypeptides are. Single amino acids as well as polypeptides are also bound by thioester links to enzyme fragments not containing pantetheine, and thus, presumably, directly to the protein. As we have pointed out in greater detail [11], this indicates that pantetheine collects the amino acids into polypeptide links in a somewhat complex mechanism which I do not want to discuss here.

I would like to summarize the results which seem to have greater general implications: 1) that the large enzyme proteins act as templates for the polymerization of the amino acids in GS [1] and in Ty [3]; 2) that polymerization involves condensation of enzyme-S~amino acid esters to a growing polypeptide chain which remains thioester-linked to protein until completion. If one divides the molecular weight of the large enzymes of Ty by three and six respectively, one obtains an average molecular weight of about 75,000 Daltons for every amino acid activated. This is a relatively small protein unit which, in the case of phenylalanine, also includes a racemase and transacts aminoacyl adenylation and transfer to a thioester group on the enzyme. In addition, there must be a trans-peptidation enzyme which is connected with the function of pantetheine in every step that enlarges the peptide chain. Although the transfer from a thioester-bound amino acid to the enzyme has only been shown in the case of D-phenylalanine, it seems to indicate the possibility that from a relatively simple thioester-linked amino acid precursor the aminoacyl may be transferred to a protein-linked thiol group. As we have discussed elsewhere [5], this may serve as a model for a simple mode of polymerization of amino acids on a protein.

The results reported here seem to encourage one to continue to search for metabolic fossils, and to approach experimentally the manner by which the prebiotic and prokaryotic evolutions merged into each other. I believe we can also learn a good deal about early evolution by tracing the transition from prokaryotes to eukaryotes [12]. There seems to be a schism separating prokaryote and eukaryote cells, the analysis of which may help to understand the preceding transition from the prebiotic to prokaryotic phase. Once the eukaryote cell had developed, evolution seems to have progressed in a rather orderly fashion towards the present.

References

1 GEVERS, W.; KLEINKAUF, H., and LIPMANN, F.: Proc. nat. Acad. Sci., Wash. *63:* 1335 (1969).
2 ROSKOSKI, R., jr.; GEVERS, W.; KLEINKAUF, H., and LIPMANN, F.: Biochemistry *9:* 4839 (1970).
3 ROSKOSKI, R., jr.; KLEINKAUF, H.; GEVERS, W., and LIPMANN, F.: Biochemistry *9:* 4846 (1970).
4 KLEINKAUF, H.; GEVERS, W.; ROSKOSKI, R., jr., and LIPMANN, F.: Biochem. biophys. Res. Commun. *41:* 1218 (1970).
5 LIPMANN, F.: in BUVET, R. and PONNAMPERUMA, C. Chemical evolution and the origin of life p. 371 (North Holland Publ. Co., Amsterdam 1971).
6 LYNEN, F.; OESTERHELT, D.; SCHWEIZER, E., and WILLECKE, K.: in GRAN, F. C. Cellular

compartimentalization and control of fatty acid metabolism, p. 1 (Universitetsforlaget, Oslo 1968).

7 VAGELOS, P. R.; MAJERUS, P. W.; ALBERTS, A. W.; LARRABEE, A. R., and AILHAUD, G. P.: Fed. Proc. *25:* 1483 (1966).
 8 LIPMANN, F.: Science *173:* 875 (1971).
 9 ROSKOSKI, R., jr.; RYAN, G.; KLEINKAUF, H.; GEVERS, W., and LIPMANN, F.: Arch. Biochem. biophys. *143:* 485 (1971).
11 KLEINKAUF, H.; ROSKOSKI, R., jr., and LIPMANN, F.: Proc. nat. Acad. Sci., Wash. *68:* 2069 (1971).
12 COHEN, S. S.: Amer. Scientist *58:* 281 (1970).
13 LEE, S. G.; ROSKOSKI, R., jr.; BAUER, K., and LIPMANN, F.: Biochemistry (in press).

Note Added in Proof

Recent experiments [13] have shown that some of the earlier reported data on Ty biosynthesis have to be revised. The intermediate (230,000 Daltons) and the large (440,000 Daltons) enzymes, referred to as IE and LE, have been purified to near homogeneity. The pure preparations showed: 1) that IE activates and binds the three amino acids proline, phenylalanine, and D-phenylalanine, and contains 1 mole of pantetheine per enzyme mole, and 2) that LE binds the remaining six amino acids. The data indicate a subunit (enzyme/number of amino acids) of 70,000–75,000 Daltons per amino acid. This was verified by a spontaneous de-aggregation of IE and LE in crude extracts at 37° to first one half and one third molecular weight subunits, and eventually to ca. 70,000 molecular weight subunits. Also, by SDS-gel electrophoresis, purified IE and LE could be partly degraded to three and six 70,000 molecular weight subunits, respectively, each yielding in addition one subunit of ca. 17,000 molecular weight, possibly representing the pantetheine carrier unit.

Author's address: Dr. FRITZ LIPMANN, The Rockefeller University, *New York, NY 10021* (USA)

Discussions

N. J. NINIO: Your description of the synthesis of a decapeptide is a highly stimulating contribution to the knowledge of the systems able to synthesize peptides, proteins and so on. When considering the problem of the origin of a system, people invariably try to simplify it. If we consider, for instance, the synthesis of proteins on the ribosomes, can we imagine a primitive system without ribosomes or without messenger RNA or without transfer RNA? Such an experiment is, of course, inconceivable in the study of bacteria today: we cannot take out messenger RNA. In your case, you have a system synthesizing *one* decapeptide. Perhaps you can further experiment in order to simplify it. Imagine that you succeed in synthesizing the same decapeptides, but with enzymes of lower molecular weight, how far can you go in that respect? I say that if you can reach a point where decapeptides are synthesized by a system of the same complexity i.e., composed of peptides of about the same length, then you would have a system which is very interesting for the study of evolution because the complexity of the *product* is of the same order of magnitude as that of the system *producing* it. And this, I would say is a requirement of a primitive apparatus. Have you tried to perform mutagenic experiments on the system for the synthesis of decapeptides?

F. LIPMANN: No, we have not, and I agree fully with you that of course these enzyme proteins are synthesized on the ribosomal system. But that doesn't worry me much because once such an effective way of making proteins has been adopted by the cell, every protein will eventually switch over.

With regard to the limitations of the system, we are now in the process of exploring the synthesis of another antibiotic produced by the same organism, namely, gramicidin A. The earlier discovered gramicidin A is a straight-chain peptide containing 15 amino acids starting with N-formyl-valine and ending with an ethanolamine blocking the carboxyl-terminal, so actually there are 17 members in this chain. As far as we can see it is made the same way as gramicidin S and tyrocidine. Therefore, the upper limit might be around 20.

M. EIGEN: Could you imagine also a set of evolving enzymes where one enzyme puts together another enzyme. You can start, for instance, with an enzyme which makes a decapeptide; then you can have another enzyme which links two decapeptides together, and so on, until you arrive at the size of present enzymes. The requirement is that some of the enzymes have more than one function, and that the whole sequence feeds back to its origin, otherwise it would not work. What you need then is a catalytic reaction cycle which reproduces itself.

M. SELA: I believe this is the first case when a small protein, or a big peptide, has not been transcribed or translated directly from the code but through a complex multienzyme system. The multienzyme system being of, perhaps around half a million molecular weight, this means that you would have on call in the genetic code between 1 and 5 thousand amino acids in order to reach indirectly the code for 5 to 10. Do you believe that some of the hormones in, say, the mammalian system for some special reason are also made this way? Was this interrupted at some stage in the evolution or maybe it is all through. And if it goes all through does it have a special reason which we don't understand yet?

F. LIPMANN: We went out to see if there is an enzyme that can synthesize a polypeptide without RNA template. The one we found, and have described, is admittedly a complex system and I can only repeat that once the ribosomal system was invented, then everything switched over. I don't want to go into details but would like to mention an analogy for a switch-over: in mitochondria one finds an organelle, probably a descendant of bacteria, that contains an obviously deficient DNA of a size that can only account for a small fraction of mitochondrial proteins. This DNA presumably lost more and more specific functions, which were then taken over by the nucleus. The nucleus now makes enzyme proteins and cytochromes that are eventually transferred to the mitochondria. So here one has a transfer of function to the cell nucleus, the master system for producing mRNA. With regard to peptide hormones, e.g. in the pituitary gland, these seem to be made by ribosomal systems.

T.L. BLUNDELL: The cyclic polypeptides reported all contain a D-amino acid. Others, like thiostrepton also contain unusual amino acids. I wonder if Professor LIPMANN could tell us about the synthesis of any related cyclic polypeptides where all amino acids are those which would normally be part of a protein?

F. LIPMANN: I cannot. I want only to remind you that D-amino acids are probably part of primitive peptides such as are present in the bacterial cell wall. Some people think there is a connection between antibiotics and these cell-wall polypeptides, although there is no definite proof of it.

M. EIGEN: I want to stress the fact that whenever you have a synthesis which does not involve a polymerising enzyme such as polymerases for nucleic acids or ribosoms for protein synthesis nature uses also D-amino acids. There is no selective advantage of *not* using D-amino acids. Unique chirality is found only for those systems which really offer selective advantages, e.g. polymerising enzymes.

Proc. 3rd int. Conf. From Theoretical Physics to Biology, Versailles 1971, pp. 175–239
(Karger, Basel 1973)

The Molecular Variations of Cytochrome *c* as a Function of the Evolution of Species[1]

E. MARGOLIASH

Department of Biological Sciences, Northwestern University, Evanston, Ill.

I. Prospects of Molecular Studies of Evolution

When over a decade ago it was deduced that the amino acid sequences of protein polypeptide chains were nothing but images of segments of the genome and reflected the fine structure of the corresponding stretches of DNA according to simple and inflexible rules [see CRICK *et al.,* 1961; YANOFSKY *et al.,* 1964], it became clear that the structures of proteins were the current molecular end products of the totality of the evolutionary variations undergone by their genes. As eloquently stated by FRANCIS CRICK [1958], 'Biologists should realize that before long we shall have a subject which might be called 'protein taxonomy', the study of the amino acid sequences of proteins of an organism and the comparison of them between species. It can be argued that these sequences are the most delicate expression possible of the phenotype of an organism and that vast amounts of evolutionary information may be hidden away within them'. This prediction dating back some 14 years has since been amply confirmed. However, because of the labor involved in obtaining numerous amino acid sequences and the difficulty of gathering large enough samples of pure proteins from many less common species, the data collected in the best of cases is as yet minimal. Only a few proteins have been subjected to significant comparative studies, and in many cases vast groups of species have been totally ignored. The 1969 edition of the 'Atlas of Protein Sequence and Structure' [DAYHOFF, 1969], though already out of date, is nevertheless fairly well representative of that situation. We are still largely at the threshold of the new biological

1 Reproduced by permission from Harvey Lectures, *Series 66:* 177 (1972).

era of 'protein taxonomy'. Nevertheless, since the development of techniques for the determination of the nucleotide sequences of nucleic acids [see GILHAM, 1970], one can already see the beginnings of its even more difficult extension to the other classes of informational macromolecules.

The difficulties are largely technical. Human operators are singularly ill-suited to the sort of tedious repetitive functions which constitute sequence determinations and it has become obvious that if we are to delve systematically into the vast stores of evolutionary information encoded in the structures of the informational macromolecules of extant forms of life, the determination of sequences will have to be fully automated. In addition to speed and reliability within set limits, automation lends itself to the use of smaller and smaller amounts of material as techniques and instrumentation are perfected, an advantage which could well be crucial for many evolutionarily important groups of organisms. The successes already recorded in the development of such automation for important portions of the necessary overall system of procedures [EDMAN and BEGG, 1967; EDELMAN and GALL, 1971] augur an early solution to these problems and a consequent flowering of the various fields of evolutionary biology.

Taxonomy or phylogeny via the structures of informational macromolecules is from several points of view more satisfactory than organismal morphological taxonomy or to molecular taxonomy based on substances other than informational macromolecules. This is considered below. Nevertheless, such advantages are not the only attractive or even the most important features of molecular studies of evolution. Not only do proteins represent the most delicate phenotypic expression of an organism, but their chemistry is related in a direct and stable fashion to the fine structure of their genes. Their changes are largely, if not entirely, simple expressions of the fundamental unit changes of evolution, namely, the evolutionary fixation of single nucleotide substitutions in the corresponding segments of the genome. Definite constraints are, therefore, placed on protein evolutionary variations which, though necessarily present, cannot possibly be deciphered in morphological evolutionary changes representing complicated changes often encompassing numerous genes. Similarly, changes of molecules other than informational macromolecules, which are products of whole assembly lines of enzymes, are not much less complex from the genetic standpoint than morphological evolutionary events [ZUCKERKANDL and PAULING, 1965b]. Protein evolutionary changes, on the other hand, are, in an observable fashion, restricted by both the genetic code and the phenotypic requirement of maintaining a functional structure every step of the variation. This second constraint can conceivably be bypassed when more than one gene is used to code for the same protein. One gene can maintain function while another can undergo evolutionary changes presenting eventual advantages, but having intermediate stages which could not be tolerated if it was alone in supporting function.

In general, one can therefore expect that a thorough understanding of the structure-function relations of a protein, together with the information that can be extracted, by the statistical techniques developed over the last few years [FITCH and MARGOLIASH, 1967a, 1970; FITCH and MARKOWITZ, 1970; FITCH, 1971a), from the amino acid sequences of the protein from a suitable range of species, may well lead to a view of evolution far more precise than otherwise possible. This would entail a detailed reconstruction of the complete evolutionary tree of the protein, a delineation of its evolutionary relations to other gene products, and quite possibly an understanding of the rules and regulations which govern molecular evolution. Whether these last are in all details identical to the rules which govern organismal evolution is open to question. This is exemplified by the recent and unresolved controversy [see HARRIS, 1971] as to whether there are evolutionary neutral variations in protein structure, i.e., changes which present no selective advantage or disadvantage, when such changes have long been considered to be non-existent in organismal evolution [MAYR, 1966]. The question of possible neutral mutational changes in the evolutionary variations of cytochrome *c* is discussed below.

Just how far such developments may lead, when they are extended to numerous sets of homologous proteins and possibly to nucleic acids, is difficult to judge. The remarkable successes scored so far on the basis of the amino acid sequences of proteins from a relatively minute sample of species is most encouraging. The structures of an evolutionarily relatively slowly changing protein, such as cytochrome *c*, have permitted derivations of the phylogenetic relations of widely dispersed species, including in a single taxonomic scheme higher plants, fungi and animals [FITCH and MARGOLIASH, 1967a, 1970; MARGOLIASH, FITCH and DICKERSON, 1968] (see below). Similar approaches with a rapidly changing structure, such as the fibrinopeptides A, have led to a satisfactory classification of a much smaller group of species, namely artiodactyls [MROSS and DOOLITTLE, 1967; MARGOLIASH, FITCH and DICKERSON, 1968]. Protein taxonomies do not only approximate species phylogenies, but also gene phylogenies, as first shown in the case of the relative evolutionary relations of the genes for the various hemoglobin chains and for myoglobin [ITANO, 1957; INGRAM, 1961, 1963;

ZUCKERKANDL and PAULING, 1965; FITCH and MARGOLIASH, 1967a].
Protein structures have also revealed the evolutionary relations of genes
which cannot *a priori* be suspected of homology, as in the now classical
case of lysozyme and α-lactablumin [BREW, VANAMAN and HILL, 1967]. All
in all, one can readily foresee that the accumulation of protein amino acid
sequence information will lead to definitive solutions of old unsettled phylogenetic questions. Among them one may list the precise interrelations
of prokaryotes and eukaryotes, the taxonomic relations within groups of
prokaryotes for which morphological or metabolic criteria are unsatisfactory
and the phylogenetic relations between such groups, and in what inverte-
brate group did the vertebrates originate. Ideally, this approach will eventually yield a drawing of the complete tree of biological evolution
depicting the relations of many protein structural genes and, superimposed
on this ground picture of gene flow, the topology of the descent of species
in the course of biological evolution. Such developments, particularly when
linked to our increasingly sophisticated understanding of protein structure-
function relations, are likely to provide insights into the mechanisms of
molecular evolution well beyond our present naive and largely qualitative
views. Moreover, just as one can estimate the amino acid sequence of the
ancestral form of a protein from a statistical phylogenetic tree of the
corresponding homologous set of structures [FITCH and MARGOLIASH,
1967a] (see below), a complete description of the evolutionary relations of
numerous sets of proteins should enable the approximation of ancestral
polypeptide structures dating back to the obscure period of biological history
which saw the appearance of the early replicating structures and possibly
the establishment of the mutual relations between proteins and nucleic
acids. There is no reason, other than present technical difficulty, that a
similar approach should not eventually be possible for nucleic acid sequences, leading to an independent identification of the nucleic acid
counterparts of early protein sequences, and to tests of hypotheses concerning the development of the initial successful macromolecular machinery
of biological systems [FITCH and MARGOLIASH, 1970; MARGOLIASH and
FITCH, 1970]. With the perfection of procedures for the laboratory synthesis
of long polypeptide chains one can imagine that structure-function relations
of ancestral forms of present-day proteins will be amenable to direct
experimentation, just as proteins prepared from ordinary biological sources
are utilized today.

In summary, the approach to biological evolution through the study
of the structures of informational macromolecules is likely to lead to a

description of the process in a completeness of detail and a precision far beyond that which can be attained by more classical means. Since molecular changes are subject to genetic and functional restrictions which make them amenable to direct computational and experimental attack with the proteins of extant and possibly with those of ancestral extinct species, one may expect to develop a fundamental understanding of the mechanisms of molecular evolution, of the means and pathways which led to the present living systems, and perhaps even of the extent to which the more or less successful solutions which they represent are unique. A good proportion of the techniques needed to acquire the necessary information, chemical as well as statistical, have already been developed and we can only hope that the remainder will not present insuperable difficulties.

II. Cytochrome c Protein Structures and Function

Cytochromes of the C-type have an essentially ubiquitous biological distribution. They are found in the mitochondrial respiratory chain of all eukaryotes, in the oxidation systems of many prokaryotes, and in the photosynthetic membranes of both eukaryotes and prokaryotes, including those that do not utilize oxygen as the terminal oxidant of metabolism [KEILIN, 1925, 1966; MARGOLIASH and SCHEJTER, 1966; KAMEN and HORIO, 1970]. So far, following the determination of the primary structure of horse cytochrome c in 1960 [MARGOLIASH et al., 1961] (fig. 1), the amino acid sequences of over 50 cytochromes c of the eukaryotic or mitochondrial type from different species have been examined (see legend to fig. 2) through the continued efforts of several groups of investigators. The taxonomic coverage achieved to date is very wide, including 14 mammals, 6 birds, 2 reptiles, 1 amphibian, 5 fish, 4 insects, 6 fungi, 11 higher plants, and 2 protists, far in excess of corresponding information available for any other protein. All of these proteins are small, having chain lengths slightly over 100 residues. They are strongly basic and have extensive similarities of amino acid sequence [MARGOLIASH, NEEDLEMAN and STEWART, 1963; MARGOLIASH, 1963; SMITH et al., 1963; MARGOLIASH, 1964; SMITH and MARGOLIASH, 1964; MARGOLIASH and SMITH, 1965; MARGOLIASH and SCHEJTER, 1966; MARGOLIASH, 1966; FITCH and MARGOLIASH, 1967a; SMITH, 1967; MARGOLIASH and FITCH, 1968; NOLAN and MARGOLIASH, 1968; MARGOLIASH, FITCH and DICKERSON, 1968; SMITH, 1968; FITCH and MARGOLIASH, 1970; MARGOLIASH and FITCH, 1970]. All those tested to date

Acetyl-Gly-Asp-Val-Glu-Lys-Gly-Lys-Lys-Ile-Phe-Val-Gln-Lys-Cys-Ala-Gln-Cys-His-Thr-Val-
                          10   └─ HEME ─┘   20

Glu-Lys-Gly-Gly-Lys-His-Lys-Thr-Gly-Pro-Asn-Leu-His-Gly-Leu-Phe-Gly-Arg-Lys-Thr-
                          30   40

Gly-Gln-Ala-Pro-Gly-Phe-Thr-Tyr-Thr-Asp-Ala-Asn-Lys-Asn-Lys-Gly-Ile-Thr-Trp-Lys-
                          50   60

Glu-Glu-Thr-Leu-Met-Glu-Tyr-Leu-Glu-Asn-Pro-Lys-Lys-Tyr-Ile-Pro-Gly-Thr-Lys-Met-
                          70   80

Ile-Phe-Ala-Gly-Ile-Lys-Lys-Lys-Thr-Glu-Arg-Glu-Asp-Leu-Ile-Ala-Tyr-Leu-Lys-Lys-
                          90   100

Ala-Thr-Asn-GluCOOH
104

Fig. 1. The amino acid sequence of horse cytochrome *c*, according to MARGOLIASH *et al.* [1961]. The basic residues lysine, arginine and histidine, and the hydrophobic residues, tryptophan, tyrosine, phenylalanine, leucine, isoleucine, valine and methionine, to a large extent occur in basic and in hydrophobic clusters along the amino acid sequence. There is no obvious overall grouping of acidic residues in the amino acid sequence.

react identically in cytochrome *c*-depleted mitochondria from mammalian species [BYERS *et al.*, 1971] and with cytochrome *c* oxidase and reductase preparations from mammalian species [SMITH, NAVA and MARGOLIASH, 1972; SMITH, 1972]. They have been commonly classified as 'mammalian-type, cytochromes *c*, though the present usage of 'mitochondrial' or 'eukaryotic' cytochrome *c* seems more appropriate. Even these terms may not be fully satisfactory as border-line cases such as blue-green algae which do not have distinct nuclei (are prokaryotes) or any other cell organelles, appear to carry all three varieties of C-type cytochromes, the photosynthetic green plant type, an acidic bacterial type and even a trace of a basic C-type

the protists, *Physarum* and *Euglena*. References to the structures of these proteins are listed in MARGOLIASH, FITCH and DICKERSON [1968] [see also NOLAN and MARGOLIASH, 1968; WOJCIECH and MARGOLIASH, 1970; DAYHOFF, 1969], except for the proteins of the carp [GURTLER and HORSTMANN, 1970], the elephant seal and emu [AUGUSTEN and WEBB, 1969], the camel [SOKOLOVSKY and MOLDOVAN, 1972], the bonito [NAKAYAMA, TITANI and NARITA, 1971], tobacco hornworm moth [CHAN, 1970], *Debaromyces* [TITANI, 1970], *Ustilago sphaero-gena* [BITAR, *et al.*, 1972], castor, sesame, mung bean and sunflower [BOULTER, *et al.*, 1970], pumpkin, buckwheat and cauliflower [THOMPSON, RICHARDSON and BOULTER, 1971], *Abutilon* and cotton [THOMPSON *et al.*, 1971], ginkgo [RAMSHAW, RICHARDSON and BOULTER, D., 1971], *Physarum* and *Euglena* [LIN, 1971].

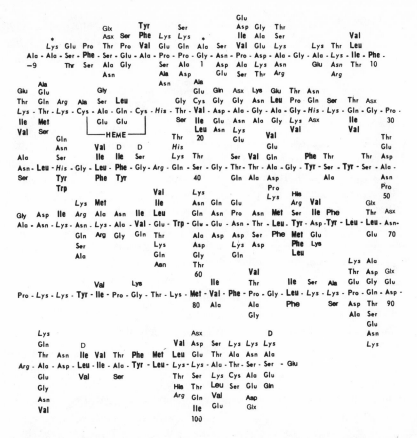

Fig. 2. Composite amino acid sequence of eukaryotic cytochromes *c*. The continuous sequence of 112 residues from residues −8 to 104 is that of the wheat germ protein. The alanyl residue at position −9 occurs in the iso-2 cytochrome *c* of baker's yeast. The glutamic acid at position 105 is found in the ginkgo protein, the only one to extend that far. The asterisk at position −8 indicates that this residues is N-acetylated in the higher plant cytochromes *c*, while the asterisk at position 1 similarly denotes the acetyl present in all vertebrate cytochromes *c*. The hydrophobic residues, valine, leucine, isoleucine, tyrosine, phenylalanine, tryptophan and methionine, are in bold face type, the basic residues, lysine, arginine and histidine, are in *italics,* and D marks the position in which a gap has been introduced in some proteins to maximize similarity. The cytochromes *c* tabulated are those of man, chimpanzee, the rhesus monkey *Macaca mulatta,* horse, donkey, cow, pig, sheep, dog, elephant seal, rabbit, California grey whale, camel, the great grey kangaroo, chicken, turkey, pigeon, duck, emu, rattlesnake, snapping turtle, bullfrog, tuna, bonito, carp, dogfish, Pacific lamprey, the moth *Samia cynthia,* the tobacco hornworm moth, the fruit fly *Drosophila melanogaster,* the screwworm fly, the fungi *Neurospora,* baker's yeast (iso-1 and iso-2 cytochromes *c*), *Debaromyces, Candida krusei, Ustilago sphaerogena,* the higher plants, wheat, mung-bean, sunflower, sesame, castor, cauliflower, buckwheat, pumpkin, *Abutilon,* cotton, ginkgo, and

cytochrome which may be of the so-called eukaryotic or mitochondrial type [HOLTON and MYERS, 1967]. The structure of none of these has yet been determined and until this is done their relationships cannot be assessed with certainty.

The properties of bacterial C-type cytochromes are much more variable than those of the eukaryotic proteins [KAMEN and HORIO, 1970], many being acidic while some are strongly basic such as the cytochromes c_3 of sulfate-reducing organisms. These have long been known to be essentially unreactive with cytochrome c oxidase preparations from mammalian sources and were, therefore, generally considered to be totally unrelated to the eukaryotic cytochromes c. More recently, as the amino acid sequences of several of these proteins were established [see KAMEN and HORIO, 1970; AMBLER, BRUSCHI and LeGALL, 1971], similarities between their structures and those of the eukaryotic proteins became obvious, to a degree which is clearly beyond what could be expected on a random basis [DUS, SLETTEN and KAMEN, 1969; FITCH and MARKOWITZ, 1970; FITCH, 1970a]. Thus, it could turn out that the C-type cytochromes of eukaryotes, prokaryotes as well as those of photosynthetic organelles are all phylogenetically related, or *homologous* proteins, namely that they all had at one time a common ancestral gene in the early stages of biological evolution. Even more, in view of the similarities of structure detected between cytochrome c and cytochrome b_5 [MARGOLIASH, FITCH and DICKERSON, 1968], the same homologous groups could include cytochromes of other than the C-type. Cytochromes could well be the proteins of choice, not only to examine the phylogenetic relationships of various groups of prokaryotes, but to determine unambiguously the evolutionary relations between prokaryotes and eukaryotes and possibly to obtain an insight into the history of that most important of evolutionary transitions, the organization of intracellular organelles, events which lead to the enormous diversification of life on this planet.

These are the tasks of the immediate future. At present the relative paucity of amino acid sequence data for the prokaryotic cytochromes c precludes any significant consideration of these problems. The following discussion will be limited to the evolutionary information that has accrued from a study of the structure of eukaryotic cytochromes c. This study, the work of numerous investigators, is not only of intrinsic biological interest, but has served as a model system to develop and test the conceptual framework and the statistical and other techniques which permit the extraction of evolutionary information from structural data for a set of homologous proteins. It is likely to remain for some time to come a main basis for extensions of knowledge of the mechanism of evolution at the molecular level well beyond the present restricted horizon.

A. The Variability of Cytochrome *c* Primary Structures

The amino acid sequences of the first eukaryotic cytochromes *c* to be determined were those of the horse [MARGOLIASH *et al.*, 1961], the pig, rabbit, chicken [CHAN *et al.*, 1963; MARGOLIASH, 1963], man [MATSUBARA and SMITH, 1963], tuna [KREIL, 1963] and baker's yeast [NARITA *et al.*, 1963a, 1963b]. At that time, well before comparable evidence was obtained for other proteins, it was already clear that the primary structure of the molecule was capable of accommodating extensive variations, and that even more remarkably the degree of molecular variability, in terms of numbers of residue differences, was demonstrably related to the phylogenetic distance between the species carrying the protein [MARGOLIASH, NEEDLEMAN and STEWART, 1963; MARGOLIASH, 1963, 1964]. It was noticed that the cytochromes *c* of mammals were roughly equally different from the cytochromes *c* of birds, that the proteins of mammals and birds together were equally different from those of fish, that those of all vertebrates were equally different from those of insects and that the cytochromes *c* of all animals, vertebrates and invertebrates, were roughly equally different from those of fungi [MARGOLIASH, 1963; MARGOLIASH and SMITH, 1965]. Thus, time elapsed since the divergence of lines of evolutionary descent seemed to be the main parameter which decided by how many residues the cytochromes *c* of any two species differed, implying that residue changes in the protein were fixed in the course of evolution at a roughly constant rate [MARGOLIASH, 1963]. Therefore, using as an internal standard the known time of divergence of the line of descent which led to birds from that which led to mammals, it was possible to define the *Unit Evolutionary Period* for cytochrome *c*, namely the evolutionary time required to effect a single change in the cytochromes *c* of two diverging lines of evolutionary descent [MARGOLIASH and SMITH, 1965]. In terms of residue changes this was calculated to be 26.4 million years, and in terms of nucleotide changes in the corresponding codons the value was 21.4 million years [MARGOLIASH and FITCH, 1968; NOLAN and MARGOLIASH, 1968] (table I). One could also draw a curve relating cytochrome *c* residue or codon variations to time elapsed since the divergences of major lines of evolutionary descent (fig. 3)

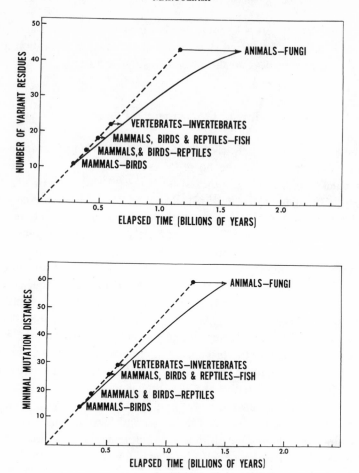

Fig. 3. Relation of number of variant residues (A) among cytochromes *c* from different classes and phyla of organisms to the time elapsed since the divergence of the corresponding lines of evolutionary descent, and (B) the similar relation of the number of variant nucleotide positions in the respective structural genes. The straight dashed lines were calculated on the basis of a value of 280 million years for the time elapsed since the divergence of the mammalian and avian lines of descent, corresponding to unit evolutionary periods of 26.4 million years for residue replacements and 21.4 million years for nucleotide replacements. The solid curved lines show the relationship corrected for multiple changes in residue (A) and nucleotide (B) positions, according to the formula, $r = n \ln n/n-\lambda$ [MARGOLIASH and SMITH, 1965], assuming a total of 76 variable residues. According to MARGOLIASH and FITCH [1968].

Table I. Average numbers of variant residues and minimal replacement distances between eukaryotic cytochromes *c*-Comparison by taxonomic groupings[1]

Groups compared	Average number of variant residues	Average of the minimal replacement distance
Mammals to birds	10.6 ± 1.5	13.6 ± 1.9
Mammals and birds to reptiles	14.5 ± 4.5	18.6 ± 6.8
Mammals, birds and reptiles to fish	17.9 ± 1.8	26.5 ± 2.4
Vertebrates to invertebrates	22.1 ± 2.0	30.4 ± 3.2
Animals to fungi	43.1 ± 2.3	59.9 ± 2.8
Unit Evolutionary Period	26.4×10^6 years	21.4×10^6 years

1   According to MARGOLIASH and FITCH [1968]. Each number is followed by its average deviation from the mean. The *Unit Evolutionary Period* [MARGOLIASH and SMITH, 1965] is defined in the text.

and correct it statistically for the chance occurrence of more than one change per residue, a situation that could not be detected by the direct comparison of amino acid sequences [MARGOLIASH and SMITH, 1965; MARGOLIASH and FITCH, 1968]. The results were satisfactory in that they accorded in general with other estimates. For example, the divergence of the chain of descent which led to fungi on the one hand, and other taxonomic groups considered on the other, was estimated at about 1.5 billion years.

These then were the first observations that demonstrated that the expected concordance between protein amino acid sequences and the phylogenetic relations of species [CRICK, 1958; ZUCKERKANDL and PAULING, 1962] was indeed a reality. Moreover, they made clear that the problems of unravelling the molecular concomitants of evolutionary transitions and of understanding the mechanisms of these changes could be attacked on a quantitative basis. They also indicated that on the basis of the genetic code, at that time only very partially determined, it would be possible to approximate the structure of ancestral forms of the protein [MARGOLIASH, 1963]. Thus, these studies laid a good part of the conceptual foundations for the more sophisticated approaches that followed.

Even though the molecular evolutionary clock concept [ZUCKERKANDL and PAULING, 1962; MARGOLIASH, 1963; MARGOLIASH and SMITH, 1965] has recently been reworked in terms of the much more extensive amino acid sequence material now available and more precise paleontological standards [DICKERSON, 1971a] leading to much the same conclusions, it can only represent an approximation, which becomes less and less satisfactory as we develop an interest in more and more accurate descriptions of molecular evolutionary history, and attempt to understand the underlying fundamental biological mechanisms.

A simple examination of the composite picture of the amino acid sequences of all the cytochromes *c* which have been examined to date (fig. 2) shows that some 26 residue positions have remained unvaried, and that the approximately 75% of the molecule which has undergone residue substitutions can be subdivided, according to evolutionary behavior, into two classes of positions [SMITH and MARGOLIASH, 1964]. There are those at which the alternative residues are so similar in structure or physico-chemical properties that one can readily believe they are capable of identical functions, constituting so-called *conservative substitutions*. Examples are given by Residue 13 which seems to be strictly confined to either of the two available strongly basic amino acids, lysine and arginine; Residue 35 which can apparently accomodate only hydrophobic amino acids such as valine, isoleucine, leucine and phenylalanine and tyrosine. Other locations seem capable of being occupied by amino acids which are so dissimilar that at first glance they appear unlikely to be importantly involved in maintaining the tertiary structure or the function of the protein. These are termed *radical substitutions* and many locations carrying such substitutions are obvious in figure 2, such as Residues 60, 89 and 92.

It must, however, not be overlooked that this simplistic approach to the significance of evolutionary residue changes in proteins assumes that all similar residues have a similar function, while chemically different residues necessarily have different functions. Either assumption can be wrong. For example, in a position at which aromaticity is the essential attribute which needs to be conserved, a tyrosine-phenylalanine substitution is conservative, but in a position at which the tyrosyl residue provides an essential hydrogen bond the same substitution is radical. As long as, at a minimum, we do not know the tertiary structure of the protein examined, or, ideally, until we understand the functional role of each residue, any classification of evolutionary variable residues can be thoroughly mislead-ing. At best, negative conclusions can be drawn. The fact that a certain residue can be substituted by others provides evidence that a function only attributable to the particular residue cannot be occurring. For example, the observation that the only constant histidine in cytochrome *c* was that at position 18, that commonly at position 33 being absent in the kangaroo [NOLAN and MARGOLIASH, 1966], snapping turtle [CHAN, TULLOSS and

MARGOLIASH, 1966] and tuna [KREIL, 1963, 1965] proteins, while that commonly found at position 26 is not present in the iso-2 cytochrome *c* of baker's yeast [STEWART, MARGOLIASH and SHERMAN, 1966] and the protein from the fungus, *Neurospora crassa* [HELLER and SMITH, 1966, 1965], provided the first definitive evidence that the cytochrome *c* hemochrome is not a diimidazole hemochrome constituted by the side chains of two histidyl residues coordinating with the heme iron from both sides of the porphyrin plane, as had long been supposed [THEORELL and ÅKESON, 1941; THEORELL, 1941; PAUL, 1951; PALEUS, EHRENBERG and TUPPY, 1955; EHRENBERG and THEORELL, 1955; HARBURY and LOACH, 1959, 1960]. This conclusion has since been amply confirmed by x-ray crystallographic [DICKERSON *et al.*, 1967, 1968, 1971] and other studies [ANDO, MATSUBARA and OKUNUKI, 1965, 1966; TSAI and WILLIAMS, 1965; SCHEJTER and GEORGE, 1965; HARBURY *et al.*, 1965; SHECHTER and SALUDJIAN, 1967; WUTHRICH, 1969; MCDONALD, PHILLIPS and VINOGRADOV, 1969; GUPTA and REDFIELD, 1970; REDFIELD and GUPTA, 1971] which demonstrate that the sulphur of methionyl residue 80 and the imidazole of histidyl residue 18 are the protein groups bound to the heme iron.

With regard to invariant positions the situation is better, in that strict invariance very likely indicates that the residues involved have a function which is compatible only with their particular structures. Moreover, when such invariance extends over a very broad taxonomic spectrum of species, as in the case of cytochrome *c*, the function subserved must be general, not one adapted only to the needs of each species, and therefore important. But, on this basis, and without independent lines of evidence, to attempt to decide what are the functions of the invariant residues is unwarranted. An excellent example is provided by the stretch of 11 residues (Residues 70 to 80), which had been found to be strictly invariant until the cytochromes *c* of 2 protists were recently examined. To note that the next longest invariant segment is only 2 residues long (see fig. 2), emphasizes the most unusual character of Residues 70 to 80. This extreme degree of conservatism was attributed either to close contact intramolecularly with the one structure in cytochrome *c* which is necessarily invariant in evolution, namely the heme, or to contact with another protein whose reaction with cytochrome *c* is functionally essential, such as the cytochrome *c* oxidase or the reductase of the mitochondrial respiratory chain, or to both types of interaction. A surface of close contact with another protein is likely to be evolutionarily conservative since amino acid substitution in one protein will require appropriately inverse substitutions in the other to

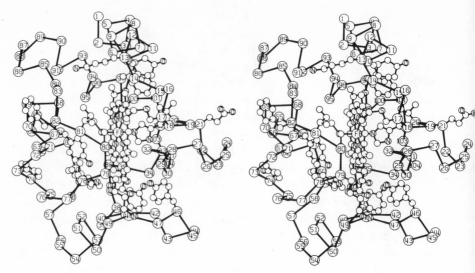

Fig. 4. Stereoscopic α-carbon diagram of tuna cytochrome *c* in the ferric form, according to DICKERSON *et al.* [1971]. The side chains that are indicated are those of Residues 10 (phenylalanine), 13 (lysine), 14 (cysteine), 17 (cysteine), 18 (histidine carrying heme iron-linked imidazole), 20 (valine), 21 (glutamic acid), 46 (tyrosine), 48 (tyrosine), 59 (tryptophan), 67 (tyrosine), 74 (tyrosine), 78 (threonine), 79 (lysine), 80 (methionine carrying heme iron-coordinated sulfur), 81 (isoleucine), 82 (phenylalanine) and 97 (tyrosine). The heme is the square structure inserted in the front of the molecule. It is seen nearly edge on, the left surface being just visible.

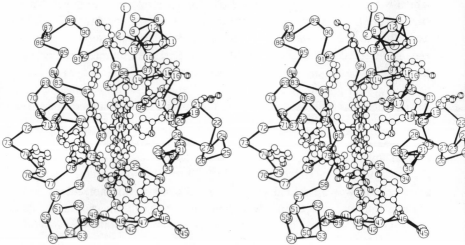

Fig. 5. Stereoscopic α-carbon diagram of tuna cytochrome *c* in the ferrous form, according to TAKANO and DICKERSON [1972] [see also TAKANO *et al.*, 1971]. The protein has been kept in the same general relative orientation as in figure 4, so that changes in conformation on oxido-reduction can be observed by direct comparisons of figures 4 and 5. The same side chains are marked as in figure 4.

maintain close contact. The x-ray crystallographic solutions of the structures of horse and bonito ferricytochrome *c* [DICKERSON *et al.*, 1971] and of tuna ferrocytochrome *c* [TAKANO *et al.*, *c*1971] have indeed shown that the carboxyl-end of the invariant stretch is near the heme, the methionyl residue at position 80 providing one of the two hemochrome-forming groups (figs. 4 and 5). Also, as long as the areas of interaction with the oxidase and the reductase have not been directly identified on cytochrome *c*, one may consider the invariant segment a likely candidate for such functions. However, an entirely novel reason for possible structural invariance was overlooked in the evolutionary argument, and subsequently beautifully brought out by the crystallographic work. The invariant segment, located at the lower left side of the protein (see figs. 4 and 5) is one which appears to be capable of considerable change of conformation, as evidenced by the preferential reaction of the internally located tyrosyl residue 67 with tetra-nitromethane [SKOV, HOFFMANN and WILLIAMS, 1969; SCHEJTER and SOKOLOVSKY, 1969; SCHEJTER, AVIRAM and SOKOLOVSKY, 1970]. The re-agent has no access to the residue unless the peptide chain comprising the invariant segment swings away from its normal position. Moreover, the immediately following segment of Residues 81 and 82 is involved in the most dramatic movement linked to oxido-reduction [TAKANO *et al.*, 1971]. This entire section is the so-called 'weak side' of the protein [DICKERSON *et al.*, 1971]. Thus, conformational mobility which is probably crucial to function, and, therefore, likely to be of a very specific variety, may be among the most stringent constraints imposing evolutionary invariance on protein structure. The invariant segment of cytochrome *c* would well represent the first case of such a mechanism for constancy.

B. The Invariance of Cytochrome *c* Tertiary Structure and Function

With respect to evolutionary considerations the amino acid sequence of a protein is operationally near the genetic end of the spectrum of biological phenomena. By itself it has no phenotypic meaning. This is acquired only when the polypeptide folds into its own spatial or tertiary structure, binds cofactors or prosthetic groups as required, and appears as a fully fledged molecular machine fit to perform its designated function in the specific cellular milieu provided, and subordinated to the various levels of organization which make up a live organism, or even a population of organisms. Since evolutionary selection operates at the phenotypic level, a most thorough understanding of the spatial structure of a protein and of its structure-function relationships in the broadest biological sense are necessary if we are ever to unravel the historical significance of molecular evolutionary changes and the mechanisms by which they occur.

For cytochrome *c*, perhaps the most striking result of the x-ray crystallographic determination of structure and of the examination of the functional activities of the cytochromes *c* from a wide variety of species is that both structure and function remain essentially the same throughout. In the case of function this statement is well documented. The first order rate constants for the oxidation of ferrocytochrome *c* by solubilized beef cytochrome *c* oxidase remains, within experimental error, the same for the protein of the cow and four other mammals, that of three birds, one fish and one insect, even though the amino acid sequences differ by as much as 27 residues out of 104 [L. SMITH, NAVA and MARGOLIASH, 1972]. The same relation is maintained over a wide range of cytochrome *c* concentrations, and an essentially identical result is obtained with the cytochrome *c* reductase system [L. SMITH, 1972]. Cytochrome *c* function can also be measured under conditions which approximate the *in vivo* situation even far better than for most proteins. It is possible to remove as much as 80 to 85% of the cytochrome *c* of mitochondria, without damaging the mitochondrial inner membrane, by procedures similar to that developed by JACOBS and SANADI [1960]. The recovery of function can then be measured on adding graded amounts of cytochrome *c*. In such cytochrome *c* repleted depleted-mitochondria, the oxidation of substrates and oxidative phosphorylation are well coupled when the repletion procedure is carried out at low ionic strength, and function can then be estimated from the rate of oxidation of substrates in the presence of ADP. Again here it is observed that with rat liver mitochondria, for example, it makes no difference whether the added cytochrome *c* is the horse, cow, rabbit, human, chicken, pigeon, snapping turtle, tuna, lamprey, moth or baker's yeast protein. Identical titration curves of functional recovery are obtained [BYERS *et al.*, 1971; BYERS, KANG and MARGOLIASH, 1972].

This is in strong contrast to the differences in the amino acid sequences of these cytochromes *c*, some of which differ by nearly half their residues. One such repletion titration curve is shown in figure 6, in which tuna cytochrome *c* is used to reactivate cytochrome *c*-depleted rat liver mitochondria. When the mitochondrial inner membrane has been damaged in the osmotic manipulations of the depletion, the titration curves do not

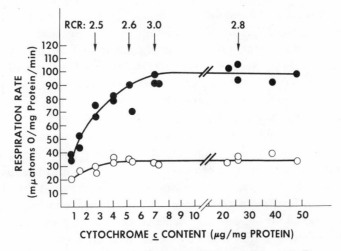

Fig. 6. Repletion of cytochrome c-depleted rat liver mitochondria with tuna cytochrome c. Respiration in the presence of ADP (filled circles) and in the absence of ADP (open circles), using succinate as substrate, is plotted against the amount of added tuna cytochrome c, in μg/mg mitochondrial protein. The numbers indicated by RCR at the top are the respiratory control ratios, estimated at the corresponding arrows. The maximal rate of O_2 uptake in the presence of ADP is the same as that obtained with the native untreated mitochondria and the amount of cytochrome c required to attain this rate is approximately that present in native mitochondria.

go to a saturation plateau, but the rate of oxidation continues to increase as more cytochrome c is added, in a manner similar to the behavior of the commonly used particle preparations from disrupted mitochondria. Interestingly, the amount of cytochrome c required to restore maximal function under coupled conditions is not very different from that present in native mitochondria. Nevertheless, the cytochrome c bound to the repleted depleted-mitochondria is most probably in equilibrium with cytochrome c in solution, as the amount of the protein needed for maximal reactivation varies with the temperature at which rates of oxidation are measured, being about twice as high at 37° as at 25° [BYERS, KANG and MARGOLIASH, 1972]. Since this is not the case for native mitochondria, and in the depletion procedure the outer membrane is damaged, these observations can be interpreted to indicate that one function of the mitochondrial outer membrane is to prevent cytochrome c from diffusing away from the organelle.

Comparisons of different cytochromes c with respect to their ability to maintain the function of the mitochondrial respiratory chain had earlier been carried out by Jacobs and Sanadi [1960] when they showed that beef cytochrome c could restore the oxidation of substrates and some 75% of the rate of oxidative phosphorylation of cytochrome c-depleted rat liver mitochondria. Mattoon and Sherman [1966] later utilized a genetic variant of baker's yeast (cy_{1-12}) which carries a chain-terminating codon at position 20 of the structural gene for cytochrome c (corresponding to position 15 of the horse cytochrome c amino acid sequence in fig. 1). This strain made mitochondria devoid of cytochrome c but otherwise normal. On repletion with either yeast or horse cytochrome c oxidative phosphorylation was recovered, the horse protein being only some 10% less effective than yeast cytochrome c. However, these cytochrome c-deficient yeast mitochondria bound many times more cytochrome c than required to saturate function, which, as noted above, carefully depleted rat liver mitochondria do not. Whether this is due to a tissue or species difference, or results from the relatively more drastic conditions required for the preparation of yeast mitochondria, cannot be decided. Moreover, the cytochrome c preparations were not exhaustively deionized and the species comparison was limited to just two preparations of the protein. For all these reasons it is difficult to judge whether these results do or do not represent an evolutionarily significant difference.

Indeed, correct functional comparisons of cytochromes c from different species were made possibly by the finding that a large variety of anions bind to ferricytochrome c while cations bind to the ferro-protein [Barlow and Margoliash, 1966; Margoliash, Barlow and Byers, 1970], and that these can be removed by electrodialysis to the isoelectric point, dialysis at the isoelectric point (pH 10) or gel filtration pH 10. Unless all cytochrome c preparations are so treated their activities in the enzymes systems or in intact mitochondria are very erratic, indicating the presence in many preparations of varying amounts of a potent inhibitor of cytochrome c function [Smith, Nava and Margoliash, 1972]. This inhibitor has not yet been identified. Observations indicating differences in the functional activities of the cytochromes c of various eukaryotes [Yamanaka and Okunuki, 1964, 1964; Estabrook, 1966; Yamanaka, 1966, 1967; Yamanaka and Okunuki, 1968], are probably vitiated by this and other types of artifact [Margoliash, 1962; Margoliash and Lustgarten, 1962].

There is much less information with regard to spatial structures. The one fully documented case is the comparison of the structures of the horse and bonito ferricytochromes c by x-ray crystallography [Dickerson et al., 1971]. Both proteins have essentially identical polypeptide backbone conformations even though they differ by 17 residues out of 104. The only differences are the expected ones in the location of some of the side chains. For example, when a tryptophan in the bonito protein replaces the histidyl residue at position 33 in the horse protein, the indole side chain is folded back into the hydrophobic interior of the protein, while the imidazole remains on the external hydrophylic surface. In view of the quantitative identity of functional cross reactivity of the cytochromes *c* from a very wide taxonomic spread of eukaryotes noted above, it seems probable that this is the general situation. If any spatial structural differences exist at all, they are likely to be so minor that they can hardly be considered to modify the general rule of essential structural identity.

C. The Role of Evolutionary Selection

From the evolutionary point of view this situation immediately poses a fundamental problem. Evolutionary selective processes can only operate by distinguishing between the functional effects of the genes being selected for or against. When the biological results of the products of two genes are precisely identical, no selection can be effected and the mutational event which led from one gene to the other is termed a *neutral* mutation. If the evidence so far developed is taken at face value, it would appear that for cytochrome *c* the vast majority of evolutionary fixations of mutations, if not all of them, belong to this category. Indeed, the proteins of all species tested are functionally indistinguishable, and though amino acid sequence variations are as extensive as about 50% of the entire chains, functional differences, if any at all, amount to less than 5%, that being roughly the estimated error of the methods of assay discussed above. The approximately 75% of residue positions which have been observed to vary as among the cytochromes *c* of different species could, on this basis, be considered to have been subject to functionally irrelevant changes, purely the result of random events. Quite possibly the metabolic machinery of cytochrome *c*, once perfected in the early ancestral form from which eukaryotes are descended, has not required any further significant functional changes, so that outside of the limitations of structure imposed by its immutable function, the rest of the changes are selectively neutral.

The difficulty is that the above contention is in essence an argument from ignorance, namely, since we have not been able to observe any functional differences between the cytochromes *c* from different species, there are none. As such it is subject to several objections. In the first place, our tests of function may not reveal evolutionarily important differences, as natural selection could well operate on properties not necessarily related to electron transport or oxidative phosphorylation, that is, cytochrome c has biological roles other than the classical ones. Alternatively, functional differences which are quantitatively so small as to be below the limits of sensitivity of our tests could have large cumulative evolutionary effects. This second possible objection is to some extent unsatisfactory since the tests of cytochrome c function are carried out under conditions which are very close to those which obtain *in vivo* in nearly intact mitochondria, and the more artifacts of preparation or testing are eliminated [SMITH, NAVA and MARGOLIASH, 1971] and tests are made rigorously quantitative, the more the cytochromes c of different species have appeared to be functionally identical, as discussed above. In contrast, the first of the above two objections to the concept of selective neutrality in cytochrome c variations is of course very strong, bearing with it the full force and extensive background of Darwinian morphological evolutionary studies. The few cases in which evolutionary changes at the morphological level were at first considered to be the result of selectively neutral variations, have so far always been shown on further study to possess clear-cut positive selective value [MAYR, 1966].

An important test of whether the evolutionary changes which do occur in cytochrome c are preponderantly selectively neutral or not, would be to determine if suitable randomly interbreeding populations of a single species shown an appropriate degree of polymorphism in the cytochrome c gene. If no polymorphism exists one would have to assume that the evolutionary changes are actively selected for. The major complicating factor in any such test is that cytochrome c varies particularly slowly in evolution. Thus, KIMURA [1968] calculates that if all mutations fixed in a gene are selectively neutral, such mutations would be established in the population with a probability equal to their probability of occurring in an individual. On this basis one can estimate the expected degree of polymorphism according to KIMURA and CROW [1964]. For a mammal with an average generation time of 4 years, the observable neutral mutation rate (μ) in the cytochrome c gene with approximately one nucleotide substitution every 21.4 million years (see above) would be about 10^{-7} per generation per gamete. With an effective population size (N_e) of 2.5×10^4, the probability that a randomly sampled individual would be homozygous in the cytochrome c gene is:

$$p = \frac{1}{1 + 4N_e\mu} = 0.99$$

Conversely, only one individual in 100 is likely to carry an 'abnormal' cytochrome c. Thus, to test whether selective neutrality is of common occurrence with cytochrome c variations would require the determination of the complete amino acid sequences of the cytochromes c of several hundred individuals of a properly chosen population. The relatively small numbers of individually tested human and horse preparations [MARGOLIASH, FITCH and DICKERSON, 1968] are much too small to detect such an expected low level of polymorphism.

This argument as to the possibility that variations of primary structure fixed in the course of evolution in cytochrome c may be of no selective value is just one part of a multi-faceted discussion that has been proceeding vigorously during the last few years over the question of the presence or absence of selectively neutral variations in proteins in general [KING and JUKES, 1969; JUKES and KING, 1971; SMITH, 1969, 1970a, 1970b; O'DONALD, 1969; ARNHEIM and TAYLOR, 1969; KIMURA, 1968, 1969; KIMURA and OTHA, 1971a, 1971b; OTHA and KIMURA, 1971; BOYER *et al.*, 1969; CLARKE, 1970a, 1970b; RICHMOND, 1970; UZZELL and CORBIN, 1971; FITCH and MARKOWITZ, 1970; FITCH, 1971b].

Parts of the discussion have been summarized by FITCH and MARGOLIASH [1970] and more recently by HARRIS [1971]. There is no need to reproduce it here. However, how the concept of selective neutrality of protein evolutionary changes has fared in the case of cytochrome c, as evolutionary and functional information was accumulated, is an important contribution of this study to our understanding of protein evolutionary transitions in general. It will be further considered below.

III. Statistical Phylogenetic Trees

If one is to attempt to examine the mechanisms by which structures have changed one into the other, the first requirement is to obtain at least an approximation of the topology of their relationships, in biological terms a phylogenetic tree. A simple, but comprehensive, statistical procedure for obtaining such a tree from amino acid sequences was first developed in 1967 [FITCH and MARGOLIASH, 1967a] and it has since proved to be very fruitful in examining the evolutionary information content of proteins, as described below [see also FITCH and MARGOLIASH, 1970; MARGOLIASH and

FITCH, 1970; NOLAN and MARGOLIASH, 1968]. Other procedures to estimate the topology of structural relationships and similar parameters have also been used [DAYHOFF, 1969; MOORE and GOODMAN, 1968; SOKOL and SNEATH, 1963; KOHNE, 1970; SARICH and WILSON, 1966; PAULING and ZUCKERKANDL, 1963]. Of these, the better developed present both some advantages and disadvantages over the original statistical technique, none of which is crucial, and indeed the final products are mostly the same for all techniques.

However, if the derivation of the similarity relations of a set of amino acid sequences is to have any biological significance, two prior conditions must be met: (1) is must be shown that the structures considered possess a degree of similarity which it greater than random, and (2) the proteins considered must be *homologous*, namely, all descendant from a common ancestral form.

A procedure for the systematic search of significant similarity between

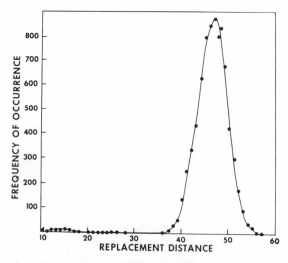

Fig. 7. Comparison of minimal replacement distance *(replacement distance)* for all possible 20-residue segments of human cytochrome *c* and baker's yeast iso-1-cytochrome *c* (7565 comparisons) by the procedure of FITCH [1966]. The numbers of times various values of minimal mutation distances occur in the comparisons are given on the ordinate *(frequency of occurrence)*. The Gaussian portion of the curve on the right represents the random part of the distribution. The extended line on the left indicates those comparisons for which the minimal mutation distances are less than expected for a random distribution and hence suggest ancestral homology. The probability that such a distribution occurred by chance is 10^{-80} [FITCH, 1970a].

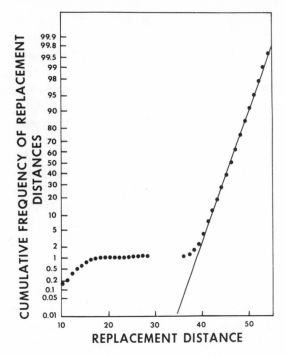

Fig. 8. Probability plot of the data from figure 7, according to FITCH [1966]. The random part of the distribution is represented by the straight line. The comparisons for which the minimal replacement distances *(replacement distance)* are less than expected for a random distribution, are represented by the points which deviate from the straight line at the lower left.

amino acid sequences proposed by FITCH in 1966 [FITCH, 1966] and more recently improved further [FITCH, 1970a], consists in comparing all possible segments of a definite length of the two chains under consideration. For each comparison the minimal mutation or replacement distance is computed. This is the minimal number of single nucleotide substitutions required to transform the gene segment coding for one protein segment into that coding for the other. The replacement distances are then plotted against the number of times each occurs in all the comparisons, to give a curve such as that in figure 7. This represents a comparison of human and baker's yeast iso-1 cytochrome *c*. The comparisons indicating random similarity fall within the confines of the Gaussian portion of the distribution curve. The summit of that portion of the curve is near a replacement distance of 45, as expected, since 30 residue segments were compared and on the average it takes near to 1.5 mutations to transform any codon to any other according to the genetic code [FITCH, 1966]. The tail of the distribution curve to the left consists of those comparisons for which the replacement distances are smaller than expected for purely random similarities, and its presence, therefore, indicates the existence of a significant degree of similarity between human and baker's yeast cytochrome c. The same data can be plotted on a probability scale, as shown in figure 8. Here, the Gaussian distribution curve becomes a straight line, and it is the deviation from that line to the lower left which denotes comparisons for which the replacement distances are smaller than for random similarities. The more recent developments of the method [FITCH, 1970a], among others, make it possible to assign a probability that the conclusion of significant similarity is in error, namely, that notwithstanding appearances, the similarity observed is a chance event. In the case of figure 7 that probability, as indicated, is less than 10^{-80}.

Evidence of significant similarity can be taken as a presumption of homology for the proteins examined. However, lack of such evidence by this test does not necessarily mean that the proteins did not at one time have a common ancestral form. It merely indicates that, even if they did enjoy such a common ancestry, the changes they have undergone since their divergence from the common line of descent are so large that it is impossible to detect any evidence of their homology in the present-day structures. As just noted, significant similarity is only presumptive evidence of homology. Indeed, structures, whether they are front limbs or polypeptide chains, may appear to be similar either because of divergence from a common ancestral form, as for the cases marked 'Recent Divergence' and 'Parallel Evolution' in figure 9, or because of so-called functional convergence from independent

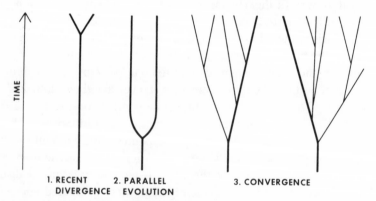

Fig. 9. Possible evolutionary reasons for similarity of present-day polypeptide chains.

evolutionary stocks, as in the third diagram of figure 9. In this last case, the proteins considered would tend to acquire the degree of similarity of structure necessitated by their similarity of function. For example, it is not inconceivable that fungal and mammalian cytochromes *c* could have had quite separate phylogenetic origins. If that were so, one would have to interpret the very considerable degree of similarity indicated by figures 7 and 8 as meaning that, in order to carry out its function, a relatively large proportion of the protein is specified within narrow limits. For this reason, even if deriving from independent lines of descent the baker's yeast and human cytochromes *c* could have tended towards structural similarity.

A final determination of whether two sets of proteins are similar because of common ancestry or through functional convergence can be made only if one can read their structures at two distinct times in the history of their evolutionary descent. This would enable one to decide whether during the course of evolutionary time they have become less similar, namely, are diverging from each other, or more similar, namely, are undergoing evolutionary convergence. To do so requires the reconstruction of ancestral sequences. Since this has been done from the statistical phylogenetic trees yet to be described, the distinction between *homologous* (similar because of common ancestry) and *analogous* (similar though of different phylogenetic origins) protein structures is discussed below.

The topology of relations described by the phylogenetic tree in figure 10 is based on the amino acid sequences of the cytochromes *c* of the species listed. Strictly the only other biological information employed in the computation is the genetic code. It is remarkable that on such a modest basis one obtains a phylogenetic tree which is so close to trees generated by more conventional morphological evolutionary criteria. The procedure is that of FITCH and MARGOLIASH [1967a]. The minimal replacement distances are calculated for all possible comparisons of the complete amino acid sequences of the cytochromes *c* employed. For n sequences there are $n(n-1)/2$ replacement distances. One can start by joining together the two proteins which show the smallest replacement distance, and calculate the average replacement distances of all other proteins to the first two, now considered as a single subset. The next protein to be joined to the tree is the one with the smallest replacement distance to the first two. It now joins their subset, and the procedure is repeated until all proteins have been linked. The final tree is merely a graphical representation of the order in which proteins were joined. The number of nucleotide substitutions required between various branch points on the tree can be calculated, and one can then reconstruct the replacement distances between any two proteins by summing the appropriate segments of the tree. This yields the so-called 'output' replacement distances, which will differ from the 'input' replacement distances —those calculated directly from the amino acid sequences of the proteins and used in the initial step of tree construction. The occurrence of these differences stems from the procedure itself in that, once two proteins are joined, only average distances from all other proteins to the two in the first subset can be used, and such averages are necessarily employed throughout the computation. Because 'input' and 'output' replacement distances do not exactly match, the initial tree produced does not necessarily provide the most effective utilization of the data. Several possible procedures can be employed to choose an optimal tree. The most useful involves calculating a percent standard deviation from the replacement distances reconstructed from the tree to those determined directly from the amino acid sequences, and examine alternative trees seeking to minimize that criterion of difference. Since the number of possible trees is very large (10^{39} for the proteins of 29 species [see FITCH and MARGOLIASH, 1968; MARGOLIASH and FITCH, 1970]), not all alternatives can be examined and common numerical taxonomic methods are used to eliminate unlikely prospects and pick trees for examination [FITCH and MARGOLIASH, 1967a; SOKOL and SNEATH, 1963; DAYHOFF, 1969]. It should, of course, not be overlooked that other criteria can also be employed for deciding which is the 'best' tree. For example, one can pick the tree for which the total number of nucleotide replacements required to derive all present-day sequences from the presumed common ancestral form is a minimum.

In the case of cytochrome c, the phylogenetic tree obtained is obviously a fair representation of the phylogeny of the species carrying the proteins employed in the computation. This is because the proteins are not only homologous but also *orthologous,* namely, that in the most recent common ancestor of all the species considered, at the topmost apex of the tree, cytochrome c was represented by a single gene. In that way a 1 to 1 gene-to-protein relationship is retained throughout, and if the evolutionary variations fixed in this particular gene are a statistically valid representation of the evolutionary changes of the species as a whole, one obtains an accurate species phylogeny. The tree in figure 10 is nevertheless not perfect, as, where the primates branch off the ancestral mammalian line before the marsupial, the turtle appears to relate more closely to the birds than to the rattlesnake, and the shark is closer to the lamprey than to the tuna. Such blemishes are likely to be eliminated as the proteins of more species

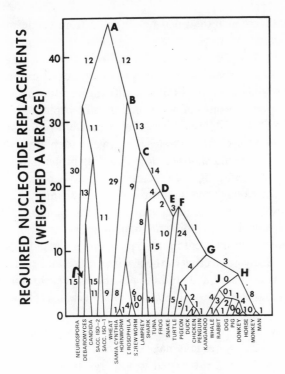

Fig. 10. Statistical phylogenetic tree based on the minimal replacement distances between the cytochromes *c* of the species listed, according to FITCH and MARGOLIASH [1967a]. Each number on the figure is the replacement distance along the line of descent as determined from the best fit of the data so far found. Each apex is placed at an ordinate value which is the average of the sums of all mutations in the lines of descent from that apex, weighting equally the two lines descending from any one apex. References to the amino acid sequences of the cytochromes *c* are given in the legend to figure 2. All the proteins in the figure are *orthologous*, with the exception of the iso-1 and iso-2 cytochromes *c* of baker's yeast, which occur in the same cells and are, therefore, *paralogous*. As this is the only such relationship, it does not introduce any errors in the rest of the tree.

are taken into account. Nevertheless, it is remarkable that as accurate a representation of phylogeny can at all be extracted from the changes of one small gene, out of the millions that constitute a total genome. Without these molecular procedures, a phylogeny as precise as this cannot be obtained even from a single morphological trait, possibly comprising the influence of many genes. One must assume that this is because an accounting of the number of mutations fixed in the course of evolution, even when restricted to a single gene, yields more precise estimates of the degree of evolutionary divergence than can be obtained from single morphological traits. In as wide a sweep of taxonomic groups as represented in the cytochrome c phylogenetic tree, it is unlikely that one can find single morphological characters that are present throughout all species considered, from man to baker's yeast. Similarly, difficulties also arise at the other extreme, when small groups of organisms are all morphologically so similar that it is difficult to distinguish between them. Both problems are readily obviated by the molecular approach. To cover widely divergent phylogenetic relations one would choose a slowly varying protein, while for a small, only slightly divergent group of species one would choose a protein which tends to fix evolutionary changes at relatively short intervals. The exciting prospects presented by this independence of the molecular approach to phylogeny from the complexities and inadequacies of morphology, together with the quantitation inherent in these procedures, the indefinitely increasing precision likely to accrue to them when more and more appropriate sets of proteins are examined, and the possibility of eventually introducing a valid time parameter in considerations of molecular evolution, have been noted above.

Homology does not necessarily imply orthology, as defined above. Indeed, homologous genes may have duplicated but remained side by side in the same species and proceeded to undergo independent evolutionary variations, both varieties being transmitted and continuing to change independently in the descendant species. Such genes have been termed *paraloguous* (from para, meaning in parallel [see FITCH and MARGOLIASH, 1970; MARGOLIASH and FITCH, 1970]). An examination of the phylogenetic relations which may be derived from the structures of the products of such genes yields not a species phylogeny, but rather a gene phylogeny. This is the case, for example, with the relations postulated by INGRAM [1963] between myoglobin and the various hemoglobin chains, or later between the two segments of the light chains and between the light and heavy chains of γG-globulins [SINGER and DOOLITTLE, 1966; HILL et al., 1966, 1967; RUTISHAUSER et al., 1968; EDELMAN et al., 1969]. In the case of hemoglobin, for example, one could, if enough information were available, generate a species phylogeny from the structures of the α chains of the proteins of mammals, and an independent phylogeny from the β chains. The most recent common ancestor of mammals already carried both α and β hemoglobin chains. These chains had by that time undergone considerable independent evolutionary divergence. If one attempted to obtain a species phylogeny by considering the α chains of some mammals and the β chains of others, one would necessarily obtain an absurd result: the species would be primarily segregated depending on whether their α or β chains had been employed in the computation.

IV. Ancestral Amino Acid Sequences

Once the topology of a set of presumably homologous proteins has been decided, the resultant statistical phylogenetic tree can be used to obtain a variety of derived items of evolutionary information [FITCH and MARGOLIASH, 1970]. Among the most informative of these are the amino acid sequences of ancestral forms of the protein at each of the branching points. Originally, as proposed by FITCH and MARGOLIASH [1967a], this was accomplished by following a series of rules, according to which each amino acid residue in an ancestral sequence was chosen so that during its phylogenetic descent the corresponding codon required the smallest number of mutations, the fewest segments of the phylogenetic tree containing multiple mutations, etc. The procedure was directed towards obtaining an approximation of actual evolutionary transitions. It permitted a complete reconstruction of the cytochrome *c* phylogenetic tree, as well as a test of the reliability of the procedure by reconstructing satisfactorily a model case, consisting of a computer derived random phylogeny [FITCH and MARGOLIASH, 1968]. From the phylogenetic tree for the cytochromes *c* of 20 species it was possible to show, for example, that the ancestral primate cytochrome *c* differed from the ancestral mammal protein at Residues 17, 18, 21, 56 and 89, while the line of descent of human cytochrome *c* differed from the *Macaca mulatta* protein line of descent by a single mutation which had occurred in the human line, not the monkey line, etc. [FITCH and MARGOLIASH, 1967a].

This was nevertheless a statistically cumbersome process, and FITCH [1970b] has more recently developed a simplified version according to which, given the topology of relations defined by the statistical phylogenetic tree, the descent of each nucleotide for each codon representing the amino acid sequences is reconstructed separately, as follows. The nucleotide, or nucleotides, assigned to each coding position at an apex is one which is common to both the immediate descendants (the intersection of the descendant nucleotide sets), or when intersection leads to an empty set—namely, there is no common nucleotide in the descendants—the apex is given to the totality of all nucleotides in the two immediate descendants (the union of the descendant sets). At a second stage, a nucleotide replacement is noted whenever an intersection would have yielded an empty set, in such a way that the minimal number of such replacements or 'mutations' are required to account for the total phylogeny at the particular coding position under consideration, from the topmost apex to all the descendants. A diagrammatic example of this process is given in figure 11. It can be extended to all coding positions to yield a complete reconstruction of the statistical phylogenetic tree with a minimal total number of nucleotide replacements, each assigned to a specific codon and a particular branch of the tree.

Whatever the technique employed, a reconstruction of the complete pathway of the evolutionary fixation of mutations will yield a codon and hence an amino acid sequence for every ancestral form of the protein located at each branch point or apex in the phylogenetic tree. Such a reconstruction of the amino acid sequence of the ancestral form at the topmost apex of the tree is given in figure 12. Presumably, this represents an approximation of the structure of the cytochrome c of the most recent common ancestor of all eukaryotes, which probably already utilized a terminal respiratory chain similar to that so generally distributed today. More than one residue marked for an amino acid position in figure 12 indicates that the total number of 'mutations' in the tree would not change whichever alternative residue is assigned to that position, making it impossible to distinguished between them according to the criterion employed for the reconstruction of ancestral forms. It is, of course, likely that when enough cytochrome c amino acid sequences will become available for computation such ambiguities will disappear. Interestingly, the ancestral sequence has a somewhat smaller preponderance of basic residues than the present-day proteins. This is a character one would expect for an ancestral protein, since the extreme basicity of extant cytochromes c is most probably an adaptation to function from an originally less specialized precursor more like the majority of proteins with near neutral isoionic points. When the remaining ambiguities in the sequence of the ancestral form will have been resolved, a particularly important investigation would be the laboratory synthesis of that protein with a view to examining its functional activities in present-day enzyme systems. Possibly, this would gain an experimental basis for the study of the functional concomitants of evolutionary molecular transitions and might serve to define and quantitate the role of selective as contrasted to possible non-selective events. These not too distant possibilities of extending the transformation of evolutionary molecular studies

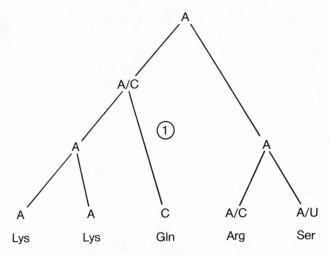

Fig. 11. Determination of the ancestral nucleotide for the first nucleotide of the codons for the amino acids marked in the bottom row, given the indicated topology of protein relationships. The procedure is described in the text. The circled 1 denotes that one mutation in the marked branch is the minimum required to derive the ancestral nucleotide.

Ala - Lys - Ser - Ala - Gly - Val - Ser - Ala - Gly - Asn - Ala - Lys - Lys - Gly - Ala - Lys - Leu - Phe - Lys
—9   10

Thr - Lys - Cys - Ala - Gln - Cys - His - Thr - Val - Glu - Gly - Gly - Gly - Thr - His - Lys - Val - Gly - Pro - Asn
        └─ HEME ─┘   20   30

Leu - His - Gly - Leu - Phe - Gly - Arg - Lys - Thr - Gly - Gln - Ala - Glu - Gly - Tyr - Ser - Tyr - Thr - Asp - Ala
                                              40   50

Asn - Lys - Lys - Lys - Gly - Val - Lys - Trp - Glu - Glu - Asn - Thr - Leu - Phe - Glu - Tyr - Leu - Glu - Asn - Pro
                          60   70

Lys - Lys - Tyr - Ile - Pro - Gly - Thr - Lys - Met - Ala - Phe - Gly - Gly - Leu - Lys - Lys - Pro - Lys - Asp - Arg
                                           80   90

Thr - Asp - Leu - Ile - Ala - Tyr - Leu - Lys - Lys - Ala - Thr - Ser - Ala
                                        100   104

Fig. 12. Amino acid sequence of the ancestral form of cytochrome *c* at the topmost apex of the phylogenetic tree. Any of the alternative amino acids shown would permit the evolution of the 29 descendent cytochromes *c* in the minimum number of 366 nucleotide replacements, assuming the topology shown in figure 10. Amino acids in brackets have not yet been observed in any present-day cytochrome *c*.

into a fully experimental science were mentioned in the introductory section, above.

Applying the newer technique of reconstructing ancestral sequences, FITCH [1970b] was able to develop a method for distinguishing analogous from homologous sets of proteins. Phylogenetic trees are independently computed for two sets of proteins. The codon sequences of the corresponding ancestral forms are obtained and examined to determine whether these are more similar or less similar to each other than the present-day sequences at each coding position. If at a certain position the ancestral forms are the same but the descendants are different, then the latter have undergone evolutionary divergence, by definition, and can be taken to be homologous. If, on the other hand, the ancestral forms are different, any similarities developed among the descendants are due to evolutionary convergence, and they can be considered to be only analogously related. Extending this to the entire ancestral nucleotide sequence gives how many positions in the two sets of proteins are divergently and how many are convergently related. An appropriate calculation performed with two sets of totally unrelated sequences yields an estimate of how many divergent and how many convergent relationships would be the expected on a purely random basis. When the two sets of proteins examined have a significantly greater number of divergently related coding positions over that resulting from chance, then one can be certain that the two sets are homologous. An excess of convergently related positions demonstrate they are analogous.

The results of a comparison of fungal and non-fungal cytochromes c by this procedure are graphed in figure 13. It is clear that the two sets of cytochromes have such a large preponderance of divergences that the probability that they are not homologously related is only 1 in 10^{10}, when a total of 24 different cytochromes c is utilized for the two phylogenetic trees in the comparison. The extensive similarities of amino acid sequence between the cytochromes c of a wide taxonomic range of species were noted very early in these studies and used to suggest a common evolutionary origin for them [MARGOLIASH, 1963; SMITH and MARGOLIASH, 1964; MARGOLIASH and SMITH, 1965]. However, these comparisons could not, in fact, distinguish structural similarity due to common ancestry from similarity resulting from evolutionary convergence. As just described, this has now been accomplished by FITCH [1970b]. His procedure provides for the first time a solid statistical demonstration that all eukaryotic cytochromes c are indeed the descendants of a common ancestral form, thereby giving a definitive proof that all living forms on Earth, as we know them, are the

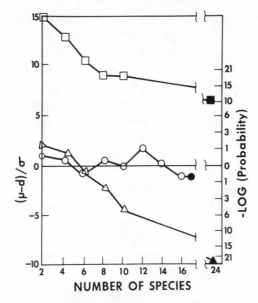

Fig. 13. Convergence and divergence as a function of the number of species. The abcissa gives the total number of sequences examined. Open symbols indicate that equal numbers of sequences were present in the two groups compared, closed symbols that they were divided unequally between the two groups. The ordinate gives the deviation (δ) from expectation (μ-d), in standard deviation units (σ) on the left, and the equivalent probability of a result being due to chance is given as negative powers of 10 on the right. Points above the zero line represent an excess of divergent comparisons, below the line, an excess of convergent comparison. Random sequences of amino acids are shown by circles ($O-O$), convergent sequences by triangles ($\triangle-\triangle$), and divergent sequences of squares ($\square-\square$). The convergent sequences were obtained by computer simulation. The divergent sequences compare fungal to non-fungal cytochromes *c*. According to Fitch [1970b].

result of a single occurrence, the basic tenet of the so-called 'unitary' theory of the origin of life.

Contrary to appearances, the technique for resolving homology from analogy in protein structure, does not contradict the principle that knowledge of any system at one point in time by itself cannot define the direction in which the system is changing as a function of time. Indeed, it cannot distinguish between pure divergence and pure convergence. In the case of the descent of the cytochromes *c* plotted in figure 13, for example, on this basis one cannot tell the difference between divergence from a single common ancestor and convergence of the 24 proteins towards an identical descendant from 24 independent evolutionary origins. What is ruled out absolutely is any mixture of divergence and convergence. If one accepts that the proteins of each of two small taxonomic groups of species are homologous within each group, one can proceed to demonstrate whether or not homology is also the case for the two groups together. Biologically, the evidence is overwhelming that at least within a well defined taxonomic group, such as mammals, a highly conservative structure such as cytochrome *c* does not have multiple phyletic origins. Moreover, from the molecular point of view to postulate an independent ancestral origin for each cytochrome *c* would make it impossible to understand how the structures of such a set of proteins could possibly yield a statistical phylogenetic tree (fig. 10) so close to that deduced from classical zoological criteria. In essence, the argument is like that between the concepts of the evolutionary origin of species and of separate creation for each living form. The latter becomes less and less tenable as more species are considered—just as pure convergence becomes less and less likely as similar results are obtained with further sets of proteins.

V. Rates of Evolutionary Change in Cytochrome c Codons

A cursory examination of the composite amino acid sequence of eukaryotic cytochromes *c* (fig. 2) and its comparison with the sequences of the individual proteins makes it obvious that the tendency to vary in the course of evolution changes over a very wide range for different positions. As already noted, some positions are unvaried, others accept few alternatives and remain unvaried over fairly extensive taxonomic groupings of species, while still others are most commonly changed whenever any amino acid substitutions whatsoever occur. Calculations of overall rates of evolutionary residue substitutions or of their corresponding nucleotide substitutions in the structural gene for the protein [MARGOLIASH and SMITH, 1965; MARGOLIASH and FITCH, 1968; DICKERSON, 1971a], can only yield averages of many rates which differ over a wide range. As described above, an approximately linear relation of cytochrome *c* variations to elapsed evolutionary time is manifested and serves to define the concept of *Unit Evolutionary Period* [MARGOLIASH and SMITH, 1965]. Nevertheless, such calculations can obviously not help in specifying why some codons change while others do not, and what are the mechanisms operating to restrict evolutionary variability, as to rate and as to acceptable alternatives, at the level of individual positions in the protein. To these ends it is first necessary to attempt to classify the codons for cytochrome *c* into classes with respect to the ease with which they undergo evolutionarily effective nucleotide substitutions and to attempt to identify the codons in each class. One would then be in a position to examine the interdependence of structure-function relations as between the corresponding residues of each class, which will hopefully lead to a complete recognition of the series of evolutionary events transforming one protein into another along the branches of the phylogenetic tree. This is the sort of historical description of molecular evolutionary phenomena likely to provide the background required for the unravelling of mechanisms operating at that level, as referred to in the first section of this paper. Only a fraction of this program has so far been achieved.

A. Hypervariable, Normally Variable and Invariant Codons

The first approach to the classification of cytochrome *c* codons according to their apparent rates of evolutionary change [FITCH and MARGOLIASH, 1967b] depended on a statistical phylogenetic tree for 20 different cytochromes *c* [FITCH and MARGOLIASH, 1967a] and the following assumptions:

a) There are particular amino acids in the protein so vital to structure-function that the probability that a strain would survive evolutionarily a mutation in the codons for these residues is essentially nil. The probability of such a variant occurring in a tree depicting the descent of the amino acid sequence of the protein is therefore also near zero. These are commonly termed 'invariant codons', and any mutation in them is defined as *malefic*, since any line carrying such a mutation will because of it eventually fail to survive.

b) The vast majority of variable amino acid positions correspond to codons which undergo mutations which are retained in the course of evolution at rates sufficiently similar that each codon can be considered to have as much chance as any other to fix the next mutation. Thus, the frequency of occurrence of such mutations in these codons will approximate a Poissons distribution. These are defined as 'normally variable codons'.

c) The relatively few remaining codons each carry so high a probability of undergoing an evolutionarily effective mutation, that the number of changes observed in the corresponding residues is large enough to permit their ready identification and exclusion from the set of normally variable codons. These are the so-called 'hypervariable codons'.

The reconstruction of the ancestral sequences from the first phylogenetic tree [FITCH and MARGOLIASH, 1967a], the 20 cytochromes c of which showed 35 unvaried positions (compare with fig. 2), identified the number of mutations that had occurred in every codon along each branch of the tree, for a total of 231. There were 5 codons showing 9 or more mutations each. These were considered to be hypervariable. Eliminating them left 182 mutations which could distribute themselves over from 105 codons, if the invariant class defined in (a) above did not exist at all, to 70, if the invariant class was as large as the 35 unvaried positions observed in the 20 cyto-chromes c. On this basis, one could identify the Poisson distribution which most closely fit the distribution of mutations observed on the reconstructed phylogenetic tree, namely, the number of codons with 1, 2, 3 etc. mutations. The best fit was found for 27 to 29 invariant codons, with the remaining 76 to 78 positions representing so-called 'normally variable' codons. This result clearly demonstrated that a very large proportion of the unvaried residues so far observed in the cytochromes c of eukaryotes were in fact invariant, namely, would never be observed to change, however many cytochromes c of different eukaryotes were examined. This is in sharp contrast to the situation with hemoglobins and myoglobins, for example, in which even with a much narrower taxonomic coverage, no more than 3 residues are found to be unvaried [see NOLAN and MARGOLIASH, 1968].

More recently, FITCH and MARKOWITZ [1970] and MARKOWITZ [1970], employing a statistical phylogenetic tree for 29 cytochromes c, have further developed this type of approach to estimates of codon variability. The reconstruction of ancestral sequences from that phylogenetic tree (see fig. 10) gave 366 mutations to be distributed over 113 codons. The improved method does not arbitrarily exclude any codons, and can deal simulta-neously with two Poisson-distributed classes. It was found that the distri-bution of codons having undergone 1, 2, 3, etc. evolutionary fixations was adequately fitted by a model which assumes 32 invariant residues, 65 'normally variable' residues and 16 'hypervariable' residues. This last class fixes mutations in the course of evolution about 3.2 times more often than the normally variable set. On the basis of the statistical phylogenetic tree employed, more than two classes of variable residues could not be di-stinguished. This, of course, does not mean that with many more amino acid sequences and a correspondingly larger tree, more classes of codon variability will not be uncovered. The descriptive limitations are merely imposed by limitations in the sequence data and in the sensitivity of the method, as well as by the degree to which the property examined is statistically well-behaved.

The above estimates of numbers of invariant residues were carried out before the more recent additions to our catalogue of cytochrome c amino acid sequences, at a time when 34 residues had not been observed to vary. However, with the amino acid sequences of the proteins from the two protists listed in the legend to figure 2, and from the ginkgo tree, variations were observed in 8 previously unvaried positions. Five are from the sequence of *Euglena* cytochrome c, one from that of the *Physarum* protein, one from both these cytochromes c and one from the ginkgo protein. Perhaps the least expected are the alanine in position 14 replacing the usual thioether heme-bonded cysteine, leaving the heme covalently attached only through one cysteinyl residue, and the variants at positions 75 and 77 in the center of the heretofore inviolate invariant segment extending from Residue 70 to Residue 80. It is particularly interesting that these changes have occurred in the cytochromes c of species outside the range of taxonomic groups previously explored, and it is obviously important to determine whether these cytochromes c do or do not have the functional characteristics common to all the other cytochromes c examined to date. If they do, then clearly the previously observed lack of variation at the positions now showing variants cannot be ascribed to requirements constant throughout the eukaryotic range. If they do not, then these cytochromes c would become prime objects of studies aimed at understanding the fundamental mechanisms of all the eukaryotic proteins. As always, the exploitation of manipulations performed by nature, when they can be recognized as such, is likely to be particularly fruitful.

B. Concomitantly Variable Codons – Covarions

The estimate of the number of evolutionarily invariant codons in the gene for the cytochromes c of eukaryotes was obtained utilizing all of the phylogenetic tree under the topmost apex in figure 10, apex A. The extent of invariance corresponds to nearly 30% of the cytochrome c gene. When the calculation is done again, excluding the cytochromes c of fungi—namely, utilizing only that portion of the tree under apex B—the degree of predicted invariance increases. It increases further and further as the taxonomic span of species is contracted by gradually excluding the proteins of groups remote from mammals. The values obtained for the set of species under apices A, B, C, etc. of figure 10, are plotted in figure 14 as a function of the weighted averages of the replacement distances for all species taken

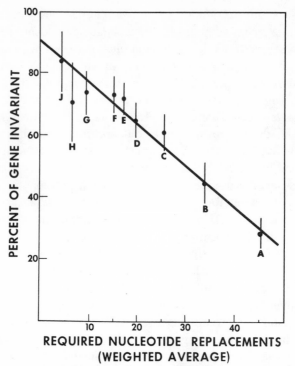

Fig. 14. Estimation of the number of concomitantly variable codons *(Covarions)* for cytochrome *c*. The percent of the gene found to be invariant is plotted as a function of the weighted average of required nucleotide replacements (heights of apices in fig. 10). Letters A to J represent the groups of cytochromes *c* indicated in figure 10. The line at each point is an estimate of the standard deviation of the ordinate value of the point. A weighted least squares fit to the results for cytochrome *c* is extrapolated to the abcissa to estimate the fraction of the cytochrome *c* gene for which all mutations are lethal or malefic [FITCH and MARGOLIASH, 1967b] in any one mammalian species. The codons that do not fall in this category are the covarions. According to FITCH and MARKOWITZ [1970].

into account for each recalculation, that is, the ordinate values at which the apices are located in figure 10. A roughly linear regression is obtained, which on extrapolation to a replacement distance of zero exhibits an evolutionary invariance of over 90% of the gene. This particularly important application of their technique for estimating the extent of invariance by FITCH and MARKOWITZ [1970] thus leads to the quite unexpected demonstration that for any one mammalian cytochrome *c* at the present time (corresponding to a replacement distance of zero), only about 10 residues are available to evolutionary amino acid substitutions. Any mutation leading to a change of amino acid in any other than these approximately 10 allowed positions is by this evidence necessarily malefic, resulting sooner or later in the extinction of the line that carries it. The calculations were carried out with a view to contracting the taxonomic coverage in the direction of mammals only because cytochromes *c* from various groups of mammals were much more generously represented than any other group in the phylogenetic tree, affording many points, and thus greater precision, near the ordinate towards which extrapolation was required. At present there are nearly enough cytochromes *c* of known amino acid sequence from other groups, such as plants, fungi and insects to make it possible to direct calculations towards them. It will, therefore, soon be possible to determine directly whether the number of residues capable of accepting evolutionary variations is the same or different for the proteins from different taxonomic groups of species. The codons corresponding to those amino acid positions, which in any one species and at any one time in the course of evolution are free to fix mutations, are termed *concomitantly variable codons,* a phrase contracted in current usage to *covarions* [FITCH and MARKOWITZ, 1970].

More recently, employing a quite different statistical approach, FITCH [1971a] has estimated that the number of covarions for cytochrome *c* is 4.5, confirming the essential previous result that the evolutionary variability of cytochrome *c* is very severely restricted. The method depends on the number of double mutations, defined as the fixation of two mutations within a single codon between two successive branch points (apices or nodes) in the statistical phylogenetic tree. The number of double mutations will depend on the number of covarions and the persistence of a codon among the set of covarions after it has fixed a first mutation. The reconstruction of the phylogenetic tree yields the total number of mutations and the number of double mutations in each internodal segment, hence the distribution of double mutations as a function of the number of mutations observed in all such segments. The number of covarions and their persistence of variability which yields a distribution most closely approximating that observed on the statistical phylogenetic tree is taken to represent the actual situation. The best fit was found for 4.5 covarions with a persistence of variability of 0.04. This indicates that, on the average, the likelihood of a covarion losing its place among the set of covarions—namely, becoming invariable as a result of the fixations of mutations among other covarions—is as high as or higher than 0.75. Thus, not only is the number of covarions small, but also their rate of turnover is high as compared to the numbers of evolutionary variations fixed in cytochrome *c*.

C. Evolutionary Significance of Covarions

Notwithstanding the remarkably small number of residues that appear
to be amenable to evolutionary substitutions at any one time in the history
of the descent of any one cytochrome c, the comparison of the amino acid
sequences of the over 50 cytochromes c of known primary structure shows
that more than 75% of the residue positions have varied during evolution
(see fig. 2). Clearly then, one must assume that when a mutation is fixed
in a particular covarion this will in many cases also change some of the
residues in the current set of covarions. Some residues previously invariable
become variable while others previously variable now lose their status of
covarions and become invariable. Over prolonged periods of evolutionary
history over two thirds of the molecule can be made to change in this
fashion. As noted above, the statistical calculation [FITCH, 1971a], does
indeed show that the turnover of covarions is remarkably rapid in cytochrome c, indicating that the variable positions in the protein are to a large
extent interdependent. A change in one tends to affect the other currently
variable positions in a large proportion of cases.

Cytochrome c has, therefore, a very low tolerance to evolutionary
change. Only some 4 to 10 residues, on the average, are available for
substitution at any one time, a picture which is very different from the
common simplistic view that since a residue has changed at some time in
the evolution of one particular cytochrome c, it is available for change
in any cytochrome c at all times. This is possibly the most clarifying
contribution that the concept of covarions has brought to our general
understanding of the evolutionary changes of protein structure. The number
of covarions can be considered an inverse expression of the stringency and
extent of structure-function requirements, and as such, this number is a
fundamental parameter expressing the quantitative effect of function on
the evolutionary behavior of a protein [MARGOLIASH and FITCH, 1970].

The rate of evolutionary change of a protein is indeed related to the
number of its covarions. Thus, the calculations [FITCH, 1971a, 1971b, 1971c;
FITCH and MARKOWITZ, 1970] which yielded 10 as the number of covarions
for cytochrome c, also showed that in fibrinopeptides A (the segment of
fibrinogen removed by the action of thrombin at the initiation of clotting),
18 of the 19 residues are covarions, in the α chain of hemoglobin there
are 50 covarions out of 141 residues, while in the β chain of hemoglobin
there are 39 covarions for 146 residues. Since in all these cases the proteins
for the horse and pig are of known amino acid sequence and had been

included in the statistical phylogenetic trees which served to estimate the number of covarions, one could calculate the rates of evolutionary fixation of mutations per codon and per covarion in these two lines of descent since their most recent common horse-pig ancestor [FITCH, 1971b]. For cytochrome c these rates turned out to be 0.048 per codon and 0.50 per covarion, for α hemoglobin 0.156 per codon and 0.44 per covarion, for β hemoglobin 0.212 per codon and 0.82 per covarion, and for fibrinopeptide A 0.684 per codon and 0.72 per covarion. The rates of evolutionary fixations per codon are highly variable, reflecting the slow changes of cytochrome c structure, the intermediate rates of change of the hemoglobin chains and the very rapid variation of fibrinopeptides. The corresponding unit evolutionary periods (see above) are about 20 millions years for cytochrome c, 5.8 million years for hemoglobin and 1.1 million years for fibrinopeptide A [DICKERSON, 1971a]. In sharp contrast to this spread, the values for residue fixations per covarion are so near each other that, within the error of the method, they may well be considered to be the same. In other words, as a function of the number of positions capable of accepting mutations without leading to the disappearance of the line of evolutionary descent, four different proteins show the same rate of evolutionary change, even though their unit evolutionary periods vary over a range of about 20-fold.

Here one comes up once more against the argument as to the presence or absence of neutral mutations in proteins in general, and in cytochromes c in particular. If a large proportion of mutations fixed in the course of evolution in proteins are selectively neutral, as maintained by KIMURA [1968], then covarions are the only sites at which mutations can manifest themselves. All other codons are prevented by selective forces from undergoing any evolutionarily effective changes whatsoever. Moreover, since neutral mutations are the result of random occurrences, then all proteins should have the same rate of neutral mutation per covarion. This is sufficiently similar to what is observed for cytochrome c, α and β chains of hemoglobin and fibrinopeptides A [FITCH, 1971b], as outlined above, that these reults can be taken to suggest that, unless a radically different explanation for the similarity of rates of protein change per covarion is forthcoming, neutral mutations may well account for a significant proportion of all protein evolutionary changes.

In summary, in the case of cytochrome c, one is left with the concept of a protein whose structure is very severely restricted by structural or functional requirements, so that in any one of its many evolutionary forms, only a small number of positions (4 to 10, approximately, out of the over

100 residues) are available for at least some possible evolutionary variations. These positions are, moreover, strongly interrelated, so that when one is changed the others are likely to be affected in such a way that some become invariable, while others that were previously invariable are now open for at least some variation. All other positions are required to carry precisely the amino acid residue found in them at the time. Selective influences are so effective that the occurrence of any other residues would inexorably lead to the extinction of the line carrying such forbidden variants. Natural selection is thus conceived only as a force which constrains the variability of the protein in evolution, eliminating the deleterious, but it is not seen providing any advantage by selecting for cytochromes c better adapted to their function or internal milieu. In support of this view, the excellent functional evidence available appears to indicate that any cytochrome c is not better adapted to its own mitochondria than to those of any other species. Moreover, the calculations of the rates of evolutionary residue substitutions per covarion, described above, yield similar rates for four different proteins, a result compatible with the idea that the majority of evolutionary changes in these proteins are selectively neutral.

VI. The Spatial Structure of Cytochrome c and Constraints on Evolutionary Variations

An understanding of protein evolutionary variations necessarily includes a rather thorough knowledge of protein function, since selection must operate via functional characteristics. In the case of cytochrome c we are fortunate that the extent of knowledge of the amino acid sequences of the protein from many species, leading to the effective statistical approaches to molecular evolutionary phenomena already outlined, is matched by satisfactory solutions of its spatial structure in both functional states—the ferric and the ferrous [DICKERSON et al., 1971; TAKANO et al., 1971]. That all this information has not yet led to a much completer and generally accepted description of the evolutionary transitions of the protein and of their mechanisms, is due in part to the fact that knowledge of spatial structure is merely a first basis upon which one may plan the experiments required to understand functional activities. Further, it is also possible that identified physiological functions, in the case of cytochrome c electron transfer in the terminal oxidase segment of the respiratory chain, may not be the only or even the most important property with respect to evolutionary selection. Other properties conceivably related to the integration and maintenance of the protein in its functional environment, to its biosynthesis or even to unknown physiological functions, etc... may play crucial evolutionary roles. Nevertheless, it is clear that present knowledge of its spatial structure has already given some insight into a few of the evolutionary aspects of cytochrome *c* variations and provided possibly useful limits in regard to others. These are described below.

A. The Structures of Ferric and Ferrous Cytochrome *c*

The structures of horse and bonito cytochromes *c* in the ferric form have been obtained to a resolution of 2.8 Å [DICKERSON *et al.*, 1971]. The molecules are prolate spheroids of about $30 \times 34 \times 34$ Å if one includes the side chains, and represent the most perfect illustration yet of the "oil drop" model of protein structure, with the hydrophobic side chains packed in the inside, and the polar residues at the surface. The heme is held in a deep crevice perpendicular to the surface termed the "front" of the molecule, and in figures 4 and 5 is seen mostly along the front edge of the heme plate (that of pyrrole ring II above and III below). In figure 5, which represents an α-carbon diagram of ferrous tuna cytochrome *c* [TAKANO *et al.*, 1971], the protein has been kept in the same relative position as in figure 4, which represents ferric tuna cytochrome *c*. However, in the conformation range attendant to oxido-reduction the heme has slightly shifted and one can now see a little less of its left surface. Cysteines 14 and 17, which are bound to the vinyl side chains of rings I and II, respectively, are at the upper right edge of the heme. The hemochrome-forming groups, the imidazole of histidyl residue 18 and the sulfur of methionyl residue 80, extend from the right and the left, respectively, to the central heme iron atom to coordinate at positions 5 and 6. The propionyl side chains of the heme point downwards to the bottom of the crevice. The interior or posterior propionic acid is buried in the hydrophobic interior of the molecule and held there by hydrogen bonds to the phenolic hydroxyl of tyrosine 48, and the indole of tryptophan 59. The front propionyl side chain is in the polar external surface. Interestingly, the imidazole of histidine 18 which is coordinated to the heme iron by its ϵ-nitrogen is also hydrogen bonded by its δ-nitrogen to the main chain carbonyl of Residue 30 at the lower front right of the molecule. This probably serves both to hold the imidazole ring in a rigid conformation for reaction with the heme iron and also to transmit any movements imparted by heme changes to the coordinated imidazole, such as in oxido-reduction, to the right half of the molecule. Indeed, the imidazole to Residue 30 hydrogen bond is one of several which appear to make the right lower part of cytochrome c the most rigid portion of the whole, relatively plastic molecule. In contrast (see Section II, 2), the left lower part of the molecule, on the side of the hemochrome-forming methionyl Residue 80, appears to be more readily deformable and capable of considerable movement.

The molecule seems to be made in roughly two halves: Residues 1 to 47, including the segment of α-helix of Residues 1–11, comprise the right side of the molceule, Residues 48 to 86, the left, while most of the remainder, Residues 87 to 102, form the second stretch of α-helix across the back of the molecule from right to left like the strap of a suitcase. There are two areas at which the hydrophobic interior seems to meet the surface. These have been called 'channels' although they present no access of the solvent to the interior. The left 'channel' is bordered by Residues 52 to 74, while the right 'channel' is surrounded by Residues 6 to 20 and the last portion of the c-terminal α-helix.

Other than the helical regions, the molecule is essentially a shell of extended chain wrapped around the heme. Hydrophobic side chains are nearly all kept in the interior of the molecule, in contact with each other or with the heme. These contacts appear to provide a good proportion of the forces stabilizing the native conformation, as the overall number of hydrogen bonds is relatively smaller than for several proteins of the same size. Such a structural arrangement also explains the early observation of evolutionary stability of the hydrophobic segments of cytochrome c [MARGOLIASH, 1963; MARGOLIASH and SMITH, 1965]. Internal residues which have to fit in a limited space, and whose close packing with other side chains or with the heme is structurally important, are likely to be preserved as long as a change in one residue (by leaving more free internal space, for example) has not made it possible for another to vary. This is very much the situation described by the existence of few covarions in cytochrome c (see above). Those positions which show a wide spectrum of acceptable substituents are external, as expected from the point of view of evolutionary selection. However, some external residues show conservative substitutions and some are invariant. Thus, the external versus internal localization is far from sufficient to explain the evolutionary variability of residues, and one must seek other parameters to explain the behavior of external residues.

Polar residues are in general external. The exceptions are few and all seem to be of functional significance. Among them is the interior propionyl side chain of the heme which is held down in the bottom of the hydrophobic crevice by hydrogen bonds. This is an energetically expensive arrangement. It is very different from that in the hemoglobins and myoglobins in which the propionyl side chains are in the external solvent, and is likely to represent an important, if still unknown, functional requirement of cytochrome c. The imidazole of histidyl Residue 18 is internal as it is coordinated to the heme iron. An interesting case is that of the evolutionarily invariant tyrosyl Residue 67. It is entirely internal. Its hydroxyl is hydrogen bonded to threonyl Residue 78 and the aromatic ring is, as it were, poised between contact with the sulfur of the heme-coordinated methionyl Residue 80 in front of it and with the ring of tyrosyl Residue 74 at the back-left surface of the protein, behind it. Such a remarkable architecture, coupled with invariance, beckons a functional or structural explanation. The hypothesis that this residue is involved in electron transfer inside the protein molecule [DICKERSON *et al.*, 1970] according to the mechanism proposed by WINFIELD [1965], and KING, LOONEY and WINFIELD [1967] does not, at this time, appear very likely. It requires the transient presence of a free radical form of tyrosyl Residue 67 in the process of cytochrome c reduction. Since acetylation of this tyrosine does not inhibit the electron transfer activity of the protein [IVANETICH *et al.*, 1971], the participation of any such free radical would appear to be ruled out. The recent observation of a phenylalanine at position 67 in a protist cytochrome c yields the same conclusion.

The strongly basic and acidic amino acid side chains are all on the exterior of the molecule and distributed in a remarkable fashion. The lysines are largely in the vicinity of the right and left "channels" described above, making the two sides of the molecule strongly cationic. In constrast, a large proportion of all acidic residues (in horse cytochrome c, for example, 9 out of 12) are located in a rather circumscribed patch at the upper half of the back of the molecule, about halfway between the positively charged regions. What is conserved in evolution is the acidic character of this area, not the acidic character of individual residues in it, all of which, except for Residue 90, can also accomodate non-acidic amino acids. Basic residues never invade the negative patch, which, at a minimum, contains 6 acidic side chains. This type of evolutionary conservatism is similar to that observed for the clusters of hydrophobic residues in the amino acid sequences of the cytochromes c of different species [MARGOLIASH, 1963; MARGOLIASH and SMITH, 1965], except that the conservation of the acidic patch occurs in regard to the spatially folded native molecule, not the primary linear structure. Here again an explanation of this architecture is needed. Whether the ideas that the basic sides are involved in reaction with the oxidase and reductase enzyme systems, and that the acidic patch binds with a mitochondrial membrane cytochrome c site [DICKERSON et al., 1971], have any validity, remains to be determined.

As the chain folds to cover the molecular surface there are several sharp bends and abrupt reversals of chain direction (see fig. 4 and 5). Some of these involve the proline residues, while others occur at bends resembling single turns of the tightly hydrogen-bonded 3_{10} helix. There are six such bends in ferricytochrome c, three of which are in one and three in the other of the two conformations such bends may take [VENKATACHALAM, 1968]. Invariant glycines occur in tight corners in which there is no place whatsoever for side chains, and in horse cytochrome c glycines occur in the positions necessary to make possible the three type II 3_{10} bends. Here again, this is insufficient to explain evolutionary invariance, as Residue 45 is a constant glycine and it is on the surface of the molecule.

The distribution of aromatic residues is also remarkable. In the ferric protein tyrosine 74 and tryptophan 59 have their rings parallel. The indole side chain points straight into the molecule from the lower third of the middle of the back and hydrogen bonds to the internal propionyl side chain of the heme, while the ring of the tyrosine is flat on the surface at the left-back corner of the protein. The ring of the completely internal tyrosine 67 is at an angle to these two aromatic rings and, as described above, seems to be bridging the gap between the two parallel rings and the heme iron-bound sulfur of methionine 80. Two other pairs of aromatic rings are also parallel and more or less in close proximity in ferricytochrome c. One of these, consisting of tyrosine 48 and tyrosine of phenylalanine 46 are at the bottom of the molecule directly below the heme, with their aromatic rings perpendicular to the heme plane. Tyrosine 48 is also hydrogen bonded to the posterior heme propionyl side chain. The other pair is not strictly parallel and slightly further apart than those already mentioned. It consists of the invariant phenylalanine 10 and the tyrosine or phenylalanine at position 97. They are located on the top right side of the molecule and form the top boundary or project into the top of the right channel. The only other invariant phenylalanine, Residue 82, is held in the external solvent in the middle of the left side, in the ferric protein. The immediately preceding Residue 81, commonly isoleucine, but also sometimes valine or alanine, is abutting the surface of the protein, but not projecting into the external medium. This leaves a sizable solvent-accessible pocket on the left side of the upper front part of the heme plane, in the vicinity of pyrrole ring II, but also possibly extending farther in and higher.

This is the area in which the most dramatic conformation change occurs on reduction (see fig. 5). Indeed, other than the aromatic pair at the upper right of the molecule, Residues 10 and 97, the other aromatic rings mentioned above are all in some way involved in the ferric $\rightleftharpoons$ ferrous functional transition of the protein. On reduction, there are large rotations in the region of Residues 80 to 82, resulting in the aromatic ring of phenylalanine 82 being moved away from the external medium and inserted into the solvent pocket near the front left top of the heme, filling it up and leaving the phenyl and the heme rings more or less parallel. At the same time Residue 81 is rotated off the surface, ending up hanging completely in the external medium, like a projecting tail. The solvent exclusion and organization effects involved in these movements and those of other hydrophobic groups are likely to be important components of the energetics of oxido-reduction in cytochrome c. The hitherto invariant segment, Residues 70 to 80, partly surrounds this area from below, and as already noted (see Section II, 2) shows evidence of having considerable mobility. This property could well be important for function, and if in addition, as previously suggested [MARGOLIASH, FITCH and DICKERSON, 1968; DICKERSON et al., 1971], this surface is one of interaction with an enzyme system; the relative evolutionary constancy of Residues 70 to 80 would appear to be explained.

The only other obvious conformation change in the ferric to ferrous transition is performed by the loop of Residues 21 to 23 near the middle of the right side of the protein. This segment moves up, so that in the reduced protein the right "channel", which is open in the ferric protein, is partly blocked. It may be noted that this movement is the one which is likely to explain the greater resistance of the reduced protein to proteolysis by an enzyme such as chymotrypsin, whose first point of attack is at phenylalanyl Residue 10, and the second at tyrosyl Residue 97, both near the top of the right channel [see MARGOLIASH and SCHEJTER, 1966].

In summary, present knowledge of the spatial structures of cytochrome c in both its oxidation states, clearly points to the protein being a highly specific piece of molecular machinery, undergoing remarkable changes in protein conformation as it transfers electrons. Satisfactory explanations of protein evolutionary variations are, therefore, certain to be closely linked to our understanding of the workings of this machine.

B. Molecular Distribution of Evolutionarily Variable Positions

From the evolutionary point of view, one of the most useful products of the solution of the structure of cytochrome c by x-ray crystallography is the possibility of examining the distribution of variable and invariant residues in the folded native molecule. In the preceding section, the structural reasons why some residues appear to be invariant were pointed out, at the same time making clear that in no case were satisfactory explanations available for the invariance of every single residue of any one type or category.

With regard to variable residues, a most interesting observation can be made, namely, that the positions at which evolutionary residue substitutions are fixed are different for different taxonomic groups of species [Margoliash et al., 1970, 1972]. Moreover, and quite strikingly, these locations are clustered into rather well defined tertiary structure regions for each such group. An example of this sort of distribution is given in figure 15A which shows the positions at which the cytochromes c from mammals vary. For this taxonomic group, residue substitutions are limited to a semicircular band in front, above and behind the heme, leaving both lateral surfaces of the molecule largely unvaried. As noted in figures 15B, 15C and 15D, the distribution of variable positions for insect cytochromes c is quite different, that for plant cytochromes c, though partially overlapping with that for insects, has yet another distinct clustered distribution, as has the distribution of variant residues in fungal cytochromes c.

That such differences are statistically significant has been demonstrated by Fitch [1971d] for an example of 4 fungal and 4 metazoan species. A previously developed technique is employed. It permits the estimation of the numbers of the evolutionarily invariant positions, of those which are variable but have not yet been observed to vary, and of the positions which have varied in a protein, from the reconstruction of the corresponding statistical phylogenetic tree [Fitch and Markowitz, 1970; Markowitz, 1970] (see above). The orthologous amino acid positions of the proteins of two groups of species are of 4 possible types with respect to the evolutionary fixation of amino acid substitutions: those which have not fixed any substitutions in either group of proteins; those which have fixed mutations in one group; those which have fixed mutations in the other, and those which have fixed mutations in both. The numbers of these 4 types of positions can be computed if one assumes a number for the positions which are invariant in common for both groups of proteins.

Computing for every admissable number of these invariant positions gives a series of results which are compared to those observed directly from the reconstructed phylogenetic tree for the cytochromes c of the two groups of species. The best fitting set of values is taken to give the best estimate

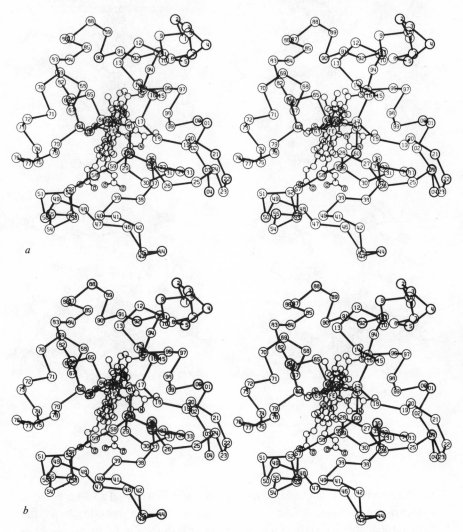

Fig. 15. Stereoscopic α-carbon diagrams of horse cytochrome c in the ferric form, according to DICKERSON *et al.* [1971]. The residues marked by heavy circles are those that vary among the cytochromes c of mammals (14 species) in figures 15a, of insects (4 species) in figure 15b, of higher plants (11 species) in figure 15c, and fungi (6 species) in figure 15d.

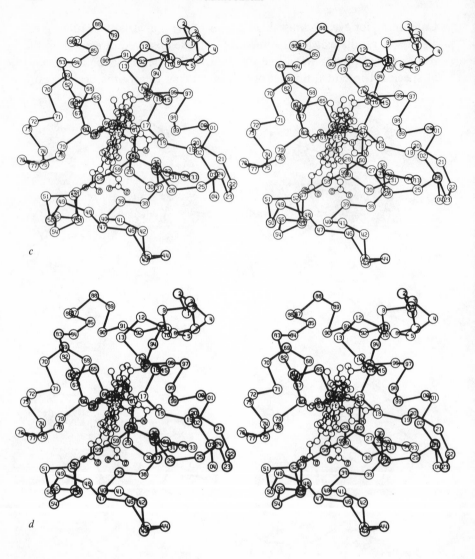

for the number of common invariant positions. For 4 fungal and 4 metazoan cytochromes *c* this was found to be 41, corresponding to about 29 for the invariable metazoan codons which are variable in the fungal proteins, about 17 for the codons variable in the metazoan proteins but invariable for the fungi, and about 23 for the codons variable in both groups. Clearly, there is a partial overlap of variable and invariant positions, but also considerable numbers of positions are variable in one group but not in the other and vice versa.

These observations make it clear that not only are selective forces operating against changes of cytochrome *c* structure to prevent any but very few residues from varying at any one time of evolutionary history in any particular species, but that these forces in fact operate, as they must, at different sets of molecular sites for different taxonomic groupings of species. Indeed, without such a situation one could not explain how a proportion of the protein of over 75% has been found to differ in the cytochromes *c* of all species examined to date, when statistical calculations of variability demonstrate that only 4 to 10 residue positions are covarions, i.e., can fix evolutionary changes in any one protein at any one time (see above). In contrast, the quite unexpected result deriving from the examination of the locations of variability on the molecular models of the protein, is that the variable positions are spatially clustered, and have distinct distributions in the cytochromes *c* of different taxonomic groups of species. This is difficult to understand if one is to maintain that all evolutionary variations in cytochrome *c* are the result of selectively neutral events. For example, why should insect cytochromes *c* not be permitted to change in the regions of the molecule in which the cytochromes *c* of mammals do, and vice versa? If all these variations had no functional biological significance whatsoever and proceeded randomly, one would expect to see, at a first approximation, a correspondingly random distribution of positions of variability in different taxonomic groups or at least a complete interdigitation of such positions.

VII. Ion-Binding Properties of Cytochrome *c* as a Basis for Species Variability

As the classical electron transport properties of cytochrome *c* give no indication of serving to provide functional distinctions between the proteins of different species, even when tested in nearly intact mitochondrial membranes [BYERS, KANG and MARGOLIASH, 1972; BYERS *et al.,* 1971; SMITH, NAVA and MARGOLIASH, 1972; SMITH, 1972], it is imperative to attempt to discover whether other functional attributes of cytochrome *c* exist which could serve as a basis for evolutionary selection. Apart from the amino acid sequences themselves, the one set of properties which has so far been shown to exhibit considerable species differences is the binding

Table II. Electrophoretic mobility of deionized horse cytochrome *c* towards the cathode in the presence of various anions[1]

Added ions	Electrophoretic mobility at 1° $(cm^2/sec/V \times 10^5)$ Ferricytochrome *c*	Ferrocytochrome *c*
None	6.1	4.6
Cl⁻	3.7; 4.2	4.6
K⁺	6.1	4.6; 5.1
Ca⁺⁺	6.1	5.1; 5.4
PO₄⁼	4.3	2.6
Citrate	1.5	1.3
α-Ketoglutarate	1.9	2.9

1   Solutions contained 0.407 to 1.017 mM cytochrome *c*, 0.10 M tris-cacodylate buffer, pH 7.2, and 0.05 M of added ions, brought to pH 7.2 with tris on cacodylic acid as required. When two electrophoretic boundaries occur, both mobilities are listed. According to MARGOLIASH, BARLOW and BYERS [1970].

of anions [BARLOW and MARGOLIASH, 1966]. Many inorganic anions were found to bind to the protein in the ferric state, including Cl^-, I^-, SO_4^{2-}, and PO_4^{3-}. In contrast, cacodylate showed no detectable binding, as did none of the cations tested, so that tris-cacodylate buffers could be used whenever a strictly non-binding medium was required. By free boundary electrophoresis, two chloride ions were shown to bind, and the relative binding affinities at the two chloride sites were roughly reflected in the ratios of the areas under the electrophoretic boundaries corresponding to one and two bound ions. This ratio was found to be 0.8, 1.8, 6, 9, and 32 for the screw worm fly, human, hog, pigeon and moth *(Samia cynthia)* proteins, respectively, implying large differences in binding affinity.

A more recent extension of this study [MARGOLIASH, BARLOW and BYERS, 1970] has led to a rather remarkable correlation between the ion-binding properties of cytochrome *c* and the ion transport properties of mitochondria [PRESSMAN, 1970; KLINGENBERG, 1970]. It was found that ions which are normally excluded from the mitochondrial matrix, typified by chloride, bind to ferricytochrome *c* but not to the ferrous protein; that those which are forcibly carried into the matrix by the use of chemical energy, such as calcium ions, bind to ferrocytochrome *c* but not to the ferric protein, while those which flow in or out of the mitochondrial matrix in response to the concentration gradient, but whose movement is carrier-mediated, such as phosphate or ADP, bind to cytochrome *c* in both oxidation states. Many examples of each class were tested and observed to follow this general rule. Some are given in table II. On this basis it was proposed that cytochrome *c* may act as a carrier for these ions in the inner membrane of mitochondria [MARGOLIASH, BARLOW and BYERS, 1970]. A similar conclusion was reached by SCHEJTER and MARGALIT [1970] from the effect of a few of these ions on the oxidation-reduction parameters of the protein. Since that time, attempts to demonstrate directly an ion-carrier function for cytochrome *c* in mitochondria have failed. This is possibly because it has not proved possible to deplete the organelles of their cytochrome *c* to an extent of more than about 80% [BYERS, KANG and MARGOLIASH, 1972], without disrupting the inner membrane. The remaining cytochrome *c* appears to be in a compartment different from the majority of the protein, as it does not reequilibrate readily with it. If the non-extractable fraction were to prove to be in the lipid phase, it could well represent the totality of the cytochrome *c* involved in some way in ion transport, hence the failure to observe any effect of ordinary cytochrome *c* depletion of mitochondria on such phenomena.

Notwithstanding these difficulties and negative results, the possible functional involvement of cytochrome *c* with ions remains an attractive possibility. If this were in regard to the transport of ions, it would very likely suffice to provide the selective forces necessary to explain how the cytochromes *c* of different species acquired their particular amino acid sequences. Indeed, the ion-transport and ion-binding properties of the mitochondria of different species have been known to vary quite extensively, even though no thorough systematic studies have been carried out so far [see PRESSMAN, 1970; LEHNINGER, CARAFOLI and ROSSI, 1967; CARAFOLI *et al.*, 1970; CHAPPELL and HAARHOFF, 1967; TULP, 1970; VAN DAM and MEYER, 1971]. The transport of ions across the inner mitochondrial membrane affects the intra- and extraorganelle concentrations of such metabolites as citrate, ADP and ATP, all known to have crucial metabolic regulatory effects [see ATKINSON, 1966; GOODWIN, 1968].

Furthermore, if the cytochrome *c* in the mitochondrial membrane were divided into pools of several interacting molecules, rather than individually strung on separate respiratory chain assemblies, ion movement via such cytochrome *c* pools could proceed by successive re-equilibrations as the ion passes from one protein molecule to another on its way through the pool [MARGOLIASH, BARLOW and BYERS, 1970]. This would result in large differences in the ability to transport various ions based on small changes in affinity, since the ratio of the rates of transport of two ions would be proportional to the ratio of their binding affinities to the carrier to the power of the number of carrier molecules encountered in one passage. Such a situation could similarly elucidate the very large temperature coefficients for the mitochondrial transport of ATP and ADP [Klingenberg, 1970].

It should also not be overlooked that the relation of cytochrome c to ions need not necessarily be as a transport carrier to account for the evolutionary phenomena. Conceivably, the protein—a macromolecular component of the inner mitochondrial membrane which is available on its outer surface and can readily move in and out of it [see Margoliash and Schejter, 1966; van Dam and Meyer, 1971]—could act to regulate the flow of ions to their transmembrane carriers, if the cytochrome c sites or pools were in some way interposed between ion carriers and the external medium. Moreover, the ion-binding properties of cytochrome c may be important in connection with its reaction with the oxidase and reductase enzyme systems, as there are large differences in the ion-binding properties of the two oxidation states of the protein (see table II). However, on such a basis it would be difficult to see how structural differences could have developed, since cytochromes c of widely differing amino acid sequences react identically with these systems, as discussed above.

Almost any physiological role that can be ascribed to the binding of ions to cytochrome c will also probably resolve the problem of the clustering of positions of variability in the proteins of different taxonomic groups, if such a role is shown to differ for the different groups of species. The mitochondria of any such group are likely to be physiologically similar, and relatively different from those of another taxonomic group. If different locations on the surface of cytochrome c bind different sets of ions, the mitochondria of the species in one taxon, because of their metabolic similarity, are likely to tolerate changes in the binding of certain ions but not in those of others, while in another taxon, required and facultative bindings will relate to different ions and hence to different surface clusters on the molecule. In any case, it is to be hoped that, whatever the fate of the hypothesis of the connection of cytochrome c to these phenomena, extensive comparative studies of the mitochondrial activities of different species will be forthcoming, to serve as a basis for the understanding of the functional variables affecting the evolutionary transformations of the molecular components of the organelle.

VIII. Concluding Summary: Character of the Evolutionary Transformations of Cytochrome c

Notwithstanding the controversy which still rages about whether evolutionary residue substitutions in proteins are selected for or are the result of so-called neutral mutational events, there is no question that the examination of the amino acid sequences of a protein in a wide taxonomic range of species has yielded considerable and often novel insight into both molecular evolutionary processes and structure-function relations. Cytochrome c remains to date the prime example of the successful application of this approach. It has served and continues to serve as a model case for the development of the statistical techniques for extracting evolutionary information from amino acid sequences.

As soon as a few cytochrome c sequences were available [MARGOLIASH, 1963], it became obvious that the numbers of variant residues tended to be constant in comparisons of the protein from large taxonomic groupings, all cytochromes c from mammals being, for example, roughly equally different from all bird cytochromes c, all vertebrate cytochromes c equally different from insect cytochromes c, etc. The protein thus appeared to vary at a constant rate in the course of evolution, permitting the definition of the *Unit Evolutionary Period* as the time required to allow, on the average, one residue substitution to occur in the cytochromes c of two diverging lines of phylogenetic descent. By comparison with the known times of major phylogenetic divergences one could then show that cytochrome c was a slowly varying protein, carrying a characteristic unit evolutionary period of about 20 million years [MARGOLIASH and SMITH, 1965]. However, such calculations could only represent rough approximations and any further advance was predicated on the computation of precise topologies of structural relationships which, in the case of an orthologous set of proteins, should correspond to a species phylogeny. Following the development of a statistical technique for distinguishing non-random from random degrees of sequence similarity [FITCH, 1966], this was accomplished by comparing amino acid sequences in terms of minimal replacement distances, the minimal number of single nucleotide changes required to transform the gene coding for one protein into that coding for another. A simple procedure yielded statistical phylogenetic trees in remarkably good accord with expectation, even though the only biological information they contained was the amino acid sequences of a number of cytochromes c and the genetic code [FITCH and MARGOLIASH, 1967a]. These trees provided the solid basis upon which the further conceptual and technical developments have rested. Among these may be listed the derivation of the amino acid sequences of ancestral forms of cytochrome *c* [FITCH and MARGOLIASH, 1967a; FITCH 1970b] which has brought within reach the synthesis of the protein from extinct and very primitive forms of life and the possible attendant clarification of the mechanisms of molecular evolutionary developments. It also became possible to demonstrate quite unequivocally that all cytochromes *c* are descendant from one ancestral form [FITCH, 1970b]. This is the first time a unique gene has been followed to very early periods of evolutionary history, providing direct proof of the concept of the unitary origin of life on this planet.

The reconstruction of a statistical phylogenetic tree gives the distribution of codons having undergone 1, 2, 3, etc. mutations, and on this basis it is possible to estimate how many codons are invariant and how many are included in Poisson-distributed variable classes. The data from the phylogenetic tree fits an invariant class of about one fourth of the protein, a normally variable and a 'hypervariable' class [FITCH and MARGOLIASH, 1967b; FITCH and MARKOWITZ, 1970]. As the taxonomic span of the species considered in the computation is gradually decreased, the invariant class increases till it comprises over 90% of the cytochrome *c* of any one species. The remaining variable codons, termed *covarions*, represent only 4 to 10 residues [FITCH and MARKOWITZ, 1970; FITCH, 1971a], and correspond to different amino acid positions in the proteins of different groups of species [FITCH, 1971d]. Their rate of turnover is high [FITCH, 1971a], so that when a covarion is changed, other covarions tend to lose their status and become invariable, while previously invariable residues tend to join the groups of covarions. This contagiousness of variability explains how more than 75% of the protein has varied over the whole span of its evolutionary history, while only 4 to 10 residues are variable at one time in any one species.

A related observation is that the distribution of variable residues on the x-ray crystallographic model of the protein [DICKERSON *et al.*, 1971] takes the form of distinct and only partially overlapping clusters for the proteins of separate taxonomic groups [MARGOLIASH *et al.*, 1970]. The surprising element is that these variable residues are clustered, rather than randomly distributed, as one would expect if precisely the same set of structure-function selective constraints were to determine the spreading of covarions during the evolutionary history of all groups of species. The number of covarions is indeed an inverse quantitative expression of the extent and stringency of selective constraints on the evolutionary variability of proteins, so that the rate of residue substitutions per covarion appears to be more or less constant for 4 proteins whose rates of change as a function of the total number of residues varies from fastest to slowest over a range of about 20-fold. Respectively, these are fibrinopeptides A, hemoglobin α and β chains and cytochromes *c* [FITCH, 1971b].

The question of the importance of selectively neutral residue changes arose from the observation of the apparent constancy of the rate of evolutionary change of the protein [MARGOLIASH, 1963; MARGOLIASH and SMITH, 1965) and the great contrast between extensive structural variability and ostensible functional identity among the cytochromes *c* of various species [see MARGOLIASH, FITCH and DICKERSON, 1968]. A possible test of the existence of selectively neutral variants—namely, the expected consequent polymorphism of the protein in a suitable population—has not been carried to the point of giving a dependable result (see above). As already discussed, the newer approaches to the study of evolutionary structural variations have also given equivocal answers. The constancy of the rate of evolutionary fixation of mutations per covarion, for four different proteins, argues for the neutralist point of view. In contrast, the observed clustering of variable positions seems to point to the proteins of different taxonomic groups of species having different functional requirements, thereby making selection possible. The hypothesis that the ion-binding properties of the protein are in some way related to the distribution or the movement of ions in mitochondria [MARGOLIASH, BARLOW and BYERS, 1970], has not been proved, though it still remains an attractive possibility as a basis for such selective effects.

Moreover, it should be noted that the apparent randomness of evolutionary changes in a group of variable codons, as evidenced from the fitting of the data by Poisson distributions, does not imply either the absence or the presence of natural selection [FITCH, 1971d]. Appropriate fixations in a group of codons could eventually lead to a selective advantage, but the order in which these fixations occur may well be partially determined by which codon first mutated and that could be a purely random event. On the other hand, had no fit to Poisson distributions been observed, one could have eliminated the possibility of neutral mutations. The precise order of a series of evolutionary changes in protein structure is at least in part determined by previous changes, so that every mutation fixed restricts the further evolutionary pathways available. This phenomenon is rather similar to developmental processes for which subsequent events depend upon prior ones having occurred. Such a situation is implicit in the existence of limited numbers of covarions having a high rate of turnover. Ordered mutation series may have simple structural backgrounds, as when particular charges have to remain neutralized, or internal space is limited so that a bulky group must be eliminated before a new one can be introduced [see FITCH and MARKOWITZ, 1970]. The same types of quasi-developmental series could conceivably even explain the clustered distributions of variable positions among the cytochromes *c* of separate taxonomic groups, shown in figure 15. If the contagiousness of variability described above is largely local in respect to the spatial structure of the protein, one would end with spatially more or less segregated clusters within which the proteins of one taxonomic group will have changed. Moreover, the older the taxon the more widespread would one expect the locations of variability to be. This last point is in agreement with the large number and relatively scattered distribution of variable positions in the 6 fungal cytochromes *c* examined to date (fig. 15D), as compared to the few and strictly clustered locations of change among 15 mammalian cytochromes *c* (fig. 15A).

Whatever the final compromise between the pure neutralist and pure selectionist positions which will be found to express reality in the evolutionary variations of cytochrome *c*, it is already quite obvious that the molecule is under extremely stringent selective pressures which prevent all but a very few residue substitutions to be acceptable for any species at any one time in its evolutionary history. Structural reasons for some of these selective pressures were apparent from the spatial structure of the protein, as noted above, but it was also clear that such explanations encompass only a fraction of the residue positions. The reasons for constancy of some residues and for variability of others could not be deciphered on this basis. A major difficulty may be that in the case of cytochrome *c* all effects of natural selection have been conceived only as eliminating undesirable variants. Positive adaptive features have so far not been detected, and their existence remains a matter of faith or disbelief.

Cytochrome *c* is an antique structure. It is recognizably present in all eukaryotes. The recent detection of amino acid sequence and possible spatial structure similarities between the cytochromes *c* of eukaryotes and those of prokaryotes [FITCH and MARKOWITZ, 1970; FITCH and MARGOLIASH, 1970; DICKERSON, 1971b] makes it likely that the family tree of the protein extends much further back than previously suspected. Even more remarkably, a direct crystallographic study has shown that exactly the same folding is present in *Rhodospirillum* cytochrome c_2 [SALEMME, FREER and KRAUT, 1972], locating the structure at the very earliest of evolutionary times even before respiratory chains of the modern type had been assembled in their full complexity. It would seem that once a successful solution to a protein structural problem is attained in evolution, the same well-tempered basic machinery is adapted to as many functions as it can satisfactorily encompass. This is certainly the case for the cytochrome *c* spatial structure. It can already be seen that a fully developed statistical phylogenetic tree for this structure will contain many different cytochromes, depict paralogous as well as orthologous relationships and cover the entire living world. The fundamental 3-dimensional pattern exemplified by cytochrome *c* is best termed the *cytochrome fold*. Such is the fabric of the expected major avenue of advance of molecular evolutionary studies discussed in the first section of this presentation. The overall number of successful basic protein structures is certain to be much smaller than the number of known proteins, which have classically been identified by their functions or physico-chemical characteristics without any knowledge of how these relate to their spatial patterns. A biological classification of proteins in terms of their fundamental types of folding will soon be necessary. This would be an essentially evolutionary classification and have all the advantages which phylogeny gives to other biological systems of classification. What is mainly conserved in the course of molecular evolution may well be the basic amino acid chain folding patterns. These are preciously maintained, in that only a relatively small number are available to undertake the myriad functions of life, and to serve them against the many hazards of the biological milieu at all its developmental stages. Thus, our difficulties of understanding the mechanisms of evolutionary transitions in cytochrome *c* may well stem from our lack of knowledge of the relation between the amino acid sequence of a protein and its spatial folding into a functional structure. Not until this much studied problem is effectively solved will it be possible to unravel the laws governing molecular evolution, a process whose main thrust seems to be the preservation of biologically successful spatial structures.

Acknowledgements

The author would like to take this occasion to thank his many colleagues, who, over the years, have borne the brunt of the amino acid sequence determination work. He is also very grateful to Dr. W. M. FITCH of the University of Wisconsin for a most pleasant and fruitful collaboration on the examination of the evolutionary information content of amino acid sequences, and to Dr. R. E. DICKERSON at the California Institute of Technology for a similar collaboration on the determination of the spatial structure of cytochrome *c*. The present work was supported by grant GM 19121 from the National Institutes of Health.

References

AMBLER, R. P.; BRUSCHI, M., and LE GALL, J.: Proc. Xth Int. Congr. Microbiol., Mexico City, Mexico; Symp. on Recent Adv. Microbiol., p. 25 (1971).

ANDO, K.; MATSUBARA, H., and OKUNUKI, K.: Proc. Japan Acad., *41:* 79 (1965).—Biochim. biophys. Acta *118:* 240, 256 (1966).

ARNHEIM, N. and TAYLOR, C. E.: Nature, Lond. *223:* 900 (1969).

ATKINSON, D. E.: Ann. Rev. Biochem. *35:* 85 (1966).

BARLOW, G. H. and MARGOLIASH, E.: J. biol. Chem. *241:* 1473 (1966).

BITAR, K.; VINOGRADOV, S. N.; NOLAN, C.; WEISS, L., and MARGOLIASH, E.: Unpublished experiments (1971).

BOULTER, D.; THOMPSON, E. W.; RAMSHAW, J. A. M., and RICHARDSON, M.: Nature, Lond. *124:* 789 (1970).

BOYER, S. H.; CROSBY, E. F.; THURMON, T. F.; NOYES, A. N.; FULLER, G. F.; LESLIE, S. E.; SHEPARD, M. K., and HERNDON, C. N.: Science *166:* 1428 (1969).

BREW, K.; VANAMAN, T. C., and HILL, R. L.: J. biol. Chem. *242:* 3747 (1967).

BYERS, V.; KANG, C. H., and MARGOLIASH, E.: Unpublished results (1972).

BYERS, V.; LAMBETH, D.; LARDY, H. A., and MARGOLIASH, E.: Fred. Proc. *30:* 1286 (1971).

CARAFOLI, E.; BALCAVAGE, W. X.; LEHNINGER, A. L., and MATTOON, J. R.: Biochim. biophys. Acta *205:* 18 (1970).

CHAN, S. K.: Biochim. biophys. Acta *221:* 497 (1970).

CHAN, S. K.; NEEDLEMAN, S. B.; STEWART, J. W.; WALASEK, O. F., and MARGOLIASH, E.: Fed. Proc. *22:* 658 (1963).

CHAN, S. K.; TULLOSS, I., and MARGOLIASH, E.: Biochemistry *5:* 2586 (1966).

CHAPPELL, J. B. and HAARHOFF, K. N.: in SLATER, E. C.; KANINGA, Z., and WOJTCZAK, L. Biochemistry of mitochondria, p. 75 (Academic Press, New York 1967).

CLARKE, B.: Science *168:* 1009 (1970a). — Nature *228:* 159 (1970b).

CRICK, F. H. C.: in the biological replication of macromolecules. Symp. Soc. exp. Biol. *12:* 138 (1958).

CRICK, F. H. C.; BARNETT, L.; BRENNER, S., and WATTS-TOBIN, R. J.: Nature Lond. *192:* 1227 (1961).

DAYHOFF, M. O.: Atlas of protein sequence and structure (Natl. Biomedical Res. Foundation, Silver Spring, Md. 1969).

DICKERSON, R. E.: J. molec. Evolut. *1:* 26 (1971a).—J. molec. Biol. *57:* 1 (1971b).

DICKERSON, R. E.; KOPKA, M. L.; WEINZIERL, J. E.; VARNUM, J. C.; EISENBERG, D., and MARGOLIASH, E.: J. biol. Chem. *242:* 3015 (1967). — in OKUNUKI, K.; KAMEN, M. D. and SEKUZU, I. Structure and function of cytochromes, p. 225. (University of Tokyo Press, Tokyo 1968).

DICKERSON, R. E.; TAKANO, T.; KALLAI, O. B., and SAMSON, L.: in Proc. Wenner-Gren Symp. on Oxidation-Reduction Enzymes, Stockholm (In press [1970]).

DUS, K.; SLETTEN, K., and KAMEN, M. D.: J. biol. Chem. *243:* 5507 (1969).

EDELMAN, G. M.; CUNNINGHAM, B. A.; GALL, W. E.; GOTTLIEB, P. D.; RUTISHAUSER, V., and WAXDAL, M. J.: Proc. nat. Acad. Sci., Wash. *63:* 78 (1969).

EDELMAN, G. M. and GALL, W. E.: Proc. nat. Acad. Sci., Wash. *68:* 1444 (1971).

EDMAN, P. and BEGG, G.: Europ. J. Biochem. *1:* 80 (1967).

EHRENBERG, A. and THEORELL, H.: Acta chem. scand. *9:* 1193 (1955).

ESTABROOK, R. W.: in CHANCE, B.; ESTABROOK, R. W., and YONETANI, T. The chemistry of hemes and hemoproteins, p. 405 (Academic Press, New York 1966).

FITCH, W. M.: J. molec. Biol. *16:* 9 (1966). — J. molec. Biol. *49:* 1 (1970a). — Systemat. Zool. *19:* 99 (1970b). — J. molec. Evolut. *1:* 84 (1971a). — Brookhaven Symp. Biol. *23* (in press p [1971b]). – in: Hämatologie und Bluttransfusion (Lehmann, München [in press] 1971c) . — Biochem. Genet. *5:* 231 (1971d).

FITCH, W. M. and MARGOLIASH, E.: Science *155:* 279 (1967a). — Biochem. Genet. *1:* 65 (1967b). — Brookhaven Symp. Biol. *21:* 217 (1968). — In DOBZHANSKY, Th.; HECHT, M. K., and STEERE, W.C. Evolutionary biology, vol. 4, p. 67 (Meredith, New York 1970).

FITCH, W. M. and MARKOWITZ, E.: Biochem. Genet. *4:* 579 (1970).

GILHAM, P. T.: Ann. Rev. Biochem. *39:* 227 (1970).

GOODWIN, T. W.: The metabolic roles of citrate (Academic Press, New York 1968).

GUPTA, R. K. and REDFIELD, A. G.: Science *169:* 1204 (1970).

GÜRTLER, L. and HORSTMANN, H. J.: Europ. J. Biochem. *12:* 48 (1970).

HARBURY, H. A. and LOACH, P. A.: Proc. nat. Acad. Sci., Wash. *45:* 1344 (1959). — J. biol. Chem. *235:* 3640 (1960).

HARBURY, H. A.; CRONIN, J. R.; FANGER, M. W.; HETTINGER, T. P.; MURPHY, A. J.; MYER, Y. P., and VINOGRADOV, S.: Proc. nat. Acad. Sci. Wash. *54:* 1658 (1965).

HARRIS, H.: J. med. Genet. *8:* 444 (1971).

HELLER, J. and SMITH, E. L.: Proc. nat. Acad. Sci., Wash. *54:* 1621 (1965). — J. biol. Chem. *241:* 3165 (1966).

HILL, R. L.; DELANEY, R.; FELLOWS, R. E., and LEBOWITZ, H. E.: Proc. nat. Acad. Sci., Wash. *56:* 1762 (1966).

HILL, R. L.; LEBOWITZ, H. E.; FELLOWS, R. E., and DELANEY, R.: in KILLANDER, J. Gamma-globulins, p. 109 (Interscience, New York 1967).

HOLTON, R. W. and MYERS, J.: Biochim. biophys. Acta *131*: 362, 375 (1967).

INGRAM, V. M.: Nature, Lond. *189*: 704 (1961).—The hemoglobins in genetics and evolution (Columbia University Press, New York 1963).

ITANO, H. A.: Adv. Protein Chem. *12:* 215 (1957).

IVANETICH, K. M.; CRONIN, J. R.; MAYNARD, J. R., and HARBURY, H. A.: Fed. Proc. *30:* 1143 (1971).

JACOBS, E. E. and SANADI, D. R.: J. biol. Chem. *235:* 531 (1960).

JUKES, T. H. and KING, J. L.: Nature, Lond. *231:* 114 (1971).

KAMEN, M. D. and HORIO, T.: Ann. Rev. Biochem. *39:* 673 (1970).

KEILIN, D.: Proc. Roy. Soc., Lond. B *98:* 312 (1925). — The history of cell respiration and cytochrome (Cambridge University Press, London 1966).

KIMURA, M.: Nature, Lond. *217:* 624 (1968).—Proc. nat. Acad. Sci., Wash. *63*: 1181 (1969).

KIMURA, M. and CROW, J. F.: Genetics *49:* 725 (1964).

KIMURA, M. and OHTA, T.: Nature, Lond. *229:* 467 (1971a). — J. molec. Evolut. *1:* 1 (1971b).

KING, J. L. and JUKES, T. H.: Science *164:* 788 (1969).

KING, N. K.; LOONEY, F. D., and WINFIELD, M. E.: Biochim. biophys. Acta *133:* 65 (1967).

KLINGENBERG, M.: FEBS Ltrs *6:* 145 (1970).

KOHNE, D. E.: Quat. Rev. Biophys. *3:* 327 (1970).

KREIL, G.: Z. Physiol. Chem. *334:* 154 (1963). — Z. Physiol. Chem. *340:* 86 (1965).

LEHNINGER, A. L.; CARAFOLI, E., and ROSSI, C. S.: Adv. Enzymol. *29:* 259 (1967).

LIN, D. K.: Ph. D. Thesis, Univ. of Wisconsin (1971).

MARGOLIASH, E.: Brookhaven Symp. Biol. *15:* 266 (1962). — Proc. nat. Acad. Sci. Wash. *50:* 672 (1963). — Canad. J. Biochem. *42:* 745 (1964). — In CHANCE, B.; ESTABROOK, R. W., and YONETANI, T. The chemistry of hemes and hemoproteins, p. 271 (Academic Press, New York 1966).

MARGOLIASH, E. and FITCH, W. M.: Proc. N.Y. Acad. Sci. *151:* 349 (1968). — in WHELAN, W. J. Homologies in enzymes and metabolic pathways, p. 33. (North-Holland Publ. Co., Amsterdam 1970).

MARGOLIASH, E. and LUSTGARTEN, J.: J. biol. Chem. *237:* 3397 (1962).

MARGOLIASH, E. and SCHEJTER, A.: Adv. Protein Chem. *21:* 113 (1966).

MARGOLIASH, E. and SMITH, E. L.: in BRYSON, V. and VOGEL, H. J. Evolving genes and proteins, p. 221 (Academic Press, New York 1965).

MARGOLIASH, E.; BARLOW, G. H., and BYERS, V.: Nature, Lond. *228:* 723 (1970).

MARGOLIASH, E.; FITCH, W. M., and DICKERSON, R. W.: Brookhaven Symp. Biol. *21:* 259 (1968).

MARGOLIASH, E.; FITCH, W. M.; MARKOWITZ, E., and DICKERSON, R. E.: Proc. Wenner-Gren Symp. on Oxidation-Reduction Enzymes, Stockholm (In press [1970]).

MARGOLIASH, E.; NEEDLEMAN, S. B., and STEWART, J. W.: Acta chem. scand. *17:* S250 (1963).

MARGOLIASH, E.; SMITH, E. L.; KREIL, G., and TUPPY, H.: Nature, Lond. *192:* 1125 (1961).

MARKOWITZ, E.: Biochem. Genet. *4:* 594 (1970).

MATTOON, J. R. and SHERMAN, F.: J. biol. Chem. *241:* 4330 (1966).

MAYR, E.: Animal species and evolution (Harvard University Press, Cambridge, Mass. 1966).

McDONALD, C. C.; PHILLIPS, W. D., and VINOGRADOV, S. N.: Biochem. biophys. Res. Commun. *36:* 442 (1969).

MOORE, G. W. and GOODMAN, M.: Bull. Mathem. Biophys. *30:* 279 (1968).

MROSS, G. A. and DOOLITTLE, R. F.: Arch. Biochem. Biophys. *122:* 674 (1967).

NAKAYAMA, T.; TITANI, K. and NARITA, K.: J. Biochem. *70:* 311 (1971).

NARITA, K.; TITANI, K.; YAOI. Y.; MURAKAMI, H.; KIMURA, M., and VANECEK, J.: Biochim. biophys. Acta *73:* 670 (1963a).

NOLAN, C. and MARGOLIASH, E.: J. biol. Chem. *241:* 1049 (1966). — Ann. Rev. Biochem. *37:* 727 (1968).

O'DONALD, P.: Nature, Lond. *221:* 815 (1969).

OHTA, T., and KIMURA, M.: J. molec. Evolut. *1:* 18 (1971).

PALEUS, S.; EHRENBERG, A., and TUPPY, H.: Acta chem. scand. *9:* 365 (1955).

PAUL, K.-G.: Acta chem. scand. *5:* 379 (1951).

PAULING, L. and ZUCKERKANDL, E.: Acta chem. scand. *17:* S9 (1963).

PRESSMAN, B. C.: in RACKER, E., Membranes of mitochondria and chloroplasts, p. 213 (Van Nostrand-Reinhold, New York 1970).

RAMSHAW, J. A. M.; RICHARDSON, M., and BOULTER, D.: Europ. J. Biochem *23:* 475 (1971).

REDFIELD, A. G. and GUPTA, R. K.: Cold Spr. Harb. Symp. quant. Biol. *36* (1971).

RICHMOND, R. C.: Nature, Lond. *225:* 1025 (1970).

RUTISHAUSER, V.; CUNNINGHAM, B. A.; BENNET, C.; KONIGSBERG, W. H., and EDELMAN, G. M.: Proc. nat. Acad. Sci., Wash. *61:* 1414 (1968).

SALEMME, R.; FREER, S. T., and KRAUT, J.: Personal commun. (1972).

SARICH, V. M. and WILSON, A. C.: Science *154:* 1563 (1966).

SCHEJTER, A.; AVIRAM, I. and SOKOLOVSKY, M.: Biochemistry *9:* 5113, 5118 (1970).

SCHEJTER, A. and GEORGE, P.: Nature, Lond. *206:* 1150 (1965).

SCHEJTER, A. and MARGALIT, R.: FEBS Ltrs *10:* 179 (1970).

SCHEJTER, A. and SOKOLOVSKY, M.: Fed. Europ. biochem. Soc. Ltrs. *4:* 269 (1969).

SHECHTER, E. and SALUDJIAN, P.: Biopolymers *5:* 788 (1967).

SINGER, S. J. and DOOLITTLE, R. F.: Science *153:* 13 (1966).

SKOV, K.; HOFMANN, T., and WILLIAMS, G. E.: Canad. J. Biochem. *47:* 750 (1969).

SMITH, E. L.: Harvey Lect. Ser. *62:* 231 (1967). – in OKUNUKI, K.; KAMEN, M. D., and SEKUZU, I. Structure and function of cytochromes, p. 282 (University of Tokyo Press, Tokyo 1968).

SMITH, E. L. and MARGOLIASH, E.: Fed. Proc. *23:* 1243 (1964).

SMITH, E. L.; MATSUBARA, H.; MCDOWELL, M. A., and ROTHFUS, J. A.: Science *140:* 385 (1963).

SMITH, J. M.: Nature, Lond. *219:* 1114 (1968). — Symp. zool. Soc. Lond. *26:* 371 (1970a) . — Amer. Naturalist *104:* 231 (1970b).

SMITH, L.: Unpubl. results (1972).

SMITH, L.; NAVA, M. E., and MARGOLIASH, E.: in KING, T. E.; MASON, H. S., and MORRISON, M. Oxidases and related redox systems. Proc. 2nd Int. Symp. (University Park Press, Baltimore [in press] 1972).

SOKAL. R. R. and SNEATH, P. H. A.: Principles of numerical taxonomy (Freeman, San Francisco 1963).

SOKOLOVSKY, M. and MOLDOVAN, M.: Biochemistry *11:* 145 (1972).

STEWART, J. W.; MARGOLIASH, E., and SHERMAN, F.: Fed. Proc. *25:* 647 (1966).

TAKANO, T. and DICKERSON, R. E.: Personal commun. (1972).

TAKANO, T.; SWANSON, R.; KALLAI, O. B. and DICKERSON, R. E.: Cold Spr. Harb. Symp. quant. biol. *36:* 397 (1971).

THEORELL, H.: J. Amer. chem. Soc. *63:* 1820 (1941).

THEORELL, H. and ÅKESON, Å.: J. Amer. chem. Soc. *63:* 1804, 1812, 1818 (1941).

THOMPSON, E. W.; RICHARDSON, M., and BOULTER, D.: Biochem. J. *124:* 779, 783 (1971).

THOMPSON, E. W.; NOTTON, B. A.; RICHARDSON, M., and BOULTER, D.: Biochem. J. *124:* 787 (1971).

TITANI, K.: Personal commun. (1970).

TSAI, H. J. and WILLIAMS, G. R.: Canad. J. Biochem. *43:* 1409, 1995 (1965).

TULP, A.: Biochem. J. *116:* 39 P (1970).

UZZELL, T. and CORBIN, K. W.: Science *172:* 1089 (1971).

VAN DAM, K. and MEYER, A. J.: Ann. Rev. Biochem. *40:* 115 (1971).

VENKATACHALAM, C. M.: Biopolymers *6:* 1425 (1968).

WINFIELD, M. E.: J. molec. Biol. *12:* 600 (1965).

WOJCIECH, R. and MARGOLIASH, E.: in SOBER, H. A. Handb. of Biochemistry, 2nd ed. p. C228 (Chemical Rubber Co., Cleveland, Ohio 1970).

WUTHRICH, K.: Proc. nat. Acad. Sci., Wash. *63:* 1071 (1969).

YAMANAKA, T.: Ann. Rep. Biol. Works Fac. Sci. Osaka Univ. *14:* 1 (1966). — Nature, Lond. *213:* 1183 (1967).

YAMANAKA, T. and OKUNUKI, K.: J. biol. Chem. *239:* 1813 (1964). — in OKUNUKI, K.; KAMEN, M. D., and SEKUZU, I. Structure and function of cytochromes, p. 390 (University of Tokyo Press, Tokyo 1968).

YANOFSKY, C.; CARLTON, B. C.; GUEST, J. R.; HELINSKI, D. R., and HENNING, V.: Proc. nat. Acad. Sci., Wash. *51:* 266 (1964).

ZUCKERKANDL, E. and PAULING, L.: in KASHA, M. and PULLMAN, B. Horizons in biochemistry, p. 198 (Academic Press, New York 1962). — In BRYSON, V. and VOGEL, H. J. Evolving genes and proteins, p. 97 (Academic Press, New York 1965a). — J. theoret. Biol. *8:* 357 (1965b).

Author's address: Dr. E. MARGOLIASH, Department of Biological Sciences, Northwestern University, *Evanston, IL 60201* (USA)

Discussions

P. O. LOWDIN: I was very much intrigued by your statement that guanine-adenine change seems to be dominating, and I wonder if you are aware that the quantum mechnical calculation we have carried out on the base pairs indicate that the change guanine-adenine is about 10 times as probable as the reverse change at 310°C.

E. MARGOLIASH: I was not aware of that.

H. C. LONGUET-HIGGINS: I enjoyed your paper very much indeed and I've been thinking about some of the problems that you've been working on. In my group we've been studying the mathematics of classification. You have based your analysis on the ultrametric hypothesis, which is essentially the view that all the species you're trying to put on the tree have got to be at the ends of branches. In fact there is a general assumption one could make and that is that, some of the objects may not in fact be at terminal nodes at all. This particular kind of problem arises for example if one has a set of medieval manuscripts and one is trying to put them in some family tree by noticing the copying errors made by scribes. My colleague, Dr. BUNEMAN has been working on just that problem. If you make that slightly more general assumption, I think one should be able to produce, not a best tree within certain rather crude limits, but the best tree of any kind.

E. MARGOLIASH: Does this not make the complexity much greater than with our assumption?

H. C. LONGUET-HIGGINS: It is slightly more complicated, but the mathematics of this is being developed and I think it may turn out that one will eventually obtain a unique solution within the slightly more general set of assumptions wherein one need not minimize the standard deviation at the end.

E. MARGOLIASH: It would of course be very pleasant to have a unique solution as it might well make it possible to test whether the hypothesis upon which the method is based fits or does not fit historical reality in so far as we know it.

J. NINIO: You conclude your talk by observing that in a sense, the information contained in cytochrome *c* is larger than the information strictly necessary for the catalytic reaction. It is obvious, of course, that any protein in a cell has information to perform other things than merely to catalyze its reaction. Part of that information is: 'do it at a rate that is compatible with the rate of the other reactions, and optimize the rate.' There is another class of instructions of a negative character: 'don't interfere with the polymerase while it is working', 'don't bind the repressor for another protein', and so on. In that respect you would expect that the allowed variability in a complex organism organism is less than the allowed variability in a lower one. In others words, take a cytochrome c from a yeast and try to obtain mutants. See to what extent yeast can afford different cytochromes and compare this situation to insects, to what extent an insect can afford different cytochromes. You would compare levels of variabilities which would really mean something for the comprehension of what *is* allowed and what is *not* allowed. I know that there are variations in the haemoglobins in men but have you any idea about the variations within a same species for the cytochromes?

E. MARGOLIASH: As far as it has been tested to date no obvious variations have been observed in a single species, except for one case in carp cytochrome c, and possibly, if I remember correctly, in one other species. There is certainly no indication that variations are any more or any less permissible in so-called 'lower' as compared to 'higher' organisms. As I showed, very few changes at all are permitted even on an evolutionary scale. As an experimentalist, I am afraid I cannot be satisfied with rather vague semi-philosophical concepts of what is supposed to happen, but rather want to know precisely what are the parameters on which evolutionary selection operates in the case of cytochrome c, in their full structural and physiological connotations. I think we may have a chance of finding out.

B. L. VALLEE: Among the many exciting possibilities implied by the discussion, I was intrigued by the comments on their pertinence to the function of the molecule in oxido-reduction. I know that Dr. MARGOLIASH didn't have time to elaborate on the background. He has, however, shown that a number of enzymes, e.g., cytochrome oxidase, exhibit markedly different affinities for oxidized and reduced cytochromes, and these two oxidation starts also vary in regard to the relative ease of chemical modification of residues such as tyrosine and methionine. Are you suggesting, perhaps, that these — thus far — unexplained observations, which imply conformational changes, may express functions of the protein over and beyond that of oxido-reduction? This would provide a common denominator of great simplicity.

E. MARGOLIASH: There clearly are differences in the conformation of the protein in the oxidized, as compared to the reduced state, and the more recent x-ray crystallographic work has pretty well delineated what these changes are. They do seem to explain most of the differences in properties that have long been known to exist between the two oxidation states of the protein, but this does not appear to provide a satisfactory solution for our problem. We are looking for some basis for evolutionary selection, and it seems quite likely that the oxido-reduction changes in conformation are the same for all cytochromes c. In fact, as was shown earlier, all react intramitochondrially in precisely the same way with the reductase and the oxidase ends of the respiratory chain, so that even if there were differences among the various cytochromes c in regard to oxido-reduction conformation changes, these do not seem to have any functional effects. As noted, the only differences so far observed among the cytochromes c of various species relate to ion-binding. I am afraid the consideration of oxido-reduction in the protein has so far not helped the evolutionary problem.

Systems of Order Recognition
and of Recognition

Proc. 3rd int. Conf. From Theoretical Physics to Biology, Versailles 1971, pp. 242–246
(Karger, Basel 1973)

Enzyme Specificity and Mechanisms

F. M. RICHARDS

Yale University, Department of Molecular Biophysics, New Haven, Conn.

Abstract

The solvent is considered to play a critical role in the structure of macromolecules and in their association with a variety of ligands. Competition between self-association and solvent-association is presumed to control the equilibria involved in macromolecular structures. From the known structures of some globular proteins quantitative estimates can be made of the changes in solvent contacts that occur during defined conformational changes.

Somewhat arbitrarily all carbon and sulfur atoms are considered nonpolar and all oxygen and nitrogen atoms polar. In comparing the accessibility to solvent of both the hypothetical fully extended chain and the native folded conformation it is found that solvent contact, or accessibility, decreases by about a factor of 3 in going from the extended to the folded conformation for both classes of atoms. This result is essentially the same for ribonuclease-S, lysozyme and myoglobin. The estimated energetic consequences of this change in geometry will depend intimately on 1) the assumptions made about the unit free energy changes for the different atom types and, 2) the extent of internal pairing of polar atoms that occur in the native structure. One such case has been examined in detail, the association of the 20 residue peptide, S-peptide, with the 104 residue protein, S-protein to give the enzymatically active complex RNase-S. The thermodynamics of this interaction have been examined in detail by direct calorimetry [HEARN et al.: Biochemistry 10: 806, 1971]. The structure of the complex is known, the structure of free S-peptide is close to a random coil. Only the structure of free S-protein is uncertain. Tentative correlations can be made between the geometrical changes and the observed thermodynamic changes.

The association of substrates or inhibitors with enzymes must follow essentially the same rules as those related to protein structure directly. Diffraction studies at the moment can only deal with stable complexes so that true Michaelis complexes cannot be observed. Substrate analogues can, however, serve to define the probable geometry of complexes related to the catalytic path. In this sense, the direct structural data can provide severe restrictions for proposed catalytic mechanisms. The complex of the dinucleotide analog UpcA with RNase-S will be used as an example.

Author's address: F. M. Richards, Chairman, Department of Molecular Biophysics and Biochemistry, Box 1937, Yale Station, *Newhaven, CT* 06520 (U.S.A.)

Discussion

H. FRÖHLICH: So you would say that you have a rather large system and it has levels, some degenerate, and you find that in actual fact the conformation is always exactly the same, is that correct? Thus, these energy levels, one would expect to be rather dense. In order to find, without exception, a single state, a single energy level must have moved away sufficiently far so as to be the only one that is always found. If this were not so, you would, as you increase the temperature slightly, find a combination with different probabilities, depending on temperature. If I understand this correctly, there is a general feature hidden, namely that a single energy level, one of the very many, is spearted out to move down. Such situations are met in physics, of course, quite often, although in physics one doesn't deal with these limited systems, but with larger systems, but they are not necessarily completely regular. Hence, if this is correct, there is an additional question to be asked: does this one state have a particular quality, one which is very selective and is such a quality to be found? The second question I wanted to ask is: has there been any measurement of the dipole moments of the whole system, are they very large or are they compensated?

F. M. RICHARDS: The dipole moment measurements on whole proteins are either large or small depending on how you look at it. An average protein molecule is of the order of 50 to 100 Å in dimensions. The dipole moments in general correspond to a charge separation of not more than one charge over that kind of distance.

H. FRÖHLICH: . . . over a *huge* distance . . .

F.M. RICHARDS: Yes, the moment is large in a sense. On the other hand there are in fact hundreds of charges, and, if they were separated to a larger extent the resulting moment would be staggering.

H. FRÖHLICH: . . . this is in terms of charges, but in terms of dipoles, because you speak of polar groups, in terms of dipoles it may mean that you have a lot of these nearly parallel.

F.M. RICHARDS: I believe that the true dipoles, the permanent dipoles, say the peptide bond and so on, tend to be averaged out over the structure.

H. FRÖHLICH: If it's like you said in terms of charge separation, then in terms of dipolar theory it is a fantastically large dipole moment. Then the interactions—I don't know how you calculate your interactions, but it would not be sufficient to take into account the interactions between neighbours only.

F. M. RICHARDS: You're saying that pairwise interaction is not good enough for this ...

H. FRÖHLICH: Yes, if you were to make a very crude approximation and smear out the dipoles, the total energy of the system would depend on its shape. You see if you would just take any system and say we have a dipole moment of whatever size, distributed continuously over it then the energy of this system would depend on its macroscopic shape. And that means that we have to take into account all the interactions.

H. C. LONGUET-HIGGINS: I wanted to ask if the following experiment has been done (which seems to me to be an interesting one): has anyone, for example, made a random polypeptide with the same overall amino-acid composition as a globular protein, but with the different amino acid residues arranged in a precise, but different, order from that in the protein, to see whether it fact in does have a unique tertiary structure? One might well imagine that very few polypeptides in fact have a unique tertiary structure, and that one of the things nature has selected for in the evolution of enzymes is this uniqueness of the tertiary configuration.

F. M. RICHARDS: I was hoping PHILLIPS would try to answer this question. I think the answer to your specific question is probably no. Only recently has it been possible to synthesize peptides of this length, and people have tried to make sure they ended up with the right answer rather than the wrong one, so far. What has been supplied, of course, are the mutants in various systems, with sometimes enormous changes in sequence, as MARGOLIASH pointed out the other day, apparently identical 3-dimensional structures, at the level of identity that we're talking about. Thus, the converse of your experiment has been done. If you make certain changes, the appropriate ones, you can end up with the same structure. But it is quite clear that if you mix the sequence up too much, you can do this by chemical modification which is not quite fair because then you end up with unnatural amino acids, but if you do that, you can certainly prevent the correct structure from forming. It's quite easy.

H. C. LONGUET-HIGGINS: It isn't a question of preventing the correct structure from forming; it's a question of preventing *any* unique structure from forming. I think this is a very important question and with the techniques that have now been developed, I should have thought that one could answer this question.

C. LEVINTHAL: I would like to comment on the general question of folding of proteins. The essential point to keep in mind is that in an ensemble of protein molecules, each molecule, or almost every one, has the same three-dimensional structure. If the protein is denatured and subsequently renatured, then it first loses its well-defined three-dimensional structure and subsequently regains it. Not all proteins can be reversably renatured but many can, and thus it is clear that there is a unique structure which can be reached starting from a random conformation. Many people have assumed (as Dr. FROHLICH did) that these renaturation experiments necessarily imply that the unique state is the one with the lowest free energy. This conclusion is not, however, necessarily correct. The unique state is certainly a meta-stable one. In energy space, it must be at the bottom of a local valley, but there is no necessary reason to believe that this valley is the deepest energy minimum possible. The unique state could be that meta-stable state which is reached most rapidly from the unfolded random conformation. Since the energy valleys seem to be very deep and there are an enormous number of possible conformations for a protein, it seems clear that all of conformation space could not be explored in the time that it actually takes to fold a protein molecule. For these reasons, I suggested several years ago that proteins reach their final unique active conformation by following a 'pathway of folding'. Recent experimental results from a number of laboratories have tended to support this general idea.

However, if pathways of folding exist, then the immediate question is: what starts the protein down its pathway and what are the driving forces? As a result of the crystallographic results, such as those reported by PHILLIPS and RICHARDS, we can begin to see what might be the nucleation regions in a polypeptide chain which could start the folding process. However, when we ask about the forces which might drive the protein along the pathway, we have much less real evidence. We have to understand both the energetics of a total protein in a solvent, and we have to understand the various kinetic processes involved. One of the major contributions to the free energy is that of the entropy terms, namely, the contribution of the protein itself and the contribution due to the organization or lack of organization of the solvent molecules. It seems to me that this is an area in which the people here who are concerned with this type of calculation could do very much better than has been done up until now. Perhaps there is work in this area with which I am not familiar because it seems to me that in a real understanding of the entropy changes in the water which take place when a protein folds, will have to be understood, and understood in some detail before we can really hope to do quantitative work on the folding of proteins. The contribution of this entropy change seems to be large but the real values and how they change with the conformation of the protein seems to be almost unknown. I hope someone here can disagree with me and knows of some work which really sheds light on this subject, but if not, it would seem to me to be an enormously fruitful area for theoretical work.

WIGNER: We theoretical physicists have learned so far an enormous amount but have contributed very little positively. I would like to make a minute contribution. We are all very impressed by what Dr. LONGUET-HIGGINS said and we agree wholeheartedly. On the other hand, we are most shocked by the 9–6 potential because our problem is always to account for the great binding energy and we know the coefficient of the attractive 6 potential. If the repulsive part is a 12 potential then only half of the attractive potential is effective at equilibrium. Because at the minimum of the 12–6 potential the 12 potential is half as great as the 6 potential and that will subtract one half. Now the 9–6 potential will subtract two-thirds of the attractive 6 potential which is very embarrassing for us who believe we know how to calculate the absolute value of the 6 potential. This is usually too small even against the repulsive 12 potential. With the 9 potential we would be quite at sea how to explain the total binding energy at equilibrium.

D.C. PHILLIPS: I wonder if I could show one slide which I had intended to include. This is just a selection of the proposed hydrogen-hydrogen nonbonded interaction potential curves taken from recent literature. All of these have been cogently argued for and it's really rather amazing as one turns from complicated protein to what one would have naively expected to be by now firm information, to find that this is the state of affairs.

K. MENDELSSOHN: Could I just ask my biological friends a question? The information given to us here, provides a consistent picture but it comes from various sources. As a physicist, one would consider some of these sources more reliable than others. If you have a protein crystal and you take an electron density diagram then your are quite happy about the structure. What do we actually know about the unfolded protein? Have we got here the same degree of reliable information on the unfolded protein? Do we know that it will be identical with what is inside coiled up? Or is there a certain amount of conjecture? Finally, one always talks about this process of folding and unfolding. Is there anything known about the intermediate state between stretched proteins and folded ones? Or do we just assume that we see one state and then the other state and we make a conjecture that something in between has happened?

F.M. RICHARDS: A very good question. The state of the unfolded protein is at the moment probably in the least satisfactory state. The hydrodynamic measurements, which is all one can usually do on these systems, are compatible with it being a random coil. But that's not a very critical decision. Obviously, small residual amounts of structure would not be seen. The stiffness of the coil, and so on, is subject to much argument among polymer chemists. I think that as far as intermediary states are concerned, I'd like DAVID PHILLIPS to speak on that because this is also a real problem.

D.C. PHILLIPS: One of the questions here of course is a technological question: what means do you employ to look at unfolded structures in solution? I suppose most people are enthused for the moment about the potential of high resolution proton NNR measurements. These would indeed seem to be very promising and they would tend to show that in a fully denatured protein in some extreme conditions, like high urea or something of that kind, the NNR signal is very well approximated by adding up the individual signals that you would expect from isolated amino acids. So it looks as though the chain is fully unfolded in this way. People are trying to push this method to study intermediate states in the denaturation process. There's a great deal of argument here and most people at the moment I think are holding to the two state idea, that there's either a folded, whatever that might mean, or an unfolded state, both of which are probably populations of states anyway. Though there are beginning to be some suggestions that there are intermediate states in which one particular conformation is favored above others that are certainly there. But this is just a subject that is developing and I think it is developing intensively in response to the technological developments of high resolution nuclear magnetic spectroscopy. Although perhaps this is not a very exciting thing for some physicists, it of course has rested on the production of super-conducting magnets and developments of this kind. This is no small contribution that physics is making at the moment.

K. MENDELSSOHN: Thank you very much, I asked the question because this is a point interesting to the physicist when the molecule for the first time does something which an ordinary organic molecule doesn't do. Something has to happen to make it uncoil in a particular way, which seems to be a pretty unlikely process unless you put in some vital force. The only alternative is that there exists some form of information already covering the whole molecule and there must be a physical process following the information.

Proc. 3rd int. Conf. From Theoretical Physics to Biology, Versailles 1971, pp. 247–285
(Karger, Basel 1973)

Stereochemistry of Cooperative Effects in Haemoglobin[1]

M. F. PERUTZ

MRC Laboratory of Molecular Biology, Cambridge

I. Haem-Haem Interaction and the Problem of Allostery

I wish to propose a stereochemical mechanism of haem–haem interaction which is based partly on a comparison of the atomic models of native deoxy and methaemoglobin of horse [1–3], and partly on an x-ray study of a haemoglobin derivative in which the transition from the quaternary structure of oxy- to that of deoxyhaemoglobin was inhibited (my unpublished results). For a description of the anatomy of the haemoglobin molecule and of the change in quaternary structure that accompanies its reaction with ligands, I refer the reader to previous publications [1, 4, 5]. The basic features of the molecule are shown in figures 1 and 2. I start by summarizing the nature of haem–haem interaction and then go on to describe relevant structural and other observations. I then explain the proposed mechanism, discuss its implications for the size of the functional unit of haemoglobin, and finally apply it to the theory of allostery. In section II I shall deal with the stereochemical mechanism of the Bohr effect and of the regulatory function 2,3-diphosphoglycerate.

Nature of Haem–Haem Interaction

This is best explained by analogy with the biblical saying 'To him who hath shall be given' [6]. Suppose we have two haemoglobin molecules A and B, A holding three molecules of oxygen already and B holding none; to

---

1   Reproduced with the kind permission of Nature 228: 726–739 (1970).

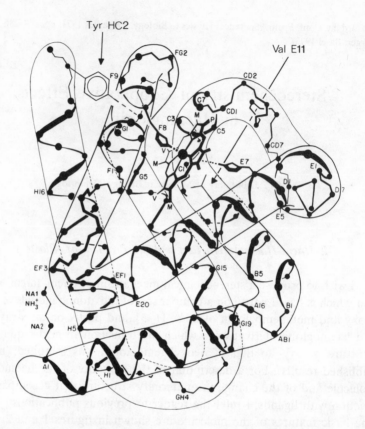

Fig. 1. Diagrammatic sketch showing the course of the polypeptide chain in the β subunit. In going from the amino to the carboxyl end, helical regions are denoted A to H; non-helical residues at the amino end NA, and at the carboxyl end HC. Non-helical regions between helices are denoted as AB, BC and so on. The haem lies in a pocket between helices E and F. The conformation of the α subunit is closely similar, except for the addition and deletion of certain residues.

which of them will an approaching free oxygen molecule stick? The chances are 70 to 1 in favour of A. If Y is the fractional saturation of haemoglobin with oxygen, p its partial pressure and K and n are constants, then the oxygen equilibrium curve can be roughly represented by Hill's equation

$$Y = \frac{Kp^n}{1 + Kp^n}$$

which corresponds to the equilibrium $Hb_n + nO_2 \rightleftharpoons Hb_n(O_2)_n$, Hb re-

Fig. 2. View of the oxy or methaemoglobin molecule showing the α-chains in white, the β-chains in black and the haems in grey. The sign in the centre indicates the two-fold symmetry axis which runs down the water-filled central cavity. The terminal regions of the α-chain are near at the top, those of the β-chains near the bottom of the molecule. In the deoxy form, the cross-links between the α-chains lie in the central cavity; those between lysine $40\alpha_1$ and histidine $146\beta_2$ lie near the letter H of the label HS in the lower centre of the molecule. The label HS indicates the position of the reactive cysteins 93β. On deoxygenation, the FG region of the α_1-chain turns away from, and that of the β_2-chain turns towards, the observer.

presenting the haemoglobin concentration in equivalents of iron. Hill's equation fits the experimental data best for a Hill's constant $n \sim 2\cdot7$; this value is independent of pH and temperature. More accurately, the oxygen equilibrium curve can be represented by ADAIR's equation which contains four separate equilibrium constants for the reactions of successive haems with oxygen [7, 8]. MONOD, WYMAN and CHANGEUX [9] have shown that on the basis of simplifying assumptions two arbitrary constants give reasonable agreement with observation, and have used haemoglobin as a model of an allosteric enzyme.

WYMAN pointed out the main features which any model of haem–haem interaction would have to satisfy [10]. These are illustrated by his logarithmic plot of the oxygen equilibrium curve which has a slope of unity at very low and very high oxygen concentrations and a slope of about three in the middle range, corresponding to Hill's constant. The total free energy interaction ΔF can be calculated from the perpendicular distance between the two asymptotes and normally lies between $2 \cdot 5$ and 3 kcal/iron atom. WYMAN writes: 'In principle, the interaction energy might equally well be interpreted as a stabilizing energy between the unliganded or the liganded haems. In the former case we might say, figuratively, that the first oxygen enters the molecule with more difficulty than the second owing to the necessity of breaking up a pre-existing partnership between the haems; in the latter case the second oxygen enters more easily owing to the decoying effect of the first' [10]. We now know that the former alternative is correct, because the oxygen affinity of isolated α and β subunits corresponds to that of the fully combined oxyhaemoglobin, showing that the low oxygen affinity of deoxyhaemoglobin must be the result of constraints imposed on this form. This is confirmed by its structure.

The physiological purpose of haem–haem interaction lies not so much in ensuring an increased oxygen affinity as successive molecules of oxygen combine, as in making the affinity fall as successive oxygen molecules dissociate.

If haemoglobin had a hyperbolic oxygen equilibrium curve, not more than a small fraction of the oxygen carried by it would be drawn off, with the result that man would asphyxiate. The Bohr effect and 2,3-diphosphoglycerate both increase the efficiency of respiratory transport further by lowering the oxygen affinity of haemoglobin in the tissues.

Movements of Iron Atoms relative to the Porphyrin

X-ray studies of iron porphyrin compounds have shown that the length of the bond from the iron atom to the nitrogen atoms of the porphyrin ring varies between $2 \cdot 061$ A in high spin ferric and $1 \cdot 99$ A in low spin ferric compounds [11, 12]. In metal-free porphyrin the distance from the centre of the ring to the nitrogen atoms is $2 \cdot 01$ A, smaller than the Fe–N distance in high spin, and larger than that distance in low spin compounds. In six-coordinated low spin ferric porphyrin complexes, therefore, the iron atom tends to lie within $0 \cdot 05$ A of the plane of the four nitrogen atoms [12]. In five-coordinated high spin ferric compounds, on the other hand, the iron atom is forced to lie out of that plane by distances of between $0 \cdot 38$ and $0 \cdot 47$ A [11] (table I).

No crystal structures of ferrous porphyrins have been published, but the radius of ferrous iron in low spin compounds is believed to be the same, within $0 \cdot 02$ A, as that in low spin ferric compounds, so that the iron atom in low spin ferrous compounds should also lie within $0 \cdot 05$ Å of the plane of the nitrogen atoms. The radius of the high spin ferrous iron, on the other hand, exceeds that of the high spin ferric by $0 \cdot 12$ Å [13]. If we add this to the Fe–N distance of $2 \cdot 061$ Å in the five-coordinated high spin ferric form, we obtain an Fe–N distance of $2 \cdot 18$ Å. This would place the iron atom $0 \cdot 83$ A out of the plane of the four nitrogen atoms.

Of ferrous haemoglobin or myoglobin derivatives, the deoxy form is high spin and five-coordinated, so that we should expect the iron atoms to lie about $0 \cdot 8$ A out of the plane of the four nitrogen atoms, while oxy and carbonmonoxy forms are both low spin and six-coordinated so that the iron atoms should lie in that plane. All ferric derivatives are six-coordinated, the acid and fluoride met derivatives being high spin and the remainder low spin. In agreement with expectation, the iron atoms in (high spin) acid metmyoglobin and meterythrocruorin lie $0 \cdot 3$ A from the plane of the porphyrin ring [14, 15], while those in a (low spin) cyanomethae-moglobin of lamprey lie in the plane [16]. Three-dimensional Fourier analyses of the deoxy derivatives of single chain haemoglobins have not yet been calculated. Difference Fourier analyses of deoxy with respect to met derivatives give conflicting results which will be discussed elsewhere.

The quaternary structure of mammalian haemoglobin depends only on the coordination of the iron atom but not on its valency or spin state. The structure of the deoxy form is unique, whereas all other derivatives have the same quaternary structure in common. We therefore call the latter simply 'the liganded form'. When we published the structure of that form we called it oxy-haemoglobin [1], because this was the state of the solution from which the crystals had been prepared, but by the time the x-ray data were collected, most of the haemoglobin had been oxidized to the (high spin) acid met state, where the ligand is a water molecule. This distinction did not matter then, but has now become important.

When Dr. H. MUIRHEAD and I measured the atomic coordinates of methaemoglobin using a plumb line and an electron density map drawn on the scale of 1 Å $= 5$ mm we did not believe our resolution of $2 \cdot 8$ A to be sufficient to determine the displacement of the iron atoms out of the plane of the porphyrin ring [1]. But by redrawing the maps on a scale of 1 Å $= 2$ cm and viewing them in Dr. F. M. RICHARD's reflecting box [17] it was clear that we had underestimated the information contained in them.

Dr. W. Bolton and I had no difficulty in fitting the porphyrin exactly to the contours. On measuring the perpendicular distance from the plane of the porphyrin ring to the observed iron peak, we found it to $0 \cdot 3$ Å in both the α and β subunits, in exact agreement with the displacement of the iron atom observed in metmyoglobin[14] and meterythrocruorin[15]. Performing the same experiment with the electron density maps of deocyhaemoglobin, on the other hand, revealed a displacement of $0 \cdot 75$ Å in both the α and β subunits, in close agreement with the value of $0 \cdot 83$ Å calculated above. Both displacements are on the side of the proximal histidine (F8). In these measurements we kept the porphyrin ring flat. When we tried to make it slightly domed, as it is in haemin chloride [18], the displacement of the iron atoms from the plane of the four nitrogens came out smaller but their displacements from the mean plane of the light atoms remained the same.

According to R. J. P. Williams, theory predicts that the length of the bond linking the iron atom to Nε of the proximal histidine will also increase by $0 \cdot 1$–$0 \cdot 2$ Å in going from a low spin to a high spin ferrous derivative. According to Hoard it should remain unchanged [11]. Our x-ray data are indecisive. Depending on its exact value, the distance of the proximal histidine from the plane of the porphyrin ring should increase by $0 \cdot 75$–$0 \cdot 95$ Å. Such a shift, occurring in the transition from oxy to deoxy-haemoglobin, would be sufficient to force drastic changes in the contacts between the porphyrin ring and the tightly packed amino-acid side chains of the surrounding globin. Hoard first predicted that the reaction with oxygen should be accompanied by a substantial movement of the iron atom relative to the porphyrin ring, and that this could lead to cooperative movements in the protein framework of the haemoglobin molecule [30]. An ealier prodiction on similar lines was made by Williams [61].

Room in the Ligand Pocket

Haem is attached to globin by a covalent bond extending from the iron atom to Nε of the proximal histidine F8. On the distal side, not directly linked to the iron atom, lie histidine E7 and valine E11 (fig. 1). The ligand comes to lie between the iron atom and these two residues. Further removed from the iron atom, 60 atoms of the globin are in van der Waals contact with the porphyrin ring [fig. 2 of ref. 1].

We first select the smallest ligand that produces full haem–haem interaction. This is a hydroxyl ion, with a van der Waals radius of $1 \cdot 5$ Å, which becomes attached to the iron atom in the oxidation of deoxy to methaemoglobin at alkaline pH. A hydroxyl ion, placed at the expected distance of $2 \cdot 1$ Å from the iron atom in either the α or the β subunits of the liganded form or in the α subunit of the deoxy form, can be accommodated without making short contacts with neighbouring amino-acid side chains. In the β subunits of the deoxy form, on the other hand, the γ methyl of valine E11(67) would come to lie within $2 \cdot 5$ A of the hydroxyl ion, which is 1 Å less than the sum of the van der Waals radii, so that there is not enough room for even the smallest ligand, let alone an oxygen molecule. Direct comparison of the electron density maps of the β-chains confirmed that the distance between the porphyrin ring and valine E11 shrinks by about 1 Å in going from the met, or oxy, to the deoxy form. Similarly, the difference electron maps of deoxy minus met-BME-haemoglobin, to be described, show a striking movement of valine E11 towards the haem group. These observations demonstrate that in the deoxy form of the β-chains the valines E11 must move relative to the porphyrin ring before the iron atoms can react, and that in going to the oxy form the distance between the porphyrin and helix E widens to make room for ligands. The width of the pocket in the α-chain, on the other hand, shows no measurable change, and has room for ligands in both forms. (Leigh Anderson has now shown that deoxy can be converted to methaemoglobin without change of quaternary structure. The reaction is accompanied by a marked widening of the haem pocket in the β-subunits and a slight widening in the α-subunits [62].

Role of the C-terminal Residues

In oxy or methaemoglobin the C-terminal residues of all four chains have complete freedom of rotation, and the penultimate tyrosines have partial freedom in the sense that they spend only a small fraction of their time in their bound position, between helices F and H. The experimental evidence for this statement will be reviewed in section II. In deoxyhaemoglobin, on the other hand, each of the C-terminal residues is doubly anchored by salt-bridges. The α-carboxyl of arigine HC3(141)α_1 is linked to the α-amino group of valine NA1(1)α_2, and its guanidinium group to aspartate H9(126)α_2. The α-carboxyl of histidine HC3(146)β_1 is linked to the ε-amino group of lysine C5(40)α_2 and its imidazole to aspartate FG1(94) β_1 (fig. 5 and 6). All four penultimate tyrosines are firmly anchored in pockets between helices F and H, partly by van der Waals contacts with the helices and partly by hydrogen bonds between their OH groups and the carbonyls of valines FG5 [2, 3]. The conformation is such that the tyrosines cannot be displaced from their pockets without also displacing the C-terminal residues and rupturing their salt-bridges. I shall show this property to be vital for the entire mechanism of interaction between the subunits (fig. 3).

Enzymatic removal of the C-terminal residues affects haem–haem interaction. Amputation of histidine HC3(146)β has the smallest effect, reducing Hill's constant from 2·7 to 2·5 [19]. But Dr. J. GREER pointed out to me that this does not mean that the salt-bridge of its carboxyl group with lysine C5(40)α is unimportant, because the α-carboxyl group of the histidine which normally forms the salt-bridge can be replaced in the truncated haemoglobin by the α-carboxyl of tyrosine HC2(145). Displacement of the C-terminal histidine by N-substituted maleimides, attached to cysteine F9(93)β, reduces Hill's constant to about 2·0 [31]. In these derivatives the salt-bridge with lysine C5(40)α is not made, but the remaining structure of the α- and β-chains is left undisturbed in both the oxy and deoxy forms. Removal of arginine HC3(141)α diminished Hill's constant to 1·7; removal of both the histidine and the arginine reduces Hill's constant to near unity as well as inhibiting most or all of the Bohr effect [19].

Why should removal of these residues inhibit interaction? Because they are free in oxyhaemoglobin, the clearly serve no function there. Even in deoxyhaemoglobin their positions are such that they could exert no direct influence on the conformation of the haem groups or their immediate environment. We are driven to conclude that their importance for haem–haem interaction lies in the constraining crosslinks they form between the subunit in deoxyhaemoglobin.

This is confirmed by x-ray studies of the derivatives just described which were crystallized by Dr. J. V. KILMARTIN. Human Des-His-146β and Des-Arg(141)α-deoxyhaemoglobin each crystallize isomorphously with normal human deoxyhaemoglobin. Des-His-(146)β + Des-Arg-(141)α deoxyhaemoglobin, on the other hand, crystallizes in a form closely related to normal oxyhaemoglobin. This shows that in the absence of the constraining salt-bridges the quaternary deoxy structure is unstable, even if all four chains are in the deoxy state [64].

Conformational Changes within the α and β Subunits

Before the quaternary structure of the haemoglobin molecule can change from the deoxy to the oxy form, the reaction of the iron atoms with oxygen or other ligands may cause the tertiary structure of the individual subunits to alter, so leading to the rearrangement of the subunits. These initial changes cannot easily be observed on electron density maps of the fully liganded or the fully deoxygenated forms. I therefore tried to detect them in a haemoglobin derivative whose quaternary structure is locked in the oxy form.

The lock is *bis*-maleimido-methyl ether (BME), a bifunctional reagent which inhibits all cooperative effects in haemoglobin by linking cysteine F9(93) to histidine FG4(97) in the same β-chain [20–22]. The reagent, which lies in the $\alpha_1\beta_2$ contact, also blocks the entry of the penultimate tyrosines HC2(145) of the β-chains into their pockets between helices F and H, and causes small changes in the remainder of the β-chains and the C-helix of the α-chains [22]. Difference electron density maps of met minus deoxy BME-haemoglobin should therefore show the initial changes undergone by the subunits on reaction with ligands.

Crystals of horse met and deoxy BME-haemoglobin are isomorphous with crystals of the native methaemoglobin, so that the phase angles of the x-ray reflexions should be approximately the same. The crystals had been grown in the acid met form. I reduced them, collected x-ray data to $3 \cdot 5$ A resolution and, with the help of Dr. J. GREER, calculated difference Fourier syntheses. When acid methaemoglobin is reduced, the haem-linked water molecule is removed and shows up as a positive peak in a difference map of met minus deoxy. The strength of that peak is a measure of the fraction of the iron atoms in the crystal that have been reduced. I first used ferrous citrate and found it had reduced most of the iron atoms in the β subunits, but very few of those in the α subunits. I next used $Na_2S_2O_4$ which reduced all the iron atoms in the β subunits and most of those in the α subunits. Ligands must be removed from the β-chains more easily than from the α-chains.

Difference electron density maps of met with respect to deoxymyoglobin and erythrocruorin reveal no significant changes in the conformation of the globin chain or movements of the haem relative to the globin. By contrast, the difference maps of met minus deoxy BME-haemoglobin show striking changes.

In the α-chains, the positive peak representing the entry of the haem-linked water molecule is matched by a negative one of the same magnitude representing the expulsion of tyrosine HC2(140) from its pocket between helices F and H. There is a further negative peak in the position of the amide group linking tyrosine HC2(140) to arginine HC3(141), corresponding to the fixed position which that amide normally takes up in native deoxyhaemoglobin. There is no positive peak corresponding to this amide, however, showing that it is rotating as freely in BME met as in native methaemoglobin. There are no peaks corresponding to either the side chain or the carboxyl group of arginine HC3(141), showing that these rotate freely im both forms of BME-haemoglobin.

There is a pair of positive and negative peaks indicating that helix F moved inwards towards the centre of the molecule, narrowing the pocket between it and helix H. This looks like the movement responsible for the expulsion of tyrosine HC2 (fig. 3). Finally, there are pairs of positive and negative peaks on either side of the propionic acid side chains of the haem group, showing that its inclination is becoming less steep in the liganded form (fig. 2).

In the β-chains the entry of tyrosine HC2(145) into its pocket between helices F and H is blocked by the BME group, but the difference Fourier indicates a similar inward movement of helix F, leading to a narrowing of the pocket between it and helix H, as in the α-chain, evidently designed to expel the tyrosine if it were there. The most prominent feature is a negative peak in the position of valine E11(67), showing the widening of the space between the porphyrin ring and helix E on ligand binding already referred to. There are also peaks near the haem group (when viewed as in fig. 18 of refs. [4, 5]) showing that its tilt becomes less steep in the liganded form, that is, its iron atom has moved away from valine E11.

The $\alpha_1\beta_2$ contacts shows signs of strain as though on dissociation of ligand the residues were trying to move towards the deoxy conformation; that is, a set of positive and negative peaks which correspond to a movement towards the observer of the β_2 subunit in fig. 2. It also looks as though the entry of the tyrosine HC2(140)α_1 into its pocket pressed on the indole ring of tryptophan C3(37)β_2, causing it to tilt over and to press on proline C2(36)β_2, and thus helping to push the β-chain in the required direction [fig. 15 of refs. 4, 5].

These results show that on binding of ligand to the α subunit, helix F moves so as to expel tyrosine HC2(140) from its pocket. A similar movement of helix F occurs in the β-chain. On removal of ligand, residues at the $\alpha_1\beta_2$ contact show signs of strain moving them towards the deoxy formation, even though the required sliding at the contact is inhibited by the BME.

I then examined the atomic models of oxy and deoxyhaemoglobin to see if the change in the width of the tyrosine pocket is detectable there, and found that, on going from deoxy to oxy, its width had shrunk by $1 \cdot 3$ Å in the α and by 2 Å in the β subunits. Model building shows that the tyrosines do not fit into the pockets of the oxy form without making short contacts with atoms of helices F and H.

At $5 \cdot 5$ Å resolution we had detected no changes in the tertiary structures of the individual chains. At $2 \cdot 8$ Å resolution various small shifts have become apparent, some distances between residues in neighbouring helices having changed by as much as 2–3 Å, which suggests that each subunit undergoes a small concerted change in tertiary structure. This possibility was first suggested by Dr. A. D. McLachlan after he had built space filling models of myoglobin and of the two haemoglobin chains. He found the side chains in the interior of the myoglobin molecule to be tightly packed, which made the model rigid, while those in the haemoglobin subunits seem more loosely packed, giving the helices some freedom of movement. The observed changes will be described elsewhere.

The differences in quaternary structure are as previously described [23], with small shifts of not more than 1 A at the pair of subunit contacts $\alpha_1\beta_2$ and $\alpha_2\beta_2$, and large relative shifts of up to 7 A at the contacts $\alpha_1\beta_2$ and $\alpha_2\beta_1$. Apart from the rupture of the salt-bridge just mentioned between histidine HC3(146)β_1 and lysine C5(40)α_2, a small increase in the number of van der Waals interactions seems to occur on oxygenation; the number of hydrogen bonds at either pair of contacts probably remains unchanged. Neither of the two forms contains any direct links between the two β-chains, but it will be shown in section II that these are furnished in the red cell by 2,3-diphosphoglycerate.

Where is the force that changes the quaternary structure applied? The evidence is overwhelmingly in favour of the contacts $\alpha_1\beta_2$. During the change of quaternary structure the contacts $\alpha_1\beta_1$ remain largely passive, whereas large movements occur at $\alpha_1\beta_2$. All but one of the mutations replacing residues at the $\alpha_1\beta_2$ contact diminish haem–haem interaction, whereas mutations at the $\alpha_1\beta_1$ contact do not. Nearly all residues at the $\alpha_1\beta_2$ contact are invariant throughout the vertebrates (except lamprey), while many of those at the $\alpha_1\beta_1$ contact are variable. The BME difference Fourier shows the beginnings of the changes that would make the contact jump from the oxy to the deoxy conformation.

Possible Intermediate Conformations

The contact $\alpha_1\beta_2$ is dovetailed, so that the CD region of one chain fits into the FG region of the other [fig. 8 of ref. 1 or fig. 14 of refs. 4, 5]. During the change of quaternary structure the two subunits rotate relative to each other so that the dovetailing of CDβ with FGα remains much the same, whereas that of CDα with FGβ changes. The change works like this: in oxyhaemoglobin a knob consisting of the side chain of threonine C3(38)α fits into a notch made up of the main chain of valine FG5(98)β. In deoxyhaemoglobin that same notch is occupied by the side chain of threonine C6(41)α—that is, the one protruding from the next turn of the C helix. At the same time the hydrogen bond linking aspartate G1(94)α to asparagine G4(102)β in oxyhaemoglobin is replaced by a hydrogen bond between tyrosine C7(42)α and aspartate G1(99)β in deoxyhaemoglobin [3]. If there were an intermediate quaternary structure, there could be no dovetailing, because knobs would fit against knobs; nor could any hydrogen bond be made to stabilize it. On stereochemical grounds its existence therefore seems unlikely. Another type of intermediate would be one where the $\alpha_1\beta_2$ contact, say, has clicked to the oxy conformation and the $\alpha_2\beta_1$ contact has remained in the deoxy conformation. In such an asymmetric intermediate all four contacts would be severely stressed, which would make it very unstable. But in view of the high speed of the reaction of haemoglobin with oxygen the occurrence of such an intermediate cannot be excluded.

Spectroscopic evidence favours the existence of an intermediate, but not necessarily one with a quaternary structure different from the oxy and deoxy forms. McCONNELL and his group find the ESR spectra of spin-labelled [24] oxy and deoxyhaemoglobin to be markedly different. Their spin lybel is attached to cysteine F9(93)β which lies near the tyrosine pocket and the $\alpha_1\beta_2$ contact, so that the label would sense changes in the conformation of the β and the α-chains as well as shifts at that contact (fig. 2). If haemoglobin existed in only two alternative conformations, then the ESR spectra at successive stages of oxygenation should exhibit common isosbestic points. Because this is not the case with one of the labels used, the authors argue that there must be an intermediate structure. I shall suggest what type of structure this might be.

We must now consider the tertiary structure of the individual subunits which changes on uptake of ligands, as the BME experiment has proved. On uptake of oxygen, the tertiary structures of the individual subunits click to the oxy conformation, but the quaternary structure may remain in the deoxy conformation until several of the subunits have reacted with oxygen; consequently, subunits in the oxy conformation will be constrained in the quaternary structure of deoxy, or vice versa. Will these constrained subunits take up an intermediate conformation?

The BME difference Fourier should tell, because it shows deoxy subunits locked in the quaternary structure of oxy. When I compared that difference Fourier with the actual differences between the native oxy and deoxy structures, I found it contained many, but not all, of the expected pairs of positive and negative peaks. This suggests but does not prove intermediate tertiary structures. Direct comparisons of the oxy and deoxy structures are still at too early a stage to judge whether intermediate tertiary structures are probable or not on stereochemical grounds. So the x-ray evidence is not yet clear.

OGAWA and SHULMAN have examined NMR spectra of mixed hybrids in which one pair of chains is kept locked in the liganded form as the cyanomet derivative, and the other pair is free to be oxygenated or deoxygenated. They find that deoxygenation of $(\alpha O_2 \beta^+ CN)_2$ in phosphate buffer did not show any effect on the paramagnetically shifted proton resonances of the β-haem and its immediate environment, while marked changes can occur in the spectrum of the α-haem on deoxygenation of $(\alpha^+ CN \beta O_2)_2$. They interpret this result to mean that the former derivative remains in the quaternary oxy structure throughout the reaction, while the latter may change to the quaternary deoxy structure, where the liganded α-chains are subject to strains extending from the subunit boundaries right to the haem groups [25]. The magnitude of these strains seems to be too small, however, to contribute significantly to the interaction energy between the haems.

In my proposed mechanism I shall assume that each subunit can take up only two alternative stable tertiary structures, the deoxy and the liganded form. But these structures may be subject to strains when subunits are confined in the 'wrong' quaternary structure.

Mechanism of the Conformational Change

I shall discuss first the conformation changes in the individual subunits that follow the binding of ligand; next, the change in quaternary structure caused by them; and finally, the probable order of the steps of the reaction with ligands.

What is the trigger that sets the conformational changes in motion? The distances between the haem groups are far too large (25–37 Å) for electromagnetic interactions to be effective, which suggests that the trigger is stereochemical. It was not clear, however, whether the changes are triggered by some deformation of the haem group or by the ligand prising the haem pocket apart. Consider first the α subunit. Because a ligand finds ample room in its haem pocket in both the oxy and the deoxy form, and the width of the pocket does not change significantly, the latter mechanism can be safely excluded. We have seen that the transition from deoxy to oxyhaemoglobin involves a movement of the proximal histidine towards the plane of the porphyrin amounting to 0·75–0·95 Å. The transition from deoxy to acid methaemoglobin involves a smaller movement, of 0·45–0·65 Å, in the same direction. Movements of this magnitude should be sufficient to set the changes in tertiary structure of the reacting subunit in motion, because the iron atom is rigidly linked to histidine F8 and the porphyrin is in contact with about sixty atoms of the globin which is flexible and can give. The way it gives can be seen in the difference electron density maps of BME deoxy versus methaemoglobin, which showed, among other things, that helix F moves towards the centre of the molecule and expels tyrosine HC2(140) from its pocket between helices F and H (fig. 3). The expelled tyrosine must pull arginine HC3(141) with it, thus breaking its salt-bridges with the opposite α subunit and releasing Bohr protons (fig. 6).

We may now turn to the β subunits. Here the trigger is clearly composite. Before the ligand can reach the iron atom, it must prise open the haem pocket (or rather, wait until thermal vibrations provide the activation energy needed to open it); reaction with the haem then moves the iron into the plane of the porphyrin ring. The accompanying changes in tertiary structure have one feature in common with those observed in the α-chain: helix F moves towards the centre of the molecule, narrowing the pocket between it and helix F and presumably expelling tyrosine HC2(145). The validity of this vital point is supported by much indirect evidence concerning the inhibitory effects of reagents which either remove that tyrosine or prevent its entry into the pocket, but it will have to be checked by direct experiment. The expelled tyrosine must pull histidine HC3(146) with it and break its salt-bridge to aspartate FG1(94) (fig. 5). The salt-bridge to lysine C5(40)α may or may not have been broken already before the β-chain could have reacted with a ligand.

We shall now examine the likely order of the steps involved in the reaction with ligands and the ensuing conformational changes (fig. 4). In

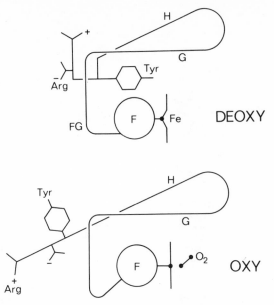

Fig. 3. Diagrammatic sketch showing the change in tertiary structure of the haemoglobin subunits on reaction with ligands. Movement of the iron atom into the plane of the porphyrin ring cuases a movement of helix F towards helix H, which expels tyrosine HC2 from its pocket between the two helices.

deoxyhaemoglobin, the α subunits have room for ligands, but the β subunits do not, so that the α subunits are likely to react first. The argument is not rigorous, for relative rates of combination with oxygen of the different subunits must depend on the activation energy needed to change their tertiary structure from the deoxy to the oxy state—which cannot be readily predicted. Low ligand affinity of the β subunits in the deoxy tetramer is not supported by kinetic data, but further work is needed to test this point. In any case the cooperative mechanism I propose would not be affected if the order of the steps turned out to be the reverse of the one given below or if it proved to be random.

Reaction of Feα_1 with oxygen causes tyrosine HC2(140)α_1 to be expelled from its pocket and the links between arginine HC3(141)α_1 and the α_2 subunit to be broken, with the release of Bohr protons. Feα_2 reacts next, followed by the expulsion of its tyrosine and the rupture of the links between the C-terminal arginine and the α_1 subunit, with further release of Bohr protons. By now four of the six salt-bridges constraining the deoxyhaemoglobin tetramer will have been broken, with a resulting change in the

262   PERUTZ

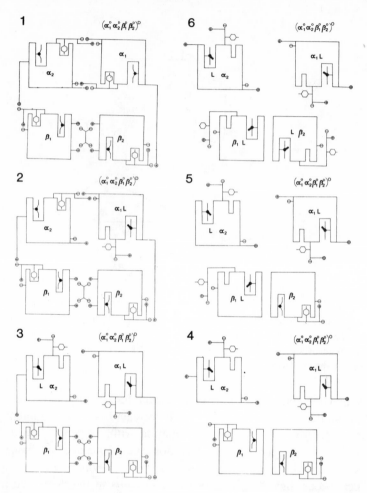

Fig. 4. Diagrammatic sketch showing one possible sequence of steps in the reaction of haemoglobin with oxygen. 1, Deoxyhaemoglobin with all salt-bridges intact and with one molecule of DPG clamped between the two β-chains. At steps 1–2 and 2–3, the α-chains are oxygenated. The tyrosine pockets are narrowed, the tyrosines expelled and the salt-bridges with the partner α-chains are broken. In step 3–4 the quaternary structure clicks to the oxy form, accompanied by the expulsion of DPG, and the breakage of the salt-bridges between subunits $\alpha_1-\beta_2$ and $\alpha_2-\beta_1$. Note that the internal salt-bridges of the β-chains are still intact and their haem pockets too narrow to admit ligands. At steps 4–5 and 5–6, the β subunits react in turn, accompanied by widening of the haem pockets, narrowing of the tyrosine pockets and rupture of the internal salt-bridges from the C-terminal histidines to aspartates 94. Note that the change of quaternary structure could in fact take place at any stage of the reaction, and the cooperative mechanism is independent of the sequence in which the individual subunits react.

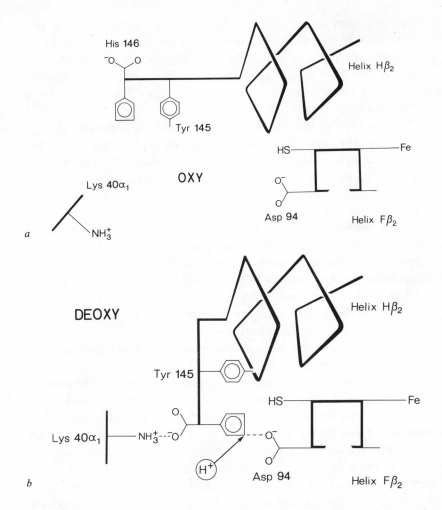

Fig. 5. Change in conformation of C-terminal residues of β-chain on deoxygenation: *a*) in oxyhaemoglobin histidines $146\beta_2$, aspartates $94\beta_2$ and lysine $40\alpha_1$ are free and far from each other. Access to the SH groups is free; *b*) in deoxyhaemoglobin, formation of a salt-bridge between aspartate 94 and the imidazole of histidine 146 leads to ionization of the latter. The salt-bridge is cemented by a further bridge between the C-terminal carboxyl of the β_2-chain and lysine $40\alpha_1$. Access to the SH groups is restricted.

allosteric equilibrium constant favour of the quaternary oxy structure. For illustrative purposes I shall assume that this is of a magnitude sufficient to change the quaternary structure to the oxy form. (In fact, the change might occur at any stage in the reaction depending on external factors such as pH and the concentrations of CO_2 and organic phosphates.) The $\alpha_1\beta_2$ and the $\alpha_2\beta_1$ contacts now give way and the tetramer clicks to the oxy form, breaking the remainder of the constraining salt-bridges, that is, the ones between lysines C5(40)α and histidine HC3(146)β, and the ones between DPG and the two β subunits. This step leads to the liberation of DPG but does not release Bohr protons.

We now have an intermediate with the quaternary structure of oxy-haemoglobin; the two-fold symmetry of the molecule is conserved; the haem and its immediate environment in the α subunit have the oxy con-formation, and those in the β-chains have the (strained) deoxy conforma-tion; that is, the tyrosines HC2(145)β are still in their pockets, keeping the ligand sites obstructed by valine E11(67). A structure of this type could be the intermediate sensed by McConnell's spin label. The change of quaternary structure has halved the activation energy needed to expel the tyrosines HC3(146)β from their pockets, because it has broken the salt-bridges between the C-terminal histidines and the α subunits, so that only the internal salt-bridges between the C-terminal histidines and aspartates of the same β-chain now remain to hold the tyrosines in place, with a resulting increase in oxygen affinity (fig. 5).

The iron atoms of the β-chains now react with ligand, each reaction being accompanied by the expulsion of a tyrosine and the rupture of one of the salt-bridges just mentioned, causing release of the last Bohr protons.

The essence of the mechanism may be summarized as follows. The haem group is so constructed that it amplifies a small change in atomic radius, undergone by the iron atom in the transition from the high spin to the low spin state, into a large movement of the haem-linked histidine relative to the porphyrin ring. This movement, and the widening of the haem pocket required in the β chain, triggers small changes in the tertiary structure of the reacting subunits, which include a movement of helix F towards helix H and the centre of the molecule. Exactly how leverage from the iron atoms pushes helix F in that direction is not yet clear, but this movement can be observed to narrow the pocket between helices F and H so that the penultimate tyrosine is expelled. The expelled tyrosine pulls the C-terminal residue with it, rupturing the salt-bridges that had held the reacting subunit to its neighbours in the deoxy tetramer. The rupture of each salt-bridge removes one of the constraints holding the molecule in the deoxy conformation and tips the equilibrium between the two alternative quaternary structures some way in favour of the oxy conformation. In this conformation the oxygen affinity is raised because the subunits are no longer constrained by the salt-bridges to maintain the tertiary deoxy structure. Section II shows that it also leads to the liberation of Bohr protons, in agreement with the finding that their release is exactly proportional to the amount of oxygen taken up, and that the two reactions are synchronous [26, 27]. I have suggested that the subunits react in the order α_1, α_2, β_1, β_2, but this is not certain and my cooperative mechanism would work equally well in whichever sequence the individual subunits reacted.

In haemoglobins where one pair of subunits, either α or β, is unreactive, the other, reactive, pair may or may not be able to actuate the change in quaternary structure: this will depend on the conformation of the unreactive pair and other factors. The literature contains instances where reaction with ligands of either the α or the β subunits alone apparently can or cannot change the quaternary structure of the tetramer. The reasons are not clear without x-ray analysis of each separate case.

Nature of Interaction Energy

Where in this large molecule should one look for that energy? Conceivably the reaction of one haem with oxygen might exercise a direct steric effect which raises the oxygen affinity of its neighbours. Nuclear magnetic resonance spectra offer an extremely sensitive probe for small changes in the environment of one haem caused by the binding of oxygen to another. Studies with mixed hybrids show no evidence that such changes occur unless there is a change in quaternary structure, which led SHULMAN and his colleagues to suggest that the energy of interaction arises at the boundaries between the subunits [25, 36].

BOLTON and PERUTZ found fewer van der Waals interactions at contacts between the α and β subunits in deoxy than in oxyhaemoglobin, and about the same number of hydrogen bonds. Consequently, neither of these is likely to provide the stabilizing interaction that constrains the subunits in the quaternary deoxy form. The most obvious constraints are the six salt-bridges made by the C-terminal residues which link the contacts $\alpha_1\beta_2$, $\alpha_1\beta_2$ and $\alpha_2\beta_1$. This is borne out by the discovery that enzymatic removal of the four C-terminal residues makes the quaternary deoxy structure unstable and inhibits haem–haem interaction. The bond energy per salt-bridge may reasonably be estimated as 1–2 kcal/M. For six salt-bridges this gives a total interaction energy of between 6 and 12 kcal/M of haemoglobin, that is, a value of the same order as the observed interaction energy of 12 kcal/M.

If it is only the salt-bridges which constrain the deoxy tetramer, then part of the energy released by the reaction of the haem groups with oxygen must be expended to break them. The change in conformation of the haem group that leads to the expulsion of the tyrosine provides the required stereochemical mechanism.

Do the changes in the free energy of the subunit contacts in going from the deoxy to the oxy tetramer account for all the interaction energy or for only part of it? This question is important for an understanding of the mechanism, because the answer should tell us whether direct interaction occurs between the β-chains, say, if they react with oxygen after the tetramer has clicked to the oxy conformation. We have seen that haem–haem interaction is absent when the transition to the deoxy structure is inhibited as in BME haemoglobin and haemoglobin from which the C-terminal residues have been removed. Another system is haemoglobin M Iwate (His F8(87)$\alpha \rightarrow$Tyr) [28] where the α-chains are unreactive and the tetramer is locked in the deoxy conformation [29]. The β-chains have a low affinity for oxygen, and their reaction with oxygen produces no haem–haem interaction or Bohr effect, showing again that these effects are absent without a change of quaternary structure. The biochemical data on other systems offer no firm evidence that haem–haem interaction ever occurs without a change of quaternary structure, which suggests that the changes in the free energy of the subunit contacts do seem to account for all the energy of interaction. On the other hand, the abnormal haemoglobins include unexplained instances of very weak haem–haem interaction where the change of quaternary structure does take place.

Dimers or Tetramers as the Functional Units

It has been suggested that $\alpha\beta$ dimers exhibit all the cooperative effects characteristic of the haemoglobin tetramer [32]. These dimers have the structure $\alpha_1\beta_1$, so that the contact $\alpha_1\beta_2$ remains open. There is no reason why an oxyhaemoglobin dimer should not function as well as a tetramer, for the terminal residues are free, and the subunits are not constrained by the $\alpha_1\beta_2$ contact. On the other hand, the deoxy form can exhibit a low oxygen affinity only in the tetrameric form, for dissociation into dimers of whatever type would remove the constraining salt-bridges on which the energy of interaction depends. This conclusion may seem to contradict the finding that haemoglobin solutions show full cooperativity in conditions where the haemoglobin is apparently split into $\alpha\beta$ dimers. KELLETT, MIDGARDEN and SCHACHMAN, however, have recently discovered that only the liganded form of haemoglobin is split, whereas the deoxy form remains tetrameric in all conditions formerly believed to produce dimers [see ref. 35]. They estimate the dissociation constants of the two forms to differ by a factor of at least 10^4. For comparison, the stabilizing free energy of either 6 or 12 kcal for the extra six salt-bridges in the deoxy form would lower the dissociation constant by a factor of 5×10^5 and 2×10^9 respectively— sufficient to make dissociation of the deoxy form unobservable. This suggests that all cooperative effects observed on binding of haem ligands take place in the tetrameric molecule while it has the quaternary deoxy structure; splitting into dimers may occur at very low haemoglobin concentrations or at high concentrations of neutral electrolytes, but only after the salt-bridges have been broken, when normally the quaternary structure would click to the oxy conformation.

Implications for Allostery

For purposes of comparison haemoglobin may be regarded as an enzyme made up of two different pairs of subunits with oxygen as the substrate, haem as the coenzyme, and H^+ and DPG as the allosteric effectors. Let α and β denote the subunits in the free conformation and α^S and β^S their conformation when substrate is bound. [ ]^T represents the unreactive and [ ]^R the reactive quaternary structure. Then for illustrative purposes, the reaction scheme of fig. 4 can be rewritten as

$$\left[\alpha_2\beta_2\right]^T \rightleftharpoons \left[\alpha_2^s\beta_2\right]^T \rightleftharpoons \left[\alpha_2^s\beta_2\right]^R \rightleftharpoons \left[\alpha_2^s\beta_2^s\right]^R$$

$$\underset{d+2s}{\quad} \underset{+2s}{\quad} \underset{+2S}{\quad} \underset{+2S}{\quad}$$

The mechanism of haem–haem interaction contains features of several of the theories that have been advanced to account for interaction between ligand sites in enzymes. MONOD's model postulates 'the existence of only two (or at least two) stable states that are accessible to allosteric oligomers. These states differ by the distribution and/or energy of intersubunit bonds, and therefore also by the conformational constraints imposed on the subunits. As a result, the affinity of one (or several) of the stereospecific sites is altered when a transition occurs to the other state' [9]. These words ring prophetically if we look at the mechanism in terms of the quaternary structure.

268

PERUTZ

MONOD *et al.* assumed all the subunits in the quaternary T-state to be in the unreactive and those in the quaternary R-state in the reactive form, no matter whether they were liganded or not. In their discussion they envisaged the possibility that this assumption might prove too simple and this is indeed the case. The haemoglobin subunits change their tertiary structure in response, not to the change in quaternary structure, but to the binding of ligand, as predicated by KOSHLAND's sequential model [34]. Yet KOSHLAND's model implies that the change in tertiary structure of each subunit directly affects the ligand affinity of its neighbours, and there is no evidence of this happening in haemoglobin. Rather, all—or nearly all—the interaction energy arises through the step by step release of the constraints on the unreactive quaternary structure, which changes the equilibrium in favour of the reactive quaternary structure and diminishes the work required to change the tertiary structure of each subunit from the unreactive to the reactive form (fig. 10).

In KOSHLAND's terms [35], the change in tertiary structure of the α subunits would be caused by induced fit in response to a change in conformation of the coenzyme on reacting with substrate; that of the β subunits in response to both the change of conformation of the coenzyme and the steric adaptation of the active site to the substrate.

How would allosteric effectors influence the reactivity? In haemoglobin both hydrogen ions and DPG lower the reactivity; the former strengthen the salt-bridges constraining the inactive or T state by increasing the fraction of molecules carrying positive charges on one α amino and two imidazole groups (see section II); DPG may increase the free energy of interaction by introducing additional salt-bridges specific for the T-state, to which it is stereochemically complementary but has either no affinity or a lower one for the R-state.

The cooperative mechanism outlined here explains most of the important properties of haemoglobin, even though stereochemical details are still lacking. It is of a kind that could not have been guessed without knowing the structure of the two forms in atomic detail and without further information which the inhibited BME-haemoglobin provided. Even then it could be worked out only by combining the structural information with many other results.

It is remarkable that there should be such an exceedingly complex, subtle and elegant instrument of respiratory transport, exploiting a difference in atomic radius of 13% between the covalent and ionic forms of iron.

II. The Bohr Effect and Combination with Organic Phosphates

Like haem–haem interaction, the Bohr effect results from the change in structure which the haemoglobin molecule undergoes on combination with oxygen (see preceding section). The stereochemical consequences of this change have been studied in atomic models derived from electron density maps of haemoglobin at high resolution [1–3]. The models show that the change in structure alters the environment of three pairs of weak bases so that they would tend to take up hydrogen ions on discharge of oxygen. The α-amino groups of valine 1α and the imidazole groups of histidines 146β are free in oxyhaemoglobin, but come close to carboxyl groups in the deoxy form [37, 38]; the imidazole groups of histidines 122α approach guanidinium groups in the oxy and carboxyl groups in the deoxy form. A survey of all other ionizable groups in the structure of oxy and deoxy-haemoglobins suggests that none of them are likely to change their state of ionization significantly on binding of oxygen near neutral pH.

Early Ideas concerning the Bohr Effect

The Bohr effect is produced on combination of O_2, CO or NO with the ferrous form of haemoglobin (deoxyhaemoglobin) or on oxidation of deoxyhaemoglobin to the ferric form (methaemoglobin) but not on combination of methaemoglobin with ligands. In other words, it is observed only in those reactions involving a change in quaternary structure from the deoxy form, in which the iron atoms are five coordinated, to the oxy or liganded form in which they are six coordinated. It is observable either as a change in pH on binding of haem ligands or as a change in affinity of the haem iron to ligands as a function of pH. At a pH above $6 \cdot 0$ haemoglobin takes up protons on release of haem ligands; this is known as the alkaline Bohr effect and reaches its maximum of $0 \cdot 7$ protons per haem ligands at physiological pH.

The alkaline Bohr effect has a two-fold function. On liberation of oxygen haemoglobin neutralizes the protons released in the blood by the uptake of CO_2, thereby facilitating its transport from the tissues to the lungs. Acid pH lowers the affinity of haemoblobin for haem ligands, which helps the release of oxygen in tissues rendered acid by bicarbonate and lactic acid. Below pH $6 \cdot 0$ the Bohr effect is reversed, protons being liberated on release of haem ligands. Because pHs below 6 are not likely to occur *in vivo*, however, this reverse, or acid, Bohr effect may have no physiological significance.

WYMAN showed that the Bohr effect could be attributed to ionizable amino-acid side chains with a pK close to $7 \cdot 0$ which might either release or take up H$^+$ as a result of oxygenation of the iron atoms [39]. He determined the heat of dissociation of the Bohr protons as 6,500 cal per equivalent, and showed that this is characteristic for histidine. Further analysis based on the variation of proton released or taken up as a function of pH on oxygenation led him to suggest that each haem interacts with two histidines, one on each side of the haem, a scheme that had already been proposed by CONANT [40]. To the histidine responsible for the alkaline Bohr effect he ascribed a pK of $7 \cdot 8$ in deoxy changing to $6 \cdot 2$ in oxyhaemoglobin; to that responsible for the acid Bohr effect a pK of $4 \cdot 9$ in deoxy and $5 \cdot 7$ in oxyhaemoglobin. CORYELL and PAULING arrived at the same conclusion on the basis of magnetic susceptibility measurements and introduced the term 'haem-linked acid groups' [41]. The environment of the haems first predicted by CONANT was verified when the structure of myoglobin was solved, but the mystery of the Bohr effect was deepened because the tertiary structures of myoglobin and of each of the haemoglobin chains proved closely similar, yet myoglobin was known to show no Bohr effect [42]. This meant that the haem-linked histidines could not be responsible for it and that the term 'haem-linked acid groups' was a misnomer. When the uptake of haem ligands was found to be accompanied by a marked change in quaternary structure [43] is seemed possible that this might alter the environment of ionizable groups other than the haem-linked ones in the required way. We were able to identify two of the groups concerned [37, 38], but it took electron density maps of both the oxy and deoxy forms at high resolution before we could see all the groups and determine the exact stereochemical mechanism which changes their state of ionization in the two forms of haemoglobin.

C-terminal Residues

In the liganded form of haemoglobin the C-terminal arginines of the α-chains and histidines of the β-chains rotate freely. The penultimate tyrosines seem to rotate freely for most of the time; they spend the remainder in a slot between helices F and H. Several pieces of evidence support this picture. The C-terminal residues are invisible in the electron density maps of oxyhaemoglobins; only traces of the penultimate tyrosines are to be seen. Because model building showed various ways in which the C-terminal groups could be held down by salt-bridges [1], I thought at first that this freedom might be an artefact, arising from the weakening of salt-bridges by the 2 M ammonium sulphate solution in which the crystals were suspended.

Fortunately, crystals of methaemoglobin can be prepared free of salt; these are isomorphous with those of oxyhaemoglobin precipitated by salt. X-ray reflexions of the salt-free crystals were collected on a diffractometer to a resolution of $3 \cdot 5$ Å and a difference Fourier synthesis of salt-free minus salty haemoglobin was calculated. If the C-terminal residues had been held down in fixed positions in salt-free crystals they should have shown up there as positive peaks, but there was no sign of these. In fact, the difference map was featureless, except for a negative peak on the dyad axis, possibly representing a negative ion trapped in the internal cavity. I had to conclude therefore that the C-termini were equally free in the absence of salt.

Other evidence supporting freedom of rotation of the penultimate tyrosines of the β-chains comes from work on a haemoglobin carrying spin labels at the reactive SH-groups 93β [44]. The spin label oscillates rapidly between two alternative conformations, in one of which it is free and in the other immobilized. X-ray analysis showed that in the immobilized conformation the spin label takes the place of the penultimate tyrosine in the pocket between helices F and H, implying that the tyrosine itself oscillates rapidly between a free and an immobilized position and that these oscillations are synchronized with those of the spin label [22].

When we proposed that the contribution of the β-chains to the alkaline Bohr effect arose through the freedom of the C-terminal histidines in oxy and the linkage of their imidazoles to aspartates 94 of the same β-chains in deoxyhaemoglobin [37], we were unable to explain why this link was maintained only in the deoxy form, even though no steric obstacles appeared to stand in its way in the oxy form. The expulsion of the penultimate tyrosines already described and the existence of an unsuspected salt-bridge of lysine C5(40)α with histidine HC3(146)β now provide the explanation. Fig. 5a shows the freedom of the C-terminal residues in oxyhaemoglobin.

Consider now the transition to the deoxy structure. In oxyhaemoglobin lysine C5(40)α_1 thrusts its side chain far out into the solution. During deoxygenation the α_1-chain turns relative to the β_2-chain by $13 \cdot 5°$. The ε-amino group of lysine C5(40)α_1, being furthest removed from the axis of rotation of the α_1-chain, is shifted by 7 A into the $\alpha_1\beta_2$ contact, where it can form a salt-bridge with the C-terminal carboxyl group of the β_2-chain. This salt-bridge keeps tyrosine 145β_2 fixed in its pocket between helices F and Hβ_2 and allows the further salt-bridge between the imidazole of histidine 146β_2 and aspartate 94β_2 to be formed (fig. 5b). The electron density maps of deoxyhaemoglobin at high resolution show the interatomic distances between the acidic and basic groups to have the correct length (~ 3 Å) for the salt-bridges [2, 3].

These findings suggest that histidine HC3(146)β contributes to the Bohr effect by a change in pK of its imidazole, arising through the shift from its free position in oxyhaemoglobin to its fixed position in deoxyhaemoglobin. In oxyhaemoglobin the pK of the imidazole must be slightly raised by the negative field of the histidine's own free carboxyl group. In deoxyhaemoglobin the field of the C-terminal carboxyl group is compensated by lysine 40α, but the imidazole becomes linked to aspartate 94β. This apparently produces a much larger increase in pK than the more distant interaction with its own carboxyl group in oxyhaemoglobin. We do not know of any comparable model situations which would allow us actually to predict the changes in pK, but it should soon be possible to titrate the histidines directly by NMR spectroscopy. The recently observed absence of half the alkaline Bohr effect in des-histidine (146)-haemoglobin confirms our scheme [45].

The first hint of the involvement of the C-terminal arginines of the α-chains was the observation that enzymatic removal diminished the alkaline Bohr effect [46], which has recently been confirmed [45]. This leaves a Hill's coefficient of $1 \cdot 8$, indicating that the change of quaternary structure still accompanies the reaction with oxygen [45]. Because the pKs of guanidinium groups and α-carboxyl groups are $12 \cdot 5$ and $3 \cdot 4$ respectively, neither group could itself lose or gain protons at neutral pH, which implies that their contribution must arise from interaction with other ionizable groups. The next advance came with the discovery that blocking of the α-amino groups of valine 1α with cyanate inhibited about one-quarter of the Bohr effect [38].

The electron density maps of oxyhaemoglobin at $2 \cdot 8$ A resolution showed the valines 1α to be free, and the maps of deoxyhaemoglobin at $5 \cdot 5$ Å resolution showed bridges connecting the C-terminus of one α-chain with the N-terminus of its partner chain. These observations led us to propose that the pKs of the α-amino groups are normal in oxyhaemoglobin, but are raised in the deoxy form through linkage of the α-amino groups of one chain to the α-carboxyl groups of their partner chains. At that stage

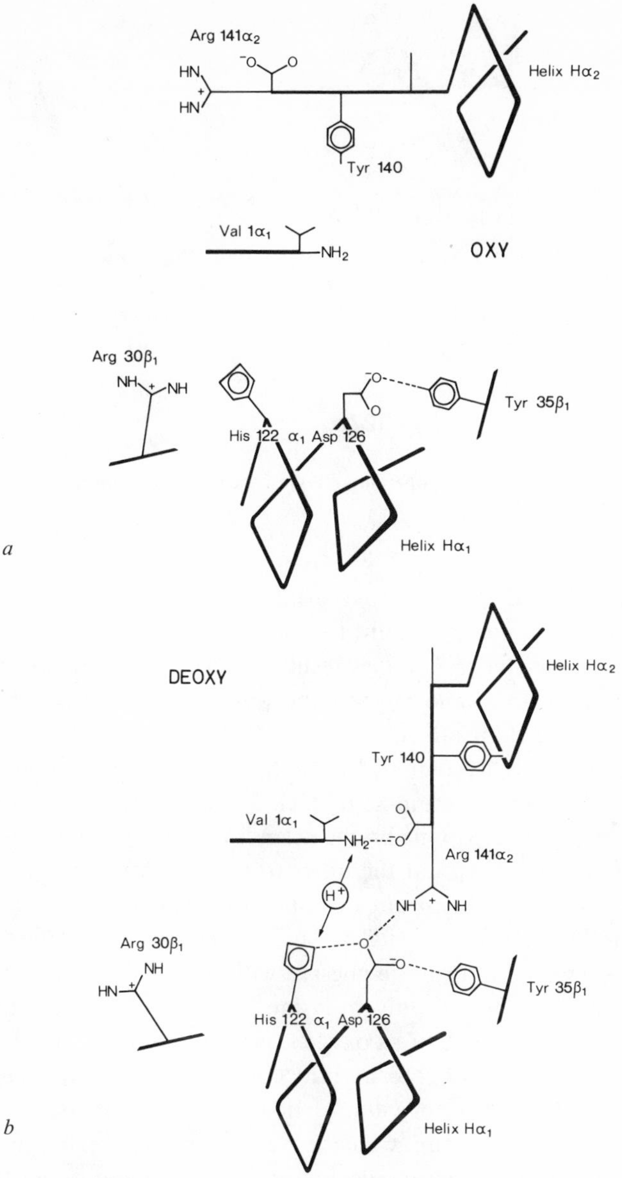

Fig. 6 Change in conformation of terminal regions of the α-chain on deoxygenation: *a*) in oxyhaemoglobin valine $1\alpha_1$ and arginine $141\alpha_2$ are free, histidine $122\alpha_1$ is close to arginine $30\beta_1$, and aspartate $122\alpha_1$ is hydrogen bonded only to tyrosine $35\beta_1$; *b*) in deoxyhaemoglobin valine $1\alpha_1$ is in contact with the carboxyl group of arginine $141\alpha_2$. Aspartate $126\alpha_1$ forms close contacts with histidine $122\alpha_1$, tyrosine $35\beta_1$ and arginine $30\alpha_2$.

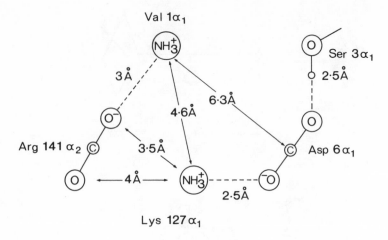

Fig. 7. Distances between members of the cluster of ionizable groups near valine 1α in deoxyhaemoglobin.

our hypothesis could not be proved, however, and in any case it was uncertain whether it could account for the entire share contributed to the Bohr effect by the α-chains. Our new results confirm the interaction between these terminal residues, and suggest the possibility of a further residue contributing to the Bohr effect.

The electron density maps of deoxyhaemoglobin at high resolution show that the C-terminal arginine of each α-chain forms two salt-bridges with its partner chain [2, 3]. One bridge extends from the guanidinium group α_2 to aspartate H9(126)α_1 and the other from the α-carboxyl group α_2 to a cluster of three ionizable groups composed of valine NA1(1), aspartate A4(6) and lysine H10(127)α_1 (fig. 6, 7). The precise constellation of this cluster is not resolved but can be guessed with the help of stereo-chemical rules. Aspartate 6 is bound only to lysine 127. The α-carboxyl group of arginine 141 must have one of its oxygens pointing towards valine 1 rather than towards lysine 127, for otherwise the O–N distance would be too short. The electron density peak for valine 1 does not resolve its isopropyl side chain from its α-amino group, which leaves one free to rotate this residue as desired. On electrostatic grounds, however, the α-amino rather than the isopropyl group would tend to face the α-carboxyl group, resulting in the arrangement of figure 7. The proximity of the positively charged lysine 127 would weaken the effect of the negatively charged carboxyl on the pK of the α-amino group of valine 1, even though the lysine's charge is compen-

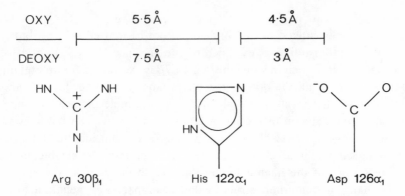

Fig. 8. Distances between histidine 122α and its neighbours in oxy and deoxyhaemo-globin.

sated by that of aspartate 6. Perhaps for this reason, the ionization of the α-amino groups may be only partial, so that it may not account for the α-chain's entire contribution to the Bohr effect.

We come now to a second possible contribution from the α-chains. Formation of links extending from the guanidinium group of the arginine HC3(141)α_2 to aspartate H9(126)α_1 causes a rearrangement in a cluster of polar groups which includes histidines H5(122)α_1. In oxyhaemoglobin the centre of the imidazole of histidine H5(122)α_1 lies within $5 \cdot 5$ Å of the guanidinium group of arginine B12(30)β_1 and is separated by $4 \cdot 5$ Å from the centre of the carboxyl group of aspartate H9(126)α_1. In deoxyhaemo-globin these distances are altered to $7 \cdot 5$ A and 3 A respectively (fig. 8). This shift would suggest that the pK of histidine 122 is lowered by the proximity of a positive charge in oxy and raised by the proximity of a negative one in deoxyhaemoglobin. On oxygenation of Feα, the expulsion of tyrosine HC2(140) would rupture the links of the C-terminal arginines with valine 1 and aspartate H9. Clearly, this would liberate the Bohr protons bound to the α-amino groups of valine 1, but it is not yet clear why this would also affect the conformation of histidine H5 relative to its two neighbours. Their movements seem to depend on the change in quaternary structure.

It may be hard to test the part played by histidine 122α chemically unless a mutant can be found in which it is replaced by a neutral residue. Similarly, the relative contributions of valine 1α and histidine 122α may prove difficult to assess exactly unless at least one of these groups can be titrated directly.

There are no changes in the conformation of the haem-linked histidines which would lead one to expect a change in their pKs. The proximal histidines F9 are linked to iron and would not be ionized; a contribution from the distal histidines is excluded by the presence of a full Bohr effect in haemoglobin Zürich in which histidine E7β is exchanged for an arginine [47]. A survey of all the other ionizable groups in oxy and deoxyhaemoglobin shows that most of the salt-bridges in the molecule are left undisturbed by the reaction of the haem group with ligands; the few that are disturbed involve ionizable groups with pKs far outside the range of physiological pH. Small changes in the general pattern of ionizable groups on the surface of the haemoglobin molecule do occur as a consequence of the change in quaternary structure that accompanies oxygenation. Some attempts are being made to use these as the basis for the interpretation of the Bohr effect, but theoretical reasons suggest that their contribution is small and experimental evidence confirms this [45]. We therefore conclude that no groups other than the ones described here contribute significantly to the alkaline Bohr effect.

The Acid Bohr Effect

The acid Bohr effect is observed below at pH 6·0 and requires ionizable groups that change their pK from about 5·7 to 4·9 in going from oxy to deoxyhaemoglobin [39]. I could not find any groups that change their environment in the way required, however, and this made me wonder if there was not a different explanation. Horse oxyhaemoglobin crystals undergo a drastic lattice change at pH 5·9, the monoclinic angle decreasing from 111° to 98° [48]. Patterson projections indicated at first that this lattice transition left the internal structure of the haemoglobin molecules unaltered [49], but isomorphous replacement with paramercuribenzoate showed that this was not true: a difference Patterson proved that the lattice transition was accompanied by a reduction of the distance between the reactive cysteines from 30 Å to 25 Å [50]. (On deoxygenation this distance increases from 30 Å to 37 Å; that is, the changes are in the opposite direction.) Such a reduction cannot occur without a change in quaternary structure which would affect the oxy and deoxy forms differently. I suggest that this is the cause of the acid Bohr effect: its details remain to be investigated. Incidentally, there is a further lattice change below pH 5·4, with a reduction of the angle from 98° to 85°, which may be the expression of a further change in quaternary structure [48].

SH Groups, the Spin Label, Invariant Residues

The reactivity of the SH groups at cysteine 93 increases in going from the deoxy to the oxy form and is linearly proportional to the degree of oxygenation [51]. This is readily accounted for by the changed conformation of the C-terminal histidines. In the oxy form they leave the SH groups accessible, while in the deoxy form they restrict access to them (fig. 6). Oxygenation of each β-chain is accompanied by expulsion of tyrosine HC2 and the rupture of the restricting salt-bridges, thus accounting for the linear relationship between the two phenomena.

A similar mechanism accounts for the changed immobilization of the spin label attached to cysteine 93β. In deoxyhaemoglobin the tyrosine pocket is occupied, so that the spin label is free. In oxyhaemoglobin the tyrosine pocket is open, for part of the time, so that the spin label can be immobilized there. The transition from the free to the imunobilized conformation is therefore proportional to the number of β chains which have taken up oxygen, as has been observed [52].

As far as can be juged from amino-acid sequences so far determined, all the residues involved in our scheme for the alkaline Bohr effect are present in all mammalian haemoglobins; they are also present in the α-chain of chicken, the only bird haemoglobin sequenced so far [53]. In the α-chains of carp the N-terminal serine is acetylated and could not contribute to the Bohr effect [54]; but histidine 122, aspartate 126 and arginine 141 are all present.

Chemical evidence bearing on our scheme will be described by KILMARTIN and WOOTTON [45]; the temporal sequence in which the different residues liberate their hydrogen ions during the reaction with oxygen was discussed in section I where it was shown that the combination of each atom of iron with ligand leads to the expulsion of the penultimate tyrosine from its pocket, and that this, in turn, breaks the salt-bridge formed by the C-terminal residue, with the liberation of a Bohr proton. This makes the liberation of Bohr protons synchronous with the uptake of haem ligands.

Model for the Binding of Organic Phosphates

2,3-Diphosphoglycerate (DPG) is present in the red cells of humans and many other species, and lowers the oxygen affinity of haemoglobin in a physiologically advantageous way [55]. It increases the free energy of

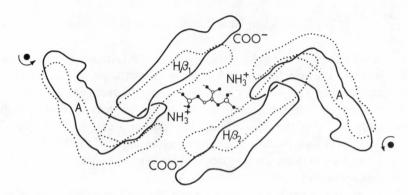

Fig. 9. Projection of the electron density maps of oxy (. . .) and deoxyhaemoglobin
(————) at 5·5 Å resolution, showing the N and C-terminal portions of the β-chains grouped
around the two-fold symmetry axis which is normal to the plane of the paper. In relation
to figure 2 of section I, the molecule is viewed from the top and the parts seen here are
near the bottom. The DPG molecule has been tentatively placed in the central cavity, centred
approximately on the two-fold axis (in fact the carboxyl group makes it unsymmetrical). The
picture shows that its phosphates are close to the N-termini of the β-chains in deoxyhaemoglobin. On oxygenation the N-termini move apart and the H helices close up, thus expelling
the DPG. The circular arrows indicate the positions of the axes around which the β subunits
rotate on deoxygenation.

haem–haem interaction but does not alter the Bohr effect. According to
several investigators it binds specifically to deoxyhaemoglobin in the ratio
of 1 M/M with a binding constant at neutral pH and physiological salt
concentration of $1·3 \times 10^5$, corresponding to a free energy of binding of
7 kcal/M. It also binds to free β-chains (β_4) in both oxy and deoxy states,
suggesting that the binding site is on the β-chains. In certain conditions
the binding is less specific and not stoichiometric, so that DPG binds to
both the oxy and the deoxy forms, but the deoxy form has the higher affinity
[56].

The affinity of DPG for haemoglobin is lowered by salt, showing that
the binding is electrostatic, and is totally inhibited at the salt concentrations
used to precipitate our crystals of deoxyhaemoglobin. For this reason we
have not yet tried to determine its binding site by x-ray analysis, but
fortunately BUNN and BRIEHL have obtained biochemical data which go
far towards pinpointing that site [57]. They measured the effect of DPG
on the oxygen affinities of normal human haemoglobin A and various other

haemoglobins, including a minor human adult component (A_{IC}) in which the N-termini of the β-chains are covalently bound to a hexose by Schiff base linkage; foetal human haemoglobin (F_{II}); and a minor component of human cord blood (F_I) in which the N-termini of the β-chains are acetylated. DPG increased the partial pressures of oxygen needed to produce half saturation with oxygen by a factor of $2 \cdot 1$ in haemoglobin A, $1 \cdot 2$ in haemoglobin A_{IC} and F_{II}, and unity in F_I, showing that the α-amino groups of the β-chains are involved in the binding of DPG. There is, however, clearly another nearby group which is responsible for the difference between haemoglobin A and haemoglobin F_{II}. A survey of the haemoglobin model and of the differences between the amino-acid sequences of the human β and γ-chains showed that of the residues which differ, 143(H21)β, which is histidine in adult and alanine in foetal haemoglobin, is the only likely one in the neighbourhood of the N-terminal amino groups.

When the atomic model of deoxyhaemoglobin was completed, Dr. W. Bolton and I placed a modell of DPG on the two-fold symmetry axis in the central cavity to see if binding to these residues would be likely on stereochemical grounds. We found that it fitted neatly so that its charged groups were within hydrogen-binding distance of the following pairs of residues of the β-chains: valine NA1(1), lysine EF6(82) and histidine H21(143). The position is shown in figure 9.

What happens on oxygenation? The distance between the α-amino groups increases from 16 A to 20 A, so that they can no longer make contact with the phosphates of the DPG, and the helices H close up so that DPG is expelled from the central cavity. In oxyhaemoglobin this particular binding site is closed, therefore, but at this stage we cannot be sure that oxyhaemoglobin does not contain an alternative site, with a weaker binding constant, to which DPG can become attached.

Sheep haemoglobin has no affinity for DPG. The explanation is this: in its β-chain residues NA2 are deleted, which would make the non-helical segment at the amino end more than 3 Å shorter, and increase the distance between the two α-amino groups to about 22 Å, so that they could not reach the two phosphate groups of DPG.

DPG is thus seen to be stereochemically complementary to a constellation of six positively charged groups in the β-chains which face the central cavity of the haemoglobin molecule. DPG has two negatively charged groups with an average pK_α of $2 \cdot 8$ and three with an average pK_α of $7 \cdot 1$, whereas the three pairs of positively charged protein groups facing them

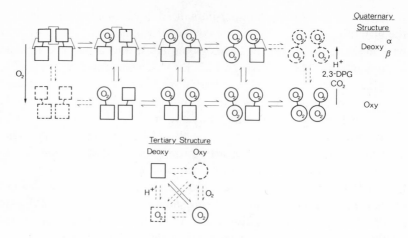

Fig. 10. Diagrammatic sketch of the allosteric mechanism of haemoglobin. The smaller subunits represent α, the larger β subunits. The clamps between them represent salt-bridges. Forms that may be too unstable to be observed have been drawn in broken lines. Note that the sequence of reaction of the individual subunits is unimportant.

have pK_{a}s of $10\cdot5$, $6\cdot8$ and $7\cdot6$ respectively. Thus the charges on the DPG could be just compensated by those of the protein. This also explains why DPG does not alter the Bohr effect. On dissociation, the number of protons bound by DPG would roughly equal the number lost by the two α-amino groups and two histidines. An explanation of this general nature for the absence of a Bohr effect has been advanced [58].

ATP, ADP and AMP have binding constants with haemoglobin lower than DPG by one, two and three orders of magnitude respectively. The steric reasons for this lie in the progressive shortening of the distances between the terminal phosphate groups which weakens their interaction between the two rigidly spaced α-amino groups of the β-chains. (For fully stretched chains the distances between the terminal phosphates are $6\cdot5$ Å in 2,3-DPG, $5\cdot5$ Å in ATP and $2\cdot8$ Å in ADP.)

Pyridoxal phosphate binds to deoxyhaemoglobin in the ratio of 1 M/M by forming a Schiff base linkage with the α-amino group of one of the β-chains [59]. Its physiological effect is similar to that of 2,3-DPG, of which it is a competitive inhibitor. Model building shows that its phosphate group would lie near the two-fold symmetry axis of the molecule, within hydrogen bonding distance of the ε-amino groups of the two lysines EF6(82)β. The proximity of the phosphate group to the two-fold axis would inhibit the binding of a second molecule of pyridoxal phosphate. The binding site coincides with that of 2,3-DPG.

Discussion

The influence of DPG may consist simply in the stabilization of the quaternary deoxy structure where it may form the equivalent of four additional salt-bridges crosslinking the β subunits, but it probably has no direct influence on the tertiary structure of the individual subunits. Hydrogen ions, on the other hand, influence the tertiary structure of the β-chains and the quaternary structure directly by stabilizing the deoxy conformation of each. This is clearly seen by reference to figures 4–6. By increasing the fraction of the α-amino groups of valine 1α and imidazoles of histidine 122α that carry positive charges, H^+ directly opposes the rupture of the salt-bridges between the α-chains; the activation energy needed to expel the tyrosines from the pockets is thereby increased and, consequently, the movement of the iron atom towards the plane of the porphyrin ring which is needed for their reaction with oxygen is opposed. Similarly, in the β-chains, H^+ increases the fraction of the C-terminal histidines which carry positive charges, thereby strengthening the salt-bridges with aspartate 94 and opposing the expulsion of the tyrosines even if the quaternary structure had already clicked to the oxy conformation. The only salt-bridges not affected by H^+ near neutral pH are those between the C-terminal carboxyl groups of the β-chains and lysine 40α and those involving DPG.

This study has perhaps conveyed a too mechanistic picture of the cooperative effects involved in respiratory transport. In fact, we should imagine a dynamic system, in which both the tertiary structure of each subunit, and the quaternary structure of the haemoglobin molecule as a whole, oscillate rapidly and continuously between the oxy and the deoxy conformations. The concentrations of the different ligands alter the equilibrium between these two conformations of the various components rather than switching any of them completely to one or the other conformation (fig. 10). This also means that the transition between the two quaternary structures could occur at any stage in the reaction, depending on the concentration of H^+, CO_2 and DPG, and on other factors, and that the mechanism would work equally well regardless of the sequence of reaction of the subunits.

Note added in proof

The preceding article is based on my proposals for the stereochemistry of the cooperative effects first published in November 1970(69). Since then further research has confirmed the general features of the mechanism I proposed there, but some of its details had to be modified or are still uncertain. The parts that are firmly established are the triggering of the change of quaternary structure, mainly by the change in the radius of the iron atom that occurs on lingand binding, and in the β-chains also by the stereochemical effect of the ligand itself (62); the parts played in the Bohr effect by the α-amino groups of Val NA1(1)α and the imidazole of His HC3(146)β; the binding site of DPG in deoxyhb (65); the changes in hydrogen-bonding at the $\alpha_1\beta_2$ interface which accompany the change in quaternary structure; the higher oxygen affinity of the α-chains in the quaternary deoxy structure, but only when this is stabilized by organic phosphates (66); the non-cooperativity of the $\alpha\beta$ dimer (67).

While it is certain that the closure of the terminal salt-bridges in deoxy- and their opening in oxyhaemoglobin in responsible for the Bohr effect, it is not certain whether the rupture of the salt-bridges accompanies the changes in tertiary structure which occur on ligand binding or the changes in quaternary structure. Also we do not know whether in oxyhaemoglobin the penultimate tyrosines are really rotating freely, or whether they merely lie more loosely in their pockets. The role of His H5(122)α in the Bohr effect is still uncertain. The terminal residues of the β-chains in oxyhaemoglobin are not free, but the α-amino group of each β-chain forms a salt-bridge with the α-carboxyl group of the other β-chain (68), as I had originally proposed (4). However the α-amino group of the β-chain need not change its state of ionization on loss of oxygen, because in deoxyhaemoglobin it forms a salt-bridge with an anion which is also coordinated to the ε-amino group of Lys EF6(82)β (65).

References

1 Perutz, M. F.; Muirhead, H.; Cox, J. M. and Goaman, L. C. G.: Nature, Lond. *219:* 131 (1968).
2 Muirhead, H. and Greer, J.: Nature, Lond. *228:* 516 (1970).
3 Bolton, W. and Perutz, M. F.: Nature, Lond. *228:* 551 (1970).
4 Perutz, M. F.: Proc. roy. Soc., B *173:* 113 (1969).
5 Perutz, M. F.: The Harvey Lectures *213:* (1967 – 68).
6 The Gospel according to Mark, chap. 4, v. 25.
7 Adair, G. S. and Adair, M. E.: Proc. roy. Soc., B *120:* 422 (1936).
8 Gibson, Q. H.: J. biol. Chem. *245:* 3285 (1970).
9 Monod, J.; Wyman, J. and Changeux, J. P.: J. molec. Biol. *12:* 88 (1965).
10 Wyman, J.: Adv. Protein Chem. *4:* 407 (1948).
11 Hoard, J. L.: in Rich, A. and Davidson, N., Structural chemistry and molecular biology (Freeman, San Francisco 1968).
12 Countryman, R.: Collins, D. M. and Hoard, J. L.: J. Amer. chem. Soc., *91:* 5166 (1969).
13 Pauling, L.: Nature of the chemical bond, 3rd ed., p. 518 (Cornell Univ. Press, Ithaca 1960).

14 KENDREW, J. C.: Science *139:* 1259 (1963).
15 HUBER, R.; EPP, O. and FORMANEK, H.: J. molec. Biol. *52:* 349 (1970).
16 LOVE, W. E.: Abst. Int. 8th Congr. Biochem., 2 (1970).
17 RICHARDS, F. M.: J. molec. Biol. *37:* 225 (1968).
18 KOENIG, D. F.: Acta Cryst. *18:* 663 (1966).
19 KILMARTIN, J. V. and WOOTTON, J.: Nature, Lond. *228:* 766 (1970).
20 SIMON, S. R.; KONIGSBERG, W. H.; BOLTON, W. and PERUTZ, M. F.: J. molec. Biol. *28:* 451 (1967).
21 SIMON, S. R.; ARNDT, D. J. and KONIGSBERG, W. H.: J. molec. Biol. (in press).
22 MOFFAT, J. K.: J. molec. Biol. (in press).
23 BOLTON, W.; COX, J. M. and PERUTZ, M. F.: J. molec. Biol. *33:* 283 (1968).
24 OGAWA, S.; MCCONNELL, H. M. and HORWITZ, A.: Proc. U.S. nat. Acad. Sci., Wash. *61:* 401 (1968).
25 OGAWA, S. and SHULMAN, R. G.: (in press).
26 ANTONINI, E.; SCHUSTER, T. M.; BRUNORI, M. and WYMAN, J.: J. biol. Chem. *240:* PC2262 (1965).
27 GRAY, R. D.: J. biol. Chem. *245:* 2914 (1970).
28 HAYASHI, N.; MOTOKAWA, Y. and KIKUCHI, G.: J. biol. Chem. *241:* 79 (1966).
29 GREER, J.: Thesis, Univ. of Cambridge (1970).
30 HOARD, J. L.: in CHANCE, B.; ESTABROOK, R. W. and YONETANI, T. Hemes and hemo-proteins, p. 9 (Academic Press, New York/London 1966).
31 SIMON, S. R.: Thesis, Rockefeller Univ., New York (1967).
32 ROSSI-FANELLI, A.; ANTONINI, E. and CAPUTO, A.: J. biol. Chem. *236:* 397 (1961).
33 KELLETT, G. C. and GUTFREUND, H.: Nature, Lond. *227:* 921 (1970).
34 KOSHLAND, D. E.; NEMETHY, G. and FILMER, D.: Biochemistry *5:* 365 (1966).
33 KOSHLAND, D. E.; NEMETHY, G. and FILMER, D.: Biochemistry *5:* 365 (1966).
35 KOSHLAND, D. E.: Proc. U.S. nat. Acad. Sci., Wash. *44:* 98 (1958).
36 SHULMAN, R. G.; OGAWA, S.; WUETHRICH, K.; YAMANE, T.; PEISACH, J. and BLUMBERG, W. E.: Science *165:* 251 (1969).
37 PERUTZ, M. F.; MUIRHEAD, H.; MAZZARELLA, L.; CROWTHER, R. A.; GREER, J. and KILMARTIN, J. V.: Nature, Lond. *222:* 1240 (1969).
38 KILMARTIN, J. V. and ROSSI-BERNARDI, L.: Nature, Lond. *222:* 1243 (1969).
39 WYMAN, J.: J. biol. Chem. *127:* 581 (1939).
40 CONANT, J. B.: Harvey Lectures *28:* 159 (1932–33).
41 CORYELL, C. D. and PAULING, L.: J. biol. Chem. *132:* 769 (1940).
42 KENDREW, J. C.; DICKERSON, R. E.; STRANDBERG, B. D.; HART, R. G.; DAVIES, D. R.; PHILLIPS, D. C. and SHORE, V. C.: Nature, Lond. *190:* 663 (1960).
43 MUIRHEAD, H. and PERUTZ, M. F.: Nature, Lond. *199:* 633 (1963).
44 MCCONNELL, H. M.; DEAL, W. and OGAWA, R. T.: Biochemistry *8:* 2580 (1969).
45 KILMARTIN, J. V. and WOOTTON, J.: Nature, Lond. *228:* 760 (1970).
46 ANTONINI, E.; WYMAN, J.; ZITO, R.; ROSSI-FANELLI, A. and CAPUTO, A.: J. biol. Chem. *236:* PC60 (1961).
47 WINTERHALTER, K. H.; ANDERSON, N. M.; AMICONI, G.; ANTONINI, E. and BRUNORI, M.: Europ. J. Biochem. *11:* 427 (1969).
48 PERUTZ, M. F.: Trans. Faraday Soc. *42B:* 187 (1946).
49 BOYES-WATSON, J.; DAVIDSON, E. and PERUTZ, M. F.: Proc. roy. Soc., A*191:* 83 (1947).
50 GREEN, D. W.; INGRAM, V. M. and PERUTZ, M. F.: Proc. roy. Soc., A*225:* 287 (1954).

51 ANTONINI, E. and BRUNORI, M.: J. biol. Chem. *244:* 3909 (1969).
52 OGAWA, S. and MCCONNELL, H. M.: Proc. U.S. nat. Acad. Sci., Wash. *58:* 19 (1967).
53 MATSUDA, G.; TAKEI, H.; WU, K. C.; MIZUNO, K. and SHIOZAWA, T.: Abst. Int. Congr. Biochem. p. 4 (1970).
54 HILSE, K. and BRAUNITZER, G.: Z. Physiol. Chem. *349:* 433 (1968).
55 BENESCH, R. and BENESCH, R. E.: Nature, Lond. *221:* 618 (1969).
56 GARBY, L.; GERBER, G. and DE VERDIER, C. H.: in Proc. Symp. Intracellular Regulation of Haemoglobin Affinity to Oxygen. Swedish J. Defence Med. *5:* 163 (1969).
57 BUNN, H. F. and BRIEHL, R. W.: J. clin. Invest. *49:* 1088 (1970).
58 BAILEY, J. E.; BEETLESTONE, J. G. and IRVINE, D. H.: J. chem. Soc., A *5:* 756 (1970).
59 BENESCH, R. E.; BENESCH, R. and CHI ING YU: Fed. Proc. *28:* 604 (1969).
60 MAZrARELLA, L. and PERUTZ, M. F.: Nature, Lond. *199:* 633 (1963).
61 WILLIAMS, R. J. P.: Fed. Proc. *20,* No 3, suppl. 10, p. 5 (1961).
62 ANDERSON, L.: J. Mol. Biol. In the press.
63 KILMARTIN, J. V. and HEWITT, J. A.: Cold Spring Harbor Symp. Quant. Biol. *36:* 311 (1971).
64 PERUTZ, M. F. and TEN EYCK, L. F.: Cold Spring Harbor Symp. Quant. Biol. *36:* 295 (1971).
65 ARNONE, A.: Nature, Lond. *237:* 146 (1972).
66 LINDSTROM, T. R. and HO, C.: Proc. Nat. Acad. Sci. USA.: *69:* 1707 (1972).
67 HEWITT, J. A.; KILMARTIN, J. V.; TEN EYCK, L. F. and PERUTZ, M. F.: Proc. Nat. Acad. Sci. USA : *69:* 203 (1972).
68 HEIDNER, E.: Unpublished.
69 PERUTZ, M. F.: Nature, Lond. *228:* 726 (1970).

Author's address: M. F. PERUTZ, MRC Laboratory of Molecular Biology, Hills Road, *Cambridge,* England

Discussion

G. LING: Dr. PHILLIP and others have shown once more how powerful a tool x-ray crystallography has become in probing with great exactness the terribly complex system of protein crystals. However, living protoplasm is even more complex than protein crystals. Living proteins, if one may so call them, are as a rule not crystalline but have the consistency of a gel. To study the interaction of metabolic products with cellular proteins, we chose a different approach. In this approach we deliberately squint our eyes, so to speak, and look for broad features.

It has long been known that the oxygen binding curve is sigmoid in shape. The data have long been analyzed with the aid of an emperical equation due to A. V. HILL:

$$\log \frac{Y}{1-y} = n \log pO_2 + n \log K_Y,$$

where y and pO_2 represent the number of complex oxygen molecules per hemoglobin molecule and the partial pressure of oxygen. The coefficient n has long been known as some sort of a measure of the 'sigmoidness' of the curve.

Figure A shows oxygen binding by hemoglobin in the presence and absence of different concentrations of 2,3-DPG [LING, Proc. nat. Acad. Sci., Wash. *67:* 296, 1970]. The experimental points are those published by BENESCH and BENESCH [Nature, Lond. *221:* 618, 1969]. The lines through them are the theoretical curves calculated on the basis of a one dimensional Ising Model which Prof. C. N. YANG and I published in 1964 [Biopolymers Symp., vol. 1, p. 91, 1964]. Within reasonable accuracy, the data and theory do agree. It also bears pointing out 1) that it gives a precise meaning to the value of n, which is equal to $\exp(\gamma/2RT)$, where γ is the nearest neighbor interaction energy. What 2,3-DPG does is to cause the displacement of affinity of hemoglobin for oxygen. It also causes some change in γ.

This approach can also be applied to an even more complex system with meaningful information in return. Figure B shows a similar group of cooperative adsorption isotherms in which case we are no longer dealing with oxygen bindings on hemoglobin but with K^+-ion binding on cellular proteins of living frog muscle cells.

The curve moves from left to right with decreasing intrinsic equilibrium constants for the K^+ ion with an increasing concentration of the drug, a cardiac glycoside, ouabain, used by clinicians to combat cardiac failures.

D. C. PHILLIPS: We've considered how protein structures are affected by interactions with substrates; and, we've heard most particularly how the protein environment can affect the reactivities of the groups in the protein. Now we're going to continue with Dr. KOSHLAND's talk about the ways in which enzymes in particular can enhance the rates of chemical reaction.

Proc. 3rd int. Conf. From Theoretical Physics to Biology, Versailles 1971, pp. 286–302
(Karger, Basel 1973)

The Contribution of Orientation to the Catalytic Power of Enzymes

D. E. KOSHLAND, jr.

Department of Biochemistry, University of California, Berkeley, Calif.

Bonds in organic compounds have a direction preference and this directional preference has an influence in reactivity [PAULING, 1931; MULLIKEN, 1932; ROBERTS, 1961; BALLHAUSEN, 1964; WOODWARD and HOFMANN, 1969]. The magnitude of this orientation dependence, however, has not been evaluated precisely and it may be of great relevance to the theories of enzyme action. If this orientation factor is large, it could be capable of explaining a large part of the existing discrepancy between the velocity of enzymatic reactions the their non-enzymatic analogs. If on the other hand it is relatively small, it can make only a small contribution to the catalytic power of enzymes.

A. Proximity Effect

The evaluation of the orientation factor in experimental situations depends on an extrapolation from a second or higher order reaction to a first order reaction on an enzyme surface. The method of calculating the contribution of the 'proximity effect' [KOSHLAND, 1962] is shown schematically in figure 1. In a bimolecular reaction, a rate constant is established for a mechanism by measuring a reaction at known concentrations of A and B and extrapolating to unit activity. Assuming the reaction proceeds by the same mechanism on the enzyme surface as in the model reaction, the enzymatic reaction rate could be calculated if the effective concentrations of A and B on the enzyme surface, A_E and B_E, could be evaluated. The situation in the upper right hand corner of figure 1, in which B is constantly juxtaposed with A on the saturated enzyme, is compared to the

Proximity

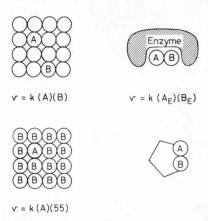

$$v = k\,(A)(B)$$

$$v = k\,(A_E)(B_E)$$

$$v = k\,(A)(55)$$

Fig. 1. Schematic illustration of proximity corrections.—Upper left: the distribution of molecules A and B in a typical bimolecular process.—Upper right: model of the active site of an enzyme where B is held in juxtaposition with A.—Lower left: a hypothetical bimolecular distribution of A entirely surrounded by 55 M B.—Lower right: intramolecular case in conformation in which B is juxtaposed with A.

situation when A is surrounded entirely by B molecules (lower left hand portion of fig. 1). The factor to be expected in such a situation therefore will be 55-fold greater than the bimolecular constant if a) the molecules of A and B are the size of water, and b) neither they, nor the solvent, have any special affinity or repulsion for each other. The factor 55 refers to the illustrative case in which only one nearest neighbor can react. In the more general case [KOSHLAND, 1962], 55 is divided by the number of nearest neighbors. This corresponds to an entropic factor of 8 entropy units. Clearly there is some similarity between intramolecular reactions in which the reactants are juxtaposed and an enzyme catalyzed reaction.

B. Orientation

The calculation of orientation factors relative to a bimolecular reaction depends on the relationship between orientation and proximity illustrated in figure 2. The assumptions made in the original derivation of the proximity correction [KOSHLAND, 1962] were: a) that molecules (or reactive atoms) A and B are the size of water molecules; b) that there is no net attraction or repulsion between A, B and/or solvent, and c) that a molecule

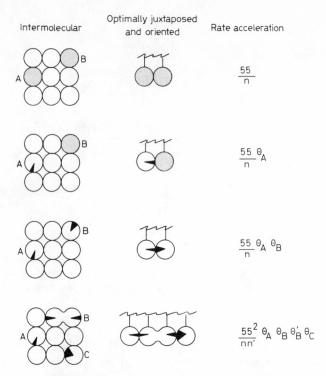

Fig. 2. Kinetic effects of proximity and orientation. The circles represent spherical molecules or functional groups about the size of water molecules. Groups reactive over their entire surface are shaded; the reactive surface area of other groups is represented in two dimensions by pie-shaped wedges. The θ factors may be considered schematically as the total surface area of the sphere divided by its reactive area and the n are corrections for the number of nearest neighbors. Note that intramolecular reactions in which only one group can have orientation preferences are unlikely in practice and that higher order reactions may involve more than one orientation parameter per group.

with no orientational preferences can react with any of its, n, nearest neighbor molecules. A molecule, A, which has an orientational preference such that it can react at $1/\theta_A$ of its solid surface has a probability of $n(B)/55\theta_A$ of being in a reactive relationship with molecule B which has no orientational preference. The rate acceleration expected therefore in a perfectly oriented and juxtaposed AB pair (as on an enzyme surface) compared to a random collision process when $(B) = 1$ M would be $(55/n)\theta_A$. The rate acceleration for two molecules, both of which have orientational requirements, would be $(55/n)\theta_A\theta_B$. As discussed earlier [KOSHLAND, 1962], for real molecules larger than water, A and B refer to the reactive groups, e.g., the OH of glucose, the phosphorus of ATP, etc., which are essentially the size of water molecules.

Theoretical calculations of the potential magnitude of various contributions to orientation seemed desirable both to ascertain what factors would be reasonable on the basis of transition state structures and for the design of future experiments. Orbital steering is defined in operational terms to describe the situation in which reactive atoms, at least one of which contains spherical asymmetry in its valence orbitals, are constrained to react along selected pathways. Such constraints can be imposed by the binding of substrates at the active site of an enzyme or by the superstructure of the molecule in an intramolecular reaction. If the constraints steer the reacting molecules along an optimal pathway, an orientation factor greater than 1 should arise; contributions to the magnitude of this factor could come from bending vibrations, stretching vibrations, non-bonded interactions, and non-productive complexing in the model systems.

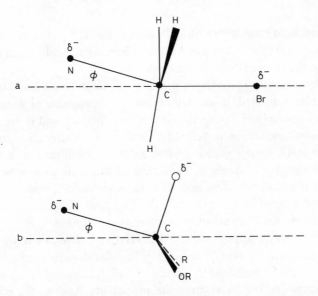

Fig. 3. Possible representations of distorted transition states results from reaction of a nucleophile N with a) methyl bromide, and b) an ester function. In a), the H–C–H plane is taken as bisecting the Br–C–N angle. In b), all bond angles except for C–N are tetrahedral. ϕ is measured from the 3-fold axis of symmetry of tetrahedral carbon.

C. Theoretical Calculations of the Contribution of Bending Vibrations to Orientation Factors

Two calculations of such orientation factors have been made and are detailed elsewhere [DAFFORN and KOSHLAND, 1971a, b]. In figure 3 are shown schematic pictures of a nucleophilic attack on an ester and on methyl bromide. If one assumes that a given deviation ϕ in any direction from the optimal angle of approach will cost the same amount of energy (i.e., that the 'reactive area' is circular), a simple set of collision equations may be obtained. The total probability P_t that a given collision on the surface of A leads to reaction can be written as

$$P_t = \int_\sigma^\pi P_1(\phi) \, P_2(\phi) \, d\phi \tag{1}$$

with $P_1(\phi)$ = probability that collision at ϕ leads to reaction

$P_1(0)$ = probability that collision at $\phi = 0°$ leads to reaction

$P_2(\phi) \, d\phi$ = probability that a given collision occurs between

ϕ and $\phi + d\phi$ (always normalized $- \int_\sigma^\pi P_2(\phi) \, d\phi = 1$)

In the idealized case where all collisions occur at $\phi = 0°$, no integration is necessary and $P_t = P_1(\phi)$. For a random bimolecular collision, it can easily be shown that $P_2(\phi) \, d\phi = \frac{1}{2} \sin \phi \, d\phi$.

The probability $P_1(\phi)$ that a collision at orientation ϕ leads to reaction must still be evaluated. If we assume that the orientation of groups A and B remains unchanged during the course of a collision and if the optimum ground state orientation is defined as that which leads to an undistorted transition state, then a collision at orientation ϕ will lead to a transition state distorted by an angle ϕ. The effect of this distortion is to raise the energy of the transition state and thus the potential difference V* between it and the ground state; as usual, the potential barrier is related to a probability by the Boltzmann or Arrhenius-type function, $e^{-V*/kT}$. In the harmonic oscillator approximation, the relative potential energy $V*(\phi)$ of any given transition state structure can be related to its configuration by a force constant F* such that $V*(\phi) = \frac{1}{2} F* \phi^2$. Effects other than orientation may be cancelled out by relating the probability $P_1(\phi)$ to the probability $P_1(0)$ for a transition state with no angle strain. $P_1(0)$ may be written $P_1(0) = Ce^{-V*(0)/kT}$ since, for a perfectly oriented collision, the probability of reaction is at least proportional to the probability that a collision is energetic enough to pass over a barrier of height $V^*(0)$. Strain in the transition state will increase the barrier height by an amount $V^*(\phi)$, so that

$$\frac{P_1(\phi)}{P_1(0)} = \frac{Ce^{-[V^*(0) + V^*(\phi)]/kT}}{Ce^{-V^*(0)/kT}} = e^{-V^*(\phi)/kT} \tag{2}$$

or

$$P_1(\phi) = P_1(0)\, e^{-F^{*(2)}\,2/2kT}$$

For the extremely simple comparison between a random bimolecular reaction and a reaction in which all collisions occur at or very near the optimum angle, the total reaction probability for one collision can now be given as:

$$P_t = P_b = \int_\delta^\pi \tfrac{1}{2}\, P_1(0)\, e^{-F^*\phi/2kT} \sin\phi\, d\phi \tag{3}$$

If the force constant is of the order of magnitude of those for ordinary vibrations, this integral is closely approximated by $P_b = P_1(0)\, kT/2F^*$. A similar expression denoted by P_b' can be obtained for the other group in a bimolecular reaction. The total rate enhancement $\theta_A \theta_B$ due to orientation is related to the inverse of $P_b P_b'$ and given by $\theta_A \theta_B = [2F^*/kT]_{\text{group A}}$ $[2F^*/kT]_{\text{group B}}$. This expression is plotted in figure 4 on the line $F = \infty$ as a function of transition state bending force constants (for the case $F_A^* = F_B^*$).

If the probability P_b (or $P_b P_b'$) is multiplied by a collision number Z to obtain a rate expression, the only exponential temperature dependence of rate appears in $P_1(0)$ and the integrals over angle-dependent terms have the effect of reducing the value of the pre-exponential factor for the bimolecular reaction; i.e., in this model system, the effect of orbital steering appears in the steric factor of ordinary collision theory. This conclusion is consistent with the usual interpretation of steric factors as 'due to improperly oriented collisions' [BENSON, 1960], as well as with the small temperature dependence of the rate ratio. It should be emphasized that this identification of orbital steering and the steric factor holds only for this simple model; for more complex reactions, particularly in solution, other effects which would tend to mix exponential and pre-exponential terms are likely. We would thus expect that measured steric factors would be a much poorer measure of orientation effects than rate ratios in which extraneous factors can be minimized.

Even in a properly oriented intramolecular or enzymatic reaction, some residual 'sloppiness' in the angle of approach is inevitable because of vibrations of the reacting groups. The correction for this effect introduces an exponential term into $P_2(\phi)$ similar to that in $P_1(\phi)$, but dependent on a ground state force constant, F, for the intramolecular reaction. Expressions similar to those above are obtained on integration and are plotted in figure 4 for three reasonable values of the ground state force constant which determines the precision of the intramolecular orientation. To further clarify the nature of F, an intramolecular reaction in which the groups are locked into place subject only to bending vibrations would have a fairly large value of F, whereas free internal rotation would lead to much smaller values. Although the optimum approach is assumed to be the most probable throughout, the calculated rate enhancement as a function of transition state

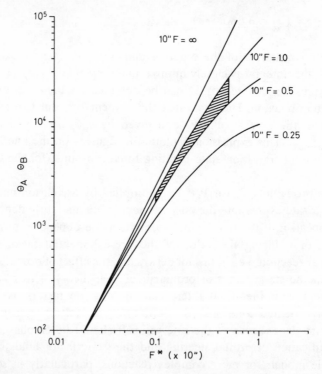

Fig. 4. Rate enhancements expected on the basis of simple collision theory when reacting groups A and B are properly oriented (if the potential walls of A and B can be described by identical force constants). $\theta_A \theta_B$, the ratio of rates of oriented reactions to bimolecular reactions, is given as a function of the transition state bending force constant F* (in erg rad^{-2}) for several values of the ground state force constant F. F describes the stiffness of the bonds maintaining the orientation. The shaded area encloses force constants for reactions in which the transition state is fairly tightly bound.

structure (F*) decreases as other angles of approach in the intramolecular reaction become more probable (as F decreases), in accord with qualitative expectations. The shaded area of figure 4 represents reasonable values of force constants for a comparison of fairly precisely oriented intramolecular reactions (no free internal rotations) and bimolecular reactions with fairly strong bonds in the transition state (nucleophilic attack on a carbonyl group as opposed to a protonation of an anion). On the basis of this model, rate enhancements of $10^3 - 10^4$ on conversion of a bimolecular to an intramolecular or enzymatic reaction appear reasonable. Such orientation factors would require an angular deviation of approximately 10 ° for each of the reacting atoms. Neither the angle of deviation nor the bending force constants are out of the range compatible with normal chemical reactions.

D. Experimental Observations

The data of table I summarize some rate constants normalized to 1 N H+ for the rates of lactonization [STORM and KOSHLAND, 1970; KOSHLAND et al., 1971]. No corrections for proximity, ring strain, non-bonded interactions, etc. are made in the values listed in that table. It is assumed that the solvation will play a small role since the reaction is relatively insensitive to the solvent and there seem to be only minor differences in the actual solvation spheres of the atoms involved in the reaction Measurement of pK's of carboxylic acid groups, IR stretching frequencies and ^{13}C NMR shifts show practically no differences in the various groups [STORM and KOSHLAND, in preparation]. Therefore, there are no apparently large abnormal factors which should change radically from one compound to another.

If the hypotheses behind orbital steering are correct, reaction velocities should be changed appreciably by changes in the angle of approach of the reacting atoms. It is difficult to devise molecules in which the angle of approach of the attacking atoms is different without introducing additional complications of strain, inductive effects, etc. One procedure previously used was to convert the attacking atom from oxygen to sulfur and use the bimolecular sulfur analog as a control [STORM and KOSHLAND, 1970].

Another approach has been to adjust the angles of approach of the hydroxyl and carboxyl groups by changing the frame to which they are attached from the bicyclo [2.2.1] heptane system to the bicyclo [2.2.2] octane system [STORM, TJIAN and KOSHLAND, 1971]. The data in the table give two examples of this. The bicyclo [2.2.2] γ-hydroxy acid analog of the 2,6 norbornane compound III reacts 1,000 times less readily than compound III itself (cf. compounds V and III, table I). The Dreiding models shows a difference in angle of 12° and a great apparent similarity in torsional strain, ring strain, solvation, etc. factors. The differing rates can be largely ascribed to the differences in angles of approach.

Further support for this position is obtained from the rate of the exo-methyl analog of compound V. Unlike the analogous substitution in the [2.2.1] system which decreases the rate by 1,000 fold, the exomethyl in the [2.2.2] system (compound VI) accelerates the rate by 100 fold relative to the H compound (compound V, table I). Again, examination of the Dreiding models shows no obvious general increases in compression strain, but does show a change in the angles of the reacting atoms.

E. Conclusions

The orientational factors in a chemical reaction arise from many factors, e.g., bending vibrations, non-bonded interactions, entropy of overall rotation, entropy of internal rotation, non-productive orientations in the bimolecular reaction, etc. In a confined system such as on an enzyme surface or in an intramolecular reaction, the reacting atoms do not have the freedom of a bimolecular collision in solution. These orientational restrictions have been called orbital steering to emphasize that the confined system is steered along confined pathways even after their juxtaposition with the reacting atoms has been allowed for. This orientation can be optimized on an enzyme surface and could, potentially, lead to large rate accelerations. The same is true of properly designed intramolecular reactions except that optimization becomes more difficult.

Experimentally, intramolecular reactions have been examined which control the direction of approach of the reacting atoms and which have revealed large factors relative to the bimolecular reaction in solution. It is, of course, not easy to design experimental molecules free of ambiguities. Alteration of structure requires chemical changes and these can cause inductive effects, compression effects, solvation effects, etc. It was precisely for this reason that a special series of compounds had to be designed for our initial studies and further compounds will be needed to establish the hypothesis. Moreover, others have indicated skepticism as to the magnitude of orientation factors in reactivity [BRUICE and TURNER, 1970; BRUICE et

Table I. Relative velocities of some intramolecular lactonizations

No.	Compound	Observed velocity	Velocity relative to bimolecular reaction
I		8.58×10^{-2}	79.5
II		7.23	6,620
III		1,120	1,030,000
IV		0.3	276
V		0.95	871
VI		107	98,200

al., 1971; PAGE and JENCKS, 1971]. Nevertheless, the accumulating evidence is beginning to be convincing. The experimental devices of a) the introduction of methyl groups which favor some orientations over others; b) the change in radius of the reacting atoms, and c) the change in the support system to alter geometry, have all indicated a sensitivity to orientation in the reacting atoms which are compatible with sizeable orbital steering factors.

References

BALLHAUSEN, C. J. and GRAY, H. B.: Molecular orbital theory (Benjamin, New York 1964).
BENSON, S. W.: The foundations of chemical kinetics; p. 277 (McGraw-Hill, New York 1960).
BRUICE, T. C.; BROWN, A., and HARRIS, D. O.: Proc. nat. Acad. Sci., Wash. *68:* 658 (1971).
BRUICE, T. C. and TURNER, A.: J. Amer. chem. Soc. *92:* 3422 (1970).
DAFFORN, G. A. and KOSHLAND, D. E., jr.: Bioorgan. Chem. (in press [1971a])—Proc. nat. Acad. Sci., Wash. (in press [1971b]).
KOSHLAND, D. E., jr.: J. theoret. Biol. *2:* 75 (1962).
KOSHLAND, D. E., jr.; CARRAWAY, K. W.; DAFFORN, G. A.; GASS, J. D., and STORM, D. R.: Cold Spr. Harb. Symp. quant. Biol. (in press).
MULLIKEN, R. S.: Physiol. Rev. *40:* 55 (1932).
PAGE, M. I. and JENCKS, W. P.: Proc. nat. Acad. Sci., Wash. *69:* 1678 (1971).
PAULING, L.: J. Amer. chem. Soc. *53:* 1367, 3225 (1931).
ROBERTS, J. D.: Molecular orbital calculations (Benjamin, New York 1961).
STORM, D. R. and KOSHLAND, D. E., jr.: Proc. nat. Acad. Sci., Wash. *66:* 445 (1970).—(In preparation).
STORM, D. R., TJIAN, R. and KOSHLAND, D. E., jr.: Chem. Commun. 854 (1971).
WOODWARD, R. B. and HOFMANN: Ang. Chem. *8:* 781 (1961).

Author's address: Dr. DANIEL E. KOSHLAND, jr., Department of Biochemistry, University of California, *Berkeley, CA 94720* (U.S.A.)

Discussion

H. FRÖHLICH: There are, of course, many other systems. Is it possible to replace some part which is not active to see whether it has any influence on the catalytic power?

D. E. KOSHLAND: Yes. Dr. AMES at Berkeley has studied mutations, not selecting for desirable or undesirable mutations, but just trying to catalog all of them. It turns out that very few mutations are actually lethal. That would make a great deal of sense with what you've heard this morning. The catalytic residues which are very specially tailored would have very little chance of being replaced in any one enzyme without complete loss of activity and essentially every one of those changes is lethal if the enzyme is essential. In a similar way a lot of the changes in the center of the molecule could occur in critical positions. Many amino acid positions in the middle of the molecule can be altered, slightly at least, and they only make the enzyme slightly less efficient. So that's a second kind of partial change, and then there are a number of changes of groups that are on the surface which cause no change at all in the activity. So that you have a mixture. If you damage the active site I think you probably turn off the enzyme. But you can make many other changes without hurting it.

P. O. LOWDIN: You mentioned that there was an unknown factor between 10^2 and 10^{10}. In chemical reactions, we are forming and breaking bonds. The strongest bond breaking power seems to come from a mechanism where one or two electrons, through transfer, are put into an antibonding orbital. Have you contemplated the fact that perhaps one action of an enzyme (or part of an enzyme) is simply to serve as an electron bridge for such a transfer?

D. E. KOSHLAND: There are some enzymes where single electrons are transferred. But in almost all of the cases the non-enzymatic reaction involves two electrons moving together. The mechanisms involve acid and base catalysis which involve electrons but are usually not called an electron bridge. In the oxidation-reduction reactions a coenzyme bound to the enzyme frequently acts as a bridge.

S. YOMOSA: Relating to FRÖHLICH's remarks and KOSHLAND's talk, I would like to represent a molecular model on the initiation of the enzymatic reaction. We consider the charge-transfer molecular compound in the local field E_L which is constituted of the Coulombic field E_C and Onsager's reaction field E_R. The field E_C is produced by the anionic and the cationic charges COO^-, NH_3^+ in the residues of amino acids and the permanent dipole moments of the peptide groups of the protein. The reaction field E_R is produced by the polarization in the surrounding medium induced by the presence of the dipole moment μ_g of the compound;

$$E_L = E_C + E_R = E_c + \lambda(\epsilon)\,\mu_g \tag{1}$$

Here μ_g represents the ground state dipole moment of the compound and ϵ represents the zero frequency dielectric constant of the medium. We assume such local field exerts on the molecular compound made up of donor D and acceptor A.

$$\overline{} \longrightarrow E_L$$
$$\text{AD}$$

Hamiltonian of the compound in the local field can be written as

$$H = H^\circ + V = H^\circ - ME_L \tag{2}$$

Here, M denotes the dipole operator of the system; $M = -\Sigma_i e r_i$. The wave function may be simply approximated by

$$\Psi = c_1\,\psi_1\,(AD) + c_2\,\psi_2\,(A - \overset{+}{D}) \tag{3}$$

For a given value of E_L, we can solve eigenvalue problem, $H\Psi = E\Psi$, and then we get the ground state dipole moment $\mu_g = (\Psi_g, M\Psi_g)$. The result obtained by this procedure can be represented as a curve (fig. 1).
On the other hand, we have a linear relation (1) between μ_g and E_L. Therefore μ_g and E_L are simultaneously determined by the intersection points M, Q, N of the curve and the line. Both states, M and N, correspond to the minimum points of free energy. When we increase

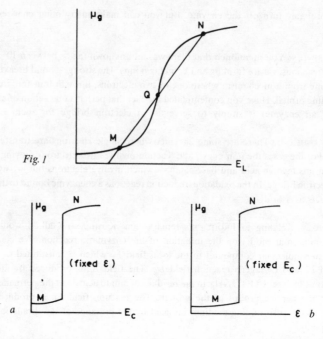

Fig. 1

Fig. 2

the value of E_C at fixed value of ε, or when we increase the value of ε at fixed value of E_C, the relation of equal area is realized at critical values of E_C or ε as shown in the figure. In this situation, the free energy of state, M, becomes equal to that of state, N, and our solution jumps from state, M, to state, N. These results are represented by figures 2a and b.

State, M, is similar to that of Mulliken complex, while state, N, is characterized as a 'super-charge-transfer-state'. These two states, M and N, may clearly be distinguished. Such strongly polarized state N is considered as the first important step of charge separation in the biochemical reaction and such abrupt polarization initiated by a small change of the Coulombic field or by the small change of the dielectric constant of the surrounding medium seems to suggest a control mechanism in the biological function. [The detailed statement may be found in the paper: Prog. theor. Phys. Suppl. No. 40, p. 249, 1967.]

D. C. PHILLIPS: Perhaps Dr. VALLEE would like to take up that kind of discussions?

B. L. VALLEE: Brevity of time no doubt precluded Dr. KOSHLAND's mentioning that he himself has made sustained and major contributions to a problem that underlies many of the points he made. Perhaps I can enlarge upon some of these in the few minutes allocated to me, since I believe them to be pertinent to Drs. PHILLIPS', RICHARDS', and KOSHLAND's discussion.

Chemical modification of enzymes has often implied that, compared with models, active site residues of enzymes are chemically hyper-reactive, even prior to the advent of the substrate. The residues that have been alluded to by the discussants as being important in catalysis have been found both to be 'hyper-reactive' to site-specific reagents and to have physico-chemical characteristics which are unusual when compared with models that have been studied. This pertains to Dr. RICHARDS' system, ribonuclease, where rates of reaction of the two catalytically essential histidyl residues toward acylating agents are unlike those of the amino acid, histidine, or histidyl peptides, or of non-functional histidines in proteins. In fact, lysine No. 41 of this enzyme is also 'hyper-reactive' toward fluorodinitrobenzene and other reagents which thereby inactivate the enzyme. In general, the chemistry underlying the inactivation of enzymes by site-specific reagents seems quite unusual, and analogous selectivity for one out of many identical residues has not been observed, thus far, in simple peptides or non-enzymatic proteins. The inactivation of certain seryl enzymes by diisopropylfluoro-phosphate, that of sulfhydryl, histidyl or lysyl enzymes by organic reagents specific for these residues under mild conditions does not seem to have been encountered elsewhere in che-mistry. The basis of such higher reactivity is currently unknown and raises questions as to the mechanisms which render certain particular groups of enzymes chemically 'hyper-reactive', a phenomenon which seems to provide a link between chemical reactivity and biological function. This problem is clearly deserving of further exploration.

However, it is not easily possible to measure 'hyper-reactivity' without adding a chemical reagent, nor to determine its relationship to catalysis without adding a substrate. An expe-rimental criterion of hyper-reactivity, independent of such experimental perturbation of the system, would be most desirable. Indeed, intrinsic components of certain enzymes, e.g., prosthetic groups, cofactors, and metals, exhibit physico-chemical properties which can serve as probes of structure and, perhaps, of catalytic potential both in the absence of such reagents and of substrates.

As a class, metal ions exhibit properties and reactivities which differ distinctly from those of amino acid side chains of proteins, and precisely this circumstance renders metallo-enzymes convenient models for the study of the mechanism of enzyme action. Catalytically essential metals of metallo-enzymes exhibit distinctive absorption, circular dichroic, magnetic circular dichroic, electron spin resonance, nuclear magnetic resonance, and Mossbauer spectra, which reflect the metal environment. Many of these spectral properties have been tabulated. In the case of iron and copper enzymes their comparison with those of the corresponding complex ions indicate that the spectra of such metallo-enzymes generally reflect irregular geometry and low symmetry of their metal atoms. In general, the physical properties of metals of metallo-enzymes seem quite unusual in the sense that unlike those of conventional complex ions of models known so far, substrates or inhibitors often markedly alter their absorption and circular dichroic spectra which then more nearly resemble those of well defined model systems. Such findings have emphasized the unusual nature of metals and their coordination in metallo-enzymes, perhaps reflecting features of their participation in catalysis. These unusual physical properties might then reflect biological capacity in some manner. The loss of such spectra on denaturation and the similarity of those of non-enzymatic metallo-proteins to known model complexes are consistent with such reasoning. The spectra of non-heme iron, copper and cobalt enzymes have been particularly revealing.

Typical of the non-heme iron enzymes, the ferredoxins and rubredoxins have at least four intense discrete absorption bands between 320 and 600 nm. Such spectra have not been seen in iron complex ions. The spectra of the numerous 'copper blue enzymes' are also quite unusual. The bands at about 450 and 750 nm particularly at 600 nm have absorptivities always much more intense than those of known models. In this regard metal atoms with poor probe qualities, e.g., zinc, have been replaced with others, e.g., cobalt, which are good probes, thereby enlarging the scope of this approach. This has revealed that the cobalt spectra of, e.g., carboxypeptidase, carbonic anydrase and alkaline phosphatase are similarly unusual as compared with cobalt complex ions examined thus far.

Dr. KOSHLAND previously summarized the factors thought to affect the catalytic potential and behavior of enzymes and to constitute the 'microscopic chemical environment' of active site residues in the tertiary structure of enzymes.

Some of these chemical features of metallo-enzymes might well denote aspects of their biological specificity and function, analogous perhaps to the unusual chemical reactivities of active-site amino acid residues, correlating to their functional properties. Unlike such hyper-reactivity, the spectral characteristics of a metal arising from its local environment can be observed in the absence of chemical reagents, substrates or inhibitors and—in analogy with hyper-reactive amino acid residues—could reflect the fact that the metal is prepared to participate in biological function perhaps by virtue of environmental factors. The term *entasis* has been introduced operationally in the sense of operational definitions as defined by BRIDGMAN in 'The Logic of Modern Physics' and in 'The Way Things Are' to designate this state of the active-site of an enzyme prior to the formation of an enzyme-substrate complex.

The state of metals in metallo-enzymes is designated *entatic* to convey an analogy of their physical and chemical properties to the hyper-reactivity of amino acid residues both arising from tertiary structure and relating to catalysis. It is conceivable that the low symmetry of the metal binding site reflects a thermodynamic state favorable to catalysis and imposed by the three-dimensional folding of the enzyme.

Such properties of metallo-enzymes can also provide a means to differentiate the structure of enzymes in solution from that in the crystalline state, a somewhat different, albeit related problem, for the discussion of which there is, unfortunately, not enough time. My main point was to emphasize that spectral characteristics of enzymes are now apparent which can relate chemical to their biological properties.

F. LIPMANN: I would like to insert a few general remarks. I was very pleased to hear Dr. PHILLIPS say that much of the contribution to enzyme activity is made nonspecifically by peptide bonds through –CO– – –HN– bonding. That paraphrases what is on my mind, namely, why is an enzyme a protein? We have heard a great deal about specific groups and the marvel of their coming together in the right coordination. But these specific groups are only a small part of the molecule compared to the large number of amino acids in the enzyme, which should contribute in some way to enzyme action and explain why the polypeptide structure of the protein is needed.

K. HEPP: I take from Prof. KOSHLAND's talk that the chemical reactivity is very sensitive with respect to changes of the tertiary structure of an enzyme. Couldn't this be used to estimate the contributions of excited states of such molecules which are weighted by a Boltzmann factor $\exp(-E/kT)$?

D. KOSHLAND: I think that would be difficult because the problem is that the protein changes shape under the influence of substrates. As Dr. RICHARDS discussed, protein forces involve the subtraction of two very large numbers to give a relatively weak net force. That's why it's so difficult to determine a unique structure. The protein was designed like a spider's web so that when you change one specific group, as Dr. PHILLIPS pointed out, the changes are often profound because of the delicate balance of forces. Therefore, absolute energies are hard to get.

H. C. LONGUET-HIGGINS: I'd like to make a remark about excited states. When you wrote something on the board about charge transfer as a possibility for the intermediate step in these reactions, one must not forget the evidence from reactions, i.e., ordinary visible spectroscopy. When there is a charge transfer complex formed, one almost always observes an absorption band in the visible region. This is a necessary consequence of the model which you proposed. I don't know, but I've never heard of any charge transfer band appearing when substrates are bound to enzymes but, I may be wrong. Of course, in haemoglobin, it might be very difficult to see such a thing because the material is colored anyway, but most enzymes are colorless. I think that if one simply wants to know where the excited states are, the simplest thing to do is to take the absorption spectrum. One other point. It's unsafe to argue, as somebody did a moment ago, that there's a mystery about enzymes simply because we haven't yet succeeded in making model systems which behave like them. This is only a matter of time. There used to be a mystery about the fixation of nitrogen. Nitrogen is fixed by certain bacteria with the aid of a molybdenum-containing enzyme. For years and years, chemists labored to construct metal complexes which would fix nitrogen. Well, they finally succeeded. The molecule which in fact does it, the first one to be discovered, is really quite small and when it's been done it's no longer a mystery. I think one must keep mystery out of this. It seems to me there are plenty of good problems in biology without inventing ones which don't really exist.

B. L. VALLEE: There was not meant to be an implication of mystery of the kind that Dr. LONGUET-HIGGINS inferred. Quite the contrary. I wanted to suggest that it would be preferable to make models that resemble enzymes rather than others aimed at predicting what they should be like, now that some of the characteristics of enzymes have become apparent.

S. FOX: Dr. KOSHLAND's introduction referred to an estimate of 3,000 enzymes for the *E. coli* cell and 30,000 for the mammalian cell. Strictly speaking isn't the 30,000 a figure that applies to the total multicellular mammalian organism? Not to make this point can yield a false picture of the contribution of processes of specialization during evolution, and thus distort the evolutionary interpretations.

D. KOSHLAND: You're right. I should have been clearer about that. I want to reassure everyone here that they are more complicated than an *E. coli* bacterium. There are more potential genes for enzymes and more actual enzymes in most mammalian cells. But some specialized cells might have even fewer enzymes than *E. coli* because only part of the DNA is read.

E. P. WIGNER: Could I ask what is known about the temperature coefficient of these reactions? That is always what interests the physical chemist or physicist first if he wants to explain or describe the differences between reaction rates.

D. KOSHLAND: If you compare the non-enzymatic with the enzymatic mechanism, let's say for the hydrolysis of an amide, the activation energy of the enzymatic reaction is generally lower. But if you look at the contribution of the rate you find that the entropy of activation, the pre-exponential factor, often goes up greatly. That pre-exponential factor in most enzymes is more important than the lowered activation energy. In fact there are some enzymatic reactions where the activation energy stays exactly the same. But both are important.

E. P. WIGNER: Could you say what is the order of magnitude of what you call exponential factor?

D. KOSHLAND: Well, in some non-enzymatic reactions which have an activation energy of 14 kcal, the enzymatic reaction may drop to as low as 5 kcal.

E. P. WIGNER: How much is the pre-exponential factor?

D. KOSHLAND: Of the order of 10^9 1/M/sec is the upper limit I believe.

D. C. PHILLIPS: Well, if there are no more pressing contributions, may I thank you for your attention, and the speakers for their presentations.

Proc. 3rd int. Conf. From Theoretical Physics to Biology, Versailles 1971, p. 303
(Karger, Basel 1973)

Antibody Heterogeneity

N. K. JERNE

Abstract

The theme of this session is the structural basis and the cellular origin of antibody specificity towards antigenic determinant patterns. It is well known that a given vertebrate animal can produce antibody molecules of a very large number of different specificities. In the present introduction, a collection of arguments will be presented that approach the question of determining the range of antibody diversity; i.e., how many different antibody molecules can an animal make? Is it possible to estimate the order of magnitude of this number?

Author's address: N. K. Jerne, Basel Institute for Immunology, 487 Grenzacherstrasse, *CH-4000 Basel* (Switzerland)

Proc. 3rd int. Conf. From Theoretical Physics to Biology, Versailles 1971, pp. 304–306
(Karger, Basel 1973)

Structural Basis of Antibody Specificity

G. M. EDELMAN

The Rockefeller University, New York, N. Y.

Abstract

Antibodies carry out two major functions in the immune response: 1) they bind to a great variety of sterically different antigen molecules (antigen-binding function or ABF); and 2) they mediate essential immune reactions within the organism, such as complement fixation (effector functions or EF). The ABF is *selective,* i.e. the information for synthesizing antibodies to a large range of chemically different antigens already exists in the organism before exposure to these antigens. Particular antigens serve only to stimulate (and thereby select) those antibody-forming cells that synthesize antibodies having appropriately complementary antigen-binding sites. This process requires that the number of sites must greatly exceed the number of different antigens. Furthermore, there must be an exceedingly efficient system for amplifying the production of specific antibodies after stimulation by injection of a particular antigen. This appears to be achieved by maturation and cell division of stimulated cells, each of which synthesizes only one kind of antibody. Division and maturation are followed by an increased synthesis of that particular antibody by the clonal progeny of the stimulated cell.

Studies of the molecular structure of antibodies or immunoglobulins have been useful in an analysis of some of the mechanisms of the selective immune response. The covalent structure of an entire γG immunoglobulin (molecular weight 150,000) has been determined and compared with portions of other immunoglobulin molecules analyzed by other workers. The light and heavy polypeptide chains of the molecule each consists of a variable, or V, region for antigen binding and a constant, or C, region for effector functions. The amino-acid sequences of V regions differ from molecule to molecule, whereas the sequence of C regions is relatively invariant. The mechanism by which diversification of amino-acid sequence is achieved in antibody V regions has not been established, but several theories on the origin of diversity have been proposed and will be reviewed.

The amino-acid sequence of γG immunoglobulin shows a linear periodic arrangement which suggests the evolutionary origin of the molecule by gene duplication. It is proposed that the molecule is arranged in successive compact domains organized about axes of symmetry and pseudosymmetry. According to this proposal, *domains* of V regions mediate antigen binding, and successive domains of the C regions mediate different effector functions. This is in accord with the hypothesis that the antibody molecule evolved from two genes, V and C, each of which underwent successive duplication.

The analysis of antibody structure suggests the possibility that a number of unusual genetic mechanisms may be required to account for the evolution of selective recognition systems. Evidence from several laboratories has led to the conclusion that each polypeptide chain is specified by a V gene and a C gene. It is suggested that the V gene may be translocated in the lymphoid cell to form a single VC gene. Translocation could account for the clonal nature of the selective response, for, although each cell may have information for many V sequences, genetic analysis indicates that it has at most only two C genes for each chain. Translocation would commit the cell to the expression of the smallest possible number of antibody chains and help to maintain the specificity and efficiency of the selective immune response.

Author's address: G. M. Edelman, The Rockefeller University, *New York, NY 10021* (USA)

Discussions

H. C. LONGUET-HIGGINS: I wasn't clear at what stage of the animal's life this somatic recombination is supposed to occur. All the time or when?

G. M. EDELMAN: It occurs in ontogeny. The process is presumed to occur in the development of an animal at the point at which immune organ systems begin to develop. At this point, this process is presumed to turn on and, therefore, the animal when born is already equipped with a reasonably competent system. Some believe that it continues throughout life, however.

J. NINIO: Do we know if *any* given component of the organism can be found inside a lymphocyte cell? Haemoglobin for instance?

G.M. EDELMAN: No. What is found in the lymphocyte is a nucleus with DNA, with chromosomes, a cytoplasm which is capable of producing ribosomes on which the messenger is read to make the protein chain, and as far as we can see, up to 50% perhaps of the ribosomes are dedicated to making the immunoglobulin product in a very active cell. In the lymphocyte this is the product made in large amounts and for which it is specialized. So there is no haemoglobin seen in the ordinary lymphocyte.

Proc. 3rd int. Conf. From Theoretical Physics to Biology, Versailles 1971, pp. 307–310
(Karger, Basel 1973)

Antigen-Antibody Relationship

M. Sela

Department of Chemical Immunology, The Weizmann Institute of Science, Rehovot

Abstract

The world of immunological phenomena may serve as a wonderful example for biological organization, as such organization—in three-dimensional space as well as in time—is typical for all its stages. This is valid for the recognition of the antigenic determinant, for the cooperation of cells in the induction of immune response, for antibody synthesis and the purely cellular immunological reaction, for the secondary response which implicates the immunological memory, for the induction and persistence of immunological tolerance, and—last but not least—for the tremendously specific interaction of the determinant group on the antigen with the combining site on the antibody. This interaction represents a uniquely specific pattern of recognition on a molecular level, and it involves two macromolecules.

The availability of synthetic antigens permitted the elucidation of the role of various molecular parameters in immunogenicity and in antigenic specificity. The parameters included composition, size, shape, accessibility, electrical charge, optical configuration and steric conformation. Antigens have been synthesized with specificity directed towards peptides and proteins, sugars, lipids, nucleic acids, drugs, etc. Investigation of the nature and size of combining sites of antibodies, making use of their interaction with hapten analogs and with affinity labeling reagents, led to their detailed mapping in terms of subsites.

The dynamics of antibody-hapten interaction were investigated by the temperature-jump chemical relaxation method on a homogeneous system: an anti-dinitrophenyl protein produced by a mouse plasma cell tumor and dinitrophenyl ligands. Only a single relaxation time was observed. More-over, no interaction between the two binding sites of the protein molecule could be detected.

Studies of synthetic models possessing either sequential or conforma-tion-dependent antigenic determinants led to the conclusion that their recognition by the biosynthetic site occurs while the immunogenic ma-cromolecule is still intact. Several examples of 'induced fit ' have been demonstrated recently in studies of antigen-antibody reactions. Thus, antibodies against a helical structure may induce helicity in a polypeptide which is largely in a random coil form when in free solution. Studies with the p-azobenzenearsonate hapten attached to either helical or random coil copolymers have demonstrated the importance of conformation, or envi-ronment, in which the hapten finds itself for its immunopotency. The role of conformation in antigenicity has been investigated also by synthesizing an immunogen capable of provoking antibodies that recognize a unique conformation-dependent region in the native hen egg-white lysozyme.

An inverse relationship exists between the net electrical charge of immunogenic macromolecules and the net charge of antibodies produced by them. Haptens will cause the formation of more acidic or more basic antibodies, depending on whether they are attached to a positively or a negatively charged carrier. The antibodies produced upon immunization with such conjugates differ in their charge, but not in their specificity or affinity. The net charge correlation has a cellular basis, as proved by making use of spleen cell fractionation on glass beads.

The capacity to provoke antibody formation to various synthetic polypeptides is controlled genetically. This genetic control in mice is determinant-specific, and is valid also for unique determinants of a complex immunogen such as the enzyme, lysozyme. It does not act at the level of the structural genes. Expression of the genetic control of inbred mouse strains to several synthetic polypeptides could be correlated with the number of antigen-sensitive precursors detected, using cells from spleen and bone marrow. In some cases the double-stranded polyadenylic-polyuridylic acid increased the antibody titers of low responder mice to the level in high responder mice.

It is clear that an antigen is much more than a determinant attached to something inert, and that the antigenic 'carrier' plays a crucial role in determining the structure and the biosynthesis of specific antibodies. Synthetic antigens have been useful, due to their simplicity, not only in elucidating the molecular basis of antigenicity, but also—in some cases

—in the detection of effects, such as genetic variations in immune response, which are not easily observable with complex natural antigens. With the advanced sophistication in molecular and cellular biology it seems that cellular immunological phenomena are now open to a molecular approach.

Author's address: M. Sela, Head of the Department of Chemical Immunology, The Weizmann Institute of Science, *Rehovot* (Israel)

Discussion

U. PALMA: As a physicist, I am perhaps allowed to ask a very simple minded question, which is in fact stimulated by Dr. EDELMAN's reply to Dr. FRÖHLICH. Have you taken into consideration the possibility that the recognition of any specific molecule might occur in a less individualized way. For instance, groups of molecules could be recognized for some common properties (e.g. symmetry) and this would cut down e.g. to 1% the 30% figure quoted by Dr. EDELMAN, which sounds impressively high.

M. SELA: It is a distinct possibility that an antibody can accomodate, even if we visualize a fixed antibody cavity, a few related antigenic determinants. Moreover, there is no reason *a priori* that such an antibody cavity could not transconform itself upon reacting with various antigenic determinants which, seemingly, are not closely related. The examples I gave were mainly to show the great changes that an antibody confers on an antigenic determinant, and not *vice versa*. As a matter of fact the few studies I described concern transconformation upon cross-reaction. For example, an antibody cavity transconforms a polymer which was not yet helical into a helical one. It is a case in which the antibody cavity causes a change within the antigenic determinants. What one would like, is to have nice examples the other way round: to see changes within the antibody combining site. The few studies which tried to show this, have shown that, if at all, the changes are very small. Undoubtedly there are some final changes, some kind of induced fit, but until now these have not been found experimentally. This does not mean, however, that they are not to be found.

U. PALMA: Perhaps this is one possibility, but the other possibility may be a real degeneracy, in that it is not necessary to accomodate for a change in shape, but just to allow for a mechanism, whereby a class of molecules can be classified as enemies so to say, without a real change in shape.

M. SELA: I don't know. At least the tremendous body of evidence in immunology, accumulated for almost 100 years, shows fantastic specificity and to assume that all this can be reduced to just a few molecule types would be very difficult.

L. COOPER: Dr. EDELMAN promised us that we would have the answer to the question of why one doesn't develop antibodies to one's own body.

M. SELA: I would say that Dr. NOSSAL can answer this question in his talk.

A. MITCHISON: I wonder whether you would accept an argument based on the reactivation of mutant enzymes by antibodies? It has been observed by POLAK and by MELCHERS that defective enzymes can be activated by combining them with antibody directed against the Wild-type enzyme. The fact that the effect works that way round, i.e. renaturing the mutant, whereas antibodies directed against the mutant do not appear to denature the Wild-type, tells us something about selection for stability. It confirms what MARGOLIASH told us this morning, namely that the conformation of an enzyme which we see as a result of evolution is much more stable than any of the alternative structures which can be reached by simple mutational steps.

Proc. 3rd int. Conf. From Theoretical Physics to Biology, Versailles 1971. pp. 311–319
(Karger, Basel 1973)

Recent Advances in Cellular Aspects
of Antibody Synthesis[1]

G. J. V. NOSSAL

The Walter and Eliza Hall Institute of Medical Research, Melbourne, Victoria

There are two key problems in cellular immunology today: the first
is, how does the genetic information for the biosynthesis of antibodies,
which must reside in the genome of the whole animal, spread itself out
amongst the individual lymphocytes of the body, which are the actual
responsive units? This is a problem in somatic genetics and cellular differentiation. The second point is, how does the antigen call forth the genetic
potential and convert that genetic potential into a synthetic reality? This
is a problem of cellular regulation. Neither of the problem areas is completely resolved, even in terms of the internal language of our own discipline. So it's clearly impossible, I think, to attempt a translation into the
language of physics or even of physical or organic chemistry. Nevertheless,
I think the skeletal outline of system design are clear by now, at least in
the language of the immunologist. So I thought it might amuse you if I
spoke to you for a few minutes about some of the actual experiments which
have brought us to our present position and then to see whether there is
anything here which may be of a wider interest. As cellular immunologists,
we have found the most useful frame of reference for our endeavours
BURNET's clonal selection hypothesis [1] put forward initially in 1957, a
modification of JERNE's [2] original views. That, simply put, says that if
a mouse has 10^9 lymphocytes, most of them are going to be different from
each other. Before the antigen comes into the picture, there is a somatic
diversification of the lymphocytes. Antigen encounters a cell pre-adapted
to react with it, and causes the cell to transform from a resting state, the

1   This work was supported by grants from the National Health and Medical Research
Council, Canberra Australia, The Australian Research Grants Committee, Canberra, Australia,
and the United States Public Health Service (AI-3958).

so-called G_0 phase of the mitotic cycle, into a more active state of cell division and differentiation. This process generates a whole population, a clone, of some hundreds of identical progeny cells that will, as well as being amplified as far as numbers are concerned, adopt an increased antibody synthetic rate at the level of a single cell. How could one come to believe in the truth in such a hypothesis? How is it that in the last 10 years the hypothesis has become so persuasive that most of us, as professionals in the game, can no longer think of the immune response in any other terms? Perhaps one started at the wrong end. One started where one could start in 1957, and that was at the level of the fully differentiated, secreting antibody forming cells [3]. Of course, the model would predict that one cell, fully turned on, would only form one antibody and that turned out to be the case [4]. But it is somewhat of an admission that it really took us a decade and by us I mean maybe 500 of us who have been engaged with this problem, to come to the definitive experiments which I now want to outline to you. The next step was to get some evidence that these lymphocytes that have not yet met the antigen have a receptor on their surface. They have something of the nature of an immunoglobulin. The evidence for immunoglobulin receptors on the lymphocytes is really threefold. First of all, it comes by a serologic trick. You make an antibody against an antibody and then use that as a chemical reagent. In that way we can show that the antireceptor antibody (using fluorescent probes or radioactive markers on the anti-antibody) specifically stains the surface of the lymphocyte [5]. So there is globulin on the surface of lymphocytes. Now of course that does not always tell you very much. The next thing you want to know is whether that globulin has any functional significance, in the regulation of the synthetic process of that cell. We can very rapidly arrive at the next experiment, which tells you that when those immuno-globulin molecules on the cell surface are covered up, with a sticky sub-stance, which can be an antiglobulin or something else, then that lymphocyte is no longer able to react to antigen. That begins to get a little closer to the idea that these immunoglobulin 'receptors' have some func-tion. The chemical identification of this lymphocyte surface immunoglo-bulin has been a matter of technical difficulty, as attempts to do chemistry on the cells is swamped a little bit by a subpopulation of differentiated, fully secreting cells. Recently, we have been able to do classical chemistry on the surface of the cells by taking the whole living lymphocyte and radioiodinating accessible surface proteins by a gentle technique that does not damage the integrity of the cell membrane [6]. By this simple trick we have been able to indeed show, by classical immunochemistry, that there is antibody on the surface of non-stimulated resting lymphocytes.

The next aspect is even more critical. One can do the very simple experiment of taking an antigen and labelling it with a radioactive isotope, say I^{125}, and mixing that in the cold, and in the presence of metabolic inhibitors, with let's say a million lymphocytes. One can smear those lymphocytes out on a slide and prepare an autoradiograph. That will give you the answer, has this lymphocyte bound to its surface the particular antigen A or not? The answer is that with any given antigen it is only a minority of the cells that has the capacity to bind it, e.g. 1 cell in 10,000 [7].

There is a tremendous heterogeneity in avidity of binding. So lymphocytes *are* heterogenous with respect to the antigen receptors that they carry on their surface. Now we can take that experiment, which again only tells you what is on the surface of the lymphocyte and does not tell you much about the function of the receptor, into a more functional phase by rendering the antigen radioactive to such a high degree that the electrons emitted during the disintegration of the isotope will kill any lymphocyte which has a sufficient amount of that particular antigen on its surface. This we call 'hot antigen suicide' [8, 9]. One can take a lymphocyte population mix it with three antigens A, B and C; A will be hot, and B and C will be cold. After 12 hours of waiting in the cold, one can test the immunological potency of that lymphocyte population. In fact, the result is a deletion of response capacity A, but the capacities B and C are left intact, just as BURNET's hypothesis would have predicted. We can do the same trick in a slightly different way and arrive at exactly the same result, namely we can use immuno-absorbant columns to physically remove from the lymphocyte population those lymphocytes with a receptor specific for antigen A, arriving at an A-potential depleted population [11]. Finally, though this has been technically a little more difficult, one has been able actually physically to enrich certain populations for cells with specific receptors, by either column or rosette techniques.

Let us restate BURNET's hypothesis in slightly more precise terms. The signal for activation of the lymphocyte is antigen meeting an antibody receptor on the surface, and it is quite clear that all of the critical events take place on the surface of the cell. By the time you have finished up with a fully switched on cell, there is no antigen inside the cell and frequently there is not even any antigen on it. The initial and key events at the cell surface are followed by de-repression, division and differentiation.

The antibody combining site, that is the variable portion of the light chain and the variable portion of the heavy chain, eventually made by that cell and by all its progeny are identical in all respects to those of the receptors on the lymphocyte before it ever saw antigen. That is what MITCHISON calls the 'accurate sample' hypothesis of lymphocyte receptors. The receptor on the surface is an accurate sample of the immunoglobulin that will be made at an immensely magnified rate by that cell after antigen activation. The cell must make an important choice; it has to make a choice at some time in its life history between being switched on, going through this business of clonal expansion and accelerated synthesis, finally secreting antibodies; or between being switched off thus achieving immunological tolerance. How can this signal discrimination between immunity and tolerance take place?

Tolerance is any condition whereby the proportion of cells capable of reacting to an antigen A is reduced from its normal level due to a pre-encounter with antigen A or a related molecule. Many people get the whole idea of tolerance wrong because they demand completeness. In fact it is rather doubtful, even in ourselves, whether tolerance is ever complete. We have probably got a few hundred or a few thousand molecules of antibody to our own red cells sitting on those red cells all the time and possibly fulfilling some functional role. So tolerance may very rarely be complete. Immunological memory is any condition whereby the proportion of reactive cells is increased to any significant degree, and is the opposite of tolerance.

Before we can attack the question of signal discrimination, the question of how antigen talks to a lymphocyte, we have got to surmount one complexity which has really only become evident in about the last three years and that is the two-fold world or universe of lymphocytes. There are really two basically different lymphocytes and two different types of adaptive immune response. On the one hand, we have what we have been talking about so far: cells secreting antibody into the blood stream, the concept of 'humoral' immunity. There is also another sort of immune response, where no antibody is secreted, at least not in ways that we can detect at the moment. We believe what is going on here is that there is a specific type of lymphocyte, which has immunoglobulin on its surface, has to make a physical encounter with antigen. Then certain metabolic changes take place in that lymphocyte, causing it to be activated and to secrete active mediators of cellular damage. This is what we call cellular immunity. Up to about three years ago, the chief interest of cellular immunity was the mechanism by which homograft or cancer cells are

Table I. Characteristics of mouse T and B lymphocytes

Characteristic	T lymphocyte	B lymphocyte
Function	Cell-mediated immunity. 'Helper' function in antibody production	Ancestors of antibody-producing cells
Origin	Thymus, after colonisation of this organ by hematogenous stem cells	Bursa of Fabricius in birds, bone marrow in mammals. Site of crucial differentiation step not known in mammals
Traffic	Extensive and relatively rapid recirculation. Large numbers found in paracortex of lymph nodes and peri-arteriolar lymphocyte sheath of spleen white pulp	Recirculation less rapid but definitely occurs. Preferred residence in primary lymphoid follicles, margin of germinal centres, medulla of lymph nodes and red pulp of spleen
Life-span	Recirculating cell has long life span measured in months	Median life span somewhat shorter. Recirculating pool takes some weeks to become labelled with ^{3}H-thymidine
Cells resulting from antigenic stimulation	'Killer cells', carrier-specific helper cells. New T cells with affinity for antigen concerned	Antibody-forming cells, including plasma cells and modified lymphocytes. New cells with affinity for antigen concerned
Immunological memory	Yes	Yes
Immunological tolerance	Yes; requires little antigen and inducible within 1–2 days	Yes; much harder to induce; requires more antigen and longer time; failure to induce has been reported in several systems
Surface Ig receptors	Difficult to demonstrate with antiglobulins though these give marginal evidence for IgM. Readily shown by surface radioiodination, suggesting 4-chained IgM molecule	Easily demonstrated by all techniques. IgM most readily shown. Adequate evidence also of α and γ chains on many cells, raising the possibility of > 1 heavy chain on some cells
Non-specific adhesiveness to glass bead columns	Low for non-activated cells. Higher for activated (larger) cells	Higher. Can be the basis for a partial separation technique
Surface markers of use for identification	Θ	MBLA
Sensitivity to PHA-induced mitosis	Yes	No, unless coupled to a solid phase

rejected. It has since become clear, and it is somewhat presumptuous of me to talk about this at all, because the two people who brought this into focus, JACQUES MILLER and AVRION MITCHISON are in the audience, that these two systems really collaborate with each other in many situations of antibody secretion.

Table I illustrates some of the properties of these two sorts of lymphocytes. T or thymus-derived lymphocytes mediate cellular immunity. B or bone marrow-derived lymphocytes are responsible for humoral immunity. In each case, antigenic drive causes a transformation to blast morphology with rapid mitotic rate and production of effector cells. Both types of cell exhibit immunological memory and immunological tolerance. And both of the types may collaborate with each other.

The basis of this collaboration rests on the need for multipoint binding of antigen to the lymphocyte surface in order to achieve activation. A *single* antigenic configuration hitting a *single* receptor, signals poorly, if at all, to the lymphocyte. Rather, a matrix or array of identical antigenic determinants must be presented to the lymphocyte to initiate the events of immune induction. Preliminary results indicate that T cells release a special type of antibody molecule (IgX), linked to antigen, which rapidly attaches to macrophages. Thus at the macrophage surface an immunogenic array of antigen is created, which in its turn activates the B cell. This is not the whole story, as stimulated T cells also release nonspecific factors that facilitate induction of B cell activation independent of antigenic specificity.

The basis of the tolerance-immunity signal switch is still controversial. We have developed the view that one important factor leading a lymphocyte to become tolerant is an oversaturation, at one local receptor patch, of antigenic determinants. This can be achieved—inefficiently—by high molar concentrations of antigen. It can also happen when the cell is presented with low concentrations of antigen polymers with closely-spaced antigenic determinants—e.g. soluble complexes of antigen and antibody in the zone of antigen excess, or haptens plastered at high density onto a polymeric carrier. We have model situations in my laboratory where the direction taken (towards immunity or towards tolerance) depends exclusively on the spacing between antigenic determinants on a polymeric carrier [12].

A final point about tolerance induction: it seems to be easier in T than in B cells, requiring less antigen and less time [10]. We are still at a loss to understand why this should be so.

References

1 NOSSAL, G. J. V.: Lymphocyte-antigen interactions and their relevance to medicine in the 1970's; in Sonderdruck der Alma mater philippina (Marburger Universitätsbund, Marburg 1972); formal acceptance oration for the Emil von Behring Prize of the Philipps-University of Marburg, October 1971.
2 JERNE, N.K.: The natural selection theory of antibody formation. Proc. nat. Acad. Sci., Wash. *41:* 849 (1955).
3 NOSSAL, G.J.V. and LEDERBERG, J.: Antibody production by single cells. Nature, Lond. *181:* 1419 (1958).
4 NOSSAL, G.J.V. and MÄKELÄ, O.: Elaboration of antibodies by single cells. Ann. Rev. Microbiol. *16:* 53 (1962).
5 RAFF, M.C.; STERNBERG, M. and TAYLOR, R.B.: Immunoglobulin determinants on the surface of mouse lymphoid cells. Nature, Lond. *225:* 553 (1970).
6 MARCHALONIS, J. J.; CONE, R. E. and ATWELL, J. L.: Isolation and partial characterization of lymphocyte surface immunoglobulins. J. exp. Med. *135:* 956 (1972).
7 NAOR, D. and SULITZEANU, D.: Binding of radioiodinated bovine serum albumin to mouse spleen cells. Nature, Lond. *214:* 687 (1967).
8 ADA, G.L. and BYRT, P.: Specific inactivation of antigen-reactive cells with ^{125}I-labelled antigen. Nature, Lond. *222:* 1291 (1969).
9 BASTEN, A.; MILLER, J.F.A.P.; WARNER, N.L. and PYE, J.: Specific inactivation of thymus-derived (T) and non-thymus-derived (B) lymphocytes by ^{125}I-labelled antigen. Nature New Biol. *231:* 104 (1971).
10 CHILLER, J. M.; HABICHT, G. S. and WEIGLE, W. O.: Kinetic differences in unresponsiveness of thymus and bone marrow cells. Science *171:* 813 (1971).
11 WIGZELL, H. and ANDERSSON, B.: Cell separation on antigen coated columns. Elimination of high rate antibody-forming cells and immunological memory cells. J. exp. Med. *129:* 23 (1969).
12 FELDMANN, M.: Induction of immunity and tolerance *in vitro* by hapten-protein conjugates. I. The relationship between the degree of hapten conjugation and the immunogenicity of DNP-polymerized flagellin. J. exp. Med. *135:* 735 (1972).

Author's address: G.J.V. Nossal, Director, The Walter and Eliza Hall, Institute of Medical Research, *Melbourne, Victoria 3050* (Australia)

Discussion

A. MITCHISON; I want to comment briefly to make it clear that some of what NOSSAL told us is generally agreed on, while other parts are not. I think that everything he told us is agreed on, until he comes to the matter of signal discrimination in the induction of tolerance. That is a much more controversial matter.

J. F. A. P. MILLER: Could Professor NOSSAL predict what might happen if a carrier was used which was a *normal* component of the body, and on this carrier was conjugated, at various substitution ratios, a determinant such as DNP. Would one get tolerance or immunity?

G.J.V. NOSSAL: It is known that this type of product can cause tolerance, and normally it does not cause immunization. However, I predict that if the 'self' component were appropriately polymerized, immunity could result. Both antigen dose and spacing would then be the determining factors.

L. COOPER: Earlier in the afternoon we spoke about the recognition of individual small groups of molecules by the potential antibody-producing molecule in the cell. Now, as I understand it, what you're saying is that what has to be recognized is a repetition of this particular small radical; not two different radicals, A and B, but a repetition of the same small one. Is that correct? So, in order for the cell on the molecules to respond, one would have essentially some nonlocal; it would have to act over distances, probably rather large ones, compared to the size of the individual radical group.

G.J.V. NOSSAL: That certainly would be my view. One is groping for words here, but words like distortion of the membrane to a certain critical degree are the sorts of words one gropes for. The concept of surface rearrangement of receptors that Dr. TAYLOR and RAFF are working on could be important in activation.

Would any of my professional colleagues here disagree with the concept that a single molecule, no matter what size, having only one antigenic determinant, doesn't speak with the cell at all, signalling neither immunity nor tolerance?

N. A. MITCHISON: I think that we all agree that a matrix of antigenic determinants stimulates a cell more effectively than do separate molecules of antigen. The question is whether matrix building is important simply in permitting multipoint binding and that is all, or alternatively whether the matrix plays a deeper role in increasing the *density* of antigenic determinants in the way that NOSSAL suggests.

E. KATCHALSKI: Is there any unequivocal proof that the receptor on the surface of a lymphocyte consists of antibody molecules: would you accept the possibility that the receptor differs in its chemical structure from that of the corresponding antibody?

G.J.V. NOSSAL: For the case of the B cell, the answer is yes. There is a chemical proof that the molecules which are sitting on the membrane are indeed antibodies. Now there is a complexity which I was hesitant to develop. The complexity is that the receptors carry a heavy chain which is not the dominant heavy chain in the serum, i.e., the μ chain. How are you going to end up with a lot of IgG being secreted when there is IgM on the surface of so many cells? It turns out not to be as grave a problem as it first appeared, because the V genes are shared. Light chains in IgM and IgG are identical, and the variable portion of the heavy chain, as far as we can tell, are also shared between γ, α and μ.

E. KATCHALSKI: Do you have any suggestions as to the mechanism by which antigen molecules trigger antibody formation after being in contact with the receptor of the cell membrane?

G.J.V. NOSSAL: The question here is whether you want facts or dreams. If you want facts, the answer is that there are none. If you want dreams, there may be an allosteric change in the receptor, with consequent release of a regulatory molecule.

E. KATCHALSKI: Are there any facts to support any one of the hypotheses which have been forwarded?

G. J. V. NOSSAL: None strictly relevant.

E. KATCHALSKI: By forming a multifunctional antigen you increase the apparent binding constant of the antigen with the receptor. Is it plausible to assume, therefore, that the stimulus for antibody formation is a function of the time of contact of the antigen with the receptor. As this triggering period is prolonged, antibody formation is enhanced.

G. J. V. NOSSAL: This is indeed not far from our views.

D. METCALF: If I could just get onto a noncontroversial subject that might interest our physicist colleagues, you recall the discussion wherein we were trying to describe the evolution of biological systems? There are two features of the immune and other blood-forming systems that are quite intriguing to speculate about. First, like many other tissues, there is a tremendous reserve of cellular capacity beyond what appears to be required and this is maybe of the order of 99%. Why is this so? What possible advantage would it be to have such a tremendous excess? This is seen to a marked degree in the population of lymphocytes. The second point is that the capacity of cells to proliferation in response to stimulation is clearly finite. Each cell has only the capacity for a limited number of divisions. Now it turns out that the cells that generate blood cells and lymphocytes, have a tremendous reserve nonetheless. so that a mouse which only lives 2 years has blood ancestral cells that are capable of keeping the mouse supplied for 200 years. It is interesting to speculate about the possible significance of this in terms of survival advantage. Why would such a system develop? It is possible that it involves an efficiency in function that we have not yet thought of. These are two other parameters of living cell systems that the physicist might like to take away with them in trying to think of how a system might organize itself.

Systems of Sensorial Analysis

Neurophysiological Aspects of Vision

Proc. 3rd int. Conf. From Theoretical Physics to Biology, Versailles 1971. pp. 322–327
(Karger, Basel 1973)

Introduction

D. M. MacKay

I think the title in our programme for this morning may be misleading. To an Englishman at least, the French suggests something like defiance between two cliques, as if the experimentalists assembled were now going to wave their flags at the theoreticians like the knights on the wall panels here, with cries of 'Charge, boys!' But of course it isn't like that at all. The situation is that all of us here, experimentalists or theoreticians, find *ourselves* challenged, by data of perhaps the most subtle and the most complex kind that we come up against in science: the data of our own experience. Our question this morning is twofold: first, what kind of challenge is this? what are the challenging data (if we've got them)? Secondly, in what ways, if at all, can we expect 'theory' in the usual sense to be applicable? What kind of theory is needed? In other words, our problem here is not merely to gain data nor to apply some standard theory, but rather to conceive of the kind of questions it would be profitable to ask about the data. This means, of course, that if the game is to go well at all, experiment and theory should combine in the same minds. So what my colleagues on this morning's program are going to do is to give a typical sample of the sorts of data which are coming up now, in one particular sensory field, the field of vision, hoping that in the discussions, those of you who are not physiologists of vision will be ready to ask, 'Why should we be interested?' And to ask any questions which will enlarge for you the understanding of the kind of challenge that this offers to the theorist. This is, if you like, a fishing expedition. But the experimentalists in vision here present are not the fishermen; they are, if you like, the hook. The fishermen are the data.

I think it may be helpful to set the scene by saying first one or two things about the problem of perception in general. Having come into the field as a physicist myself, just over 20 years ago, I can well remember some of the stumbling blocks past which I had to trip; and I think it may be good to keep a few key points in mind as landmarks when we try to relate what we are going to hear this morning to what has gone on in the rest of the week.

The central notion that focuses thinking about living organisms, for today's purpose, is the notion of *agency*. In other words, what is characteristic about a living organism for our purpose is that it *acts in view of ends,* and that it matches itself to its world by way of a sensory system. We have to think of the brain, then, not merely as a physical system, nor merely as a physiological, biochemical system, but also as an *organization.* The image of an organization (for example, a human organization) has to be kept in our minds all the time as well as images appropriate to the physical systems which we know of course to exist in the brain.

Indeed, one of the questions especially appropriate in a gathering such as this, is how we can best combine thinking about the brain as a physical system with thinking about it as an organization, so as to devise good experiments, pointed experiments, experiments with significance, whose results may bring illumination. There are many sorts of organization we can consider by way of a corrective to over-simple notions about the brain. Think of a sales organization, a military organization, a relief organization for example. The central notion in all of these is that of a *repertoire of action* (a concept that has already turned up in another context this week). This repertoire has to be used in a world which has a certain structure, certain regular, predictable features; and the problem for the organism from this formal point of view is that of selecting from the repertoire in such a way as to further the ends of the organism in the world. The sensory system can then be regarded as a surface which is in contact with that world, on which signals from the world impinge, its function being to bring up to date and keep up to date the organization (the 'conditional readiness') of the selective process in the repertoire (fig.1). Vision from this point of view, has the function of enabling such things as locomotor action and the planning of action to take place efficiently, so that obstacles are avoided, prey identified and pounced upon successfully, and so forth. The organizing system, then, can be thought of as determining implicitly (not, of course, explicitly!) an enormous conditional probability matrix—determining the probabilities that, in given circumstances, *if* the organism wants to do a particular thing, *if* the goal is such and such, *then* a particular course of

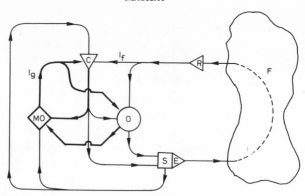

Fig. 1. A skeleton map of the organization of action in a field F from which a sensory system R extracts both feedforward and feedback information for the organizing system 0. The indicated state of the field I_f is evaluated against criteria I_g in C, and the outcome causes 0 to select subroutines via S from the effector repertoire E. MO is the 'meta-organizing' system which determines I_g and updates the 'conditional readiness' of 0 to match the demands of feedforward from R. (From "Ways of Looking at Perception" in *Models for the Perception of Speech and Visual* Form (Weiant Wathen-Dunn, ed.), 25–43, M.I.T. Press, Boston, 1967, reprinted in Perceptual Processing (P.C. Dodwell, ed.), Appleton-Century-Crofts, 1970).

action will be chosen. Now of course the repertoire is not all on one level. Just as with an organ you have the ordinary stops and then combination stops, so we can think of the effector system as extended by having many subroutines whereby individual muscles are coordinated into groups for skilled actions and so on. And then of course playing on the whole of this, the organizing system itself must develop a hierarchy of internal subroutines of a more abstract kind, operating not by selecting muscle sequences, but by concatenating subroutines which represent abstract concepts of a kind that at the human level we can express in words. This brings up an intriguing problem which might be regarded as a major 'challenge to theorists': in a hierarchic organizing system—and, again, to make this concrete, keep in mind the image of a human organization —how best can you arrange to feed in information from the sensory surface to the different levels in the organizing hierarchy? Through what kinds of processing filters and computing operations should the information go with a view to optimizing the state of readiness for action in the world in the minimum time? That's the sort of biologically adaptive arrangement of the sensory system that one would expect to find in a well-organized brain.

This leads me to one of the main points I'd like to make: that if you think about the brain from this point of view, you will be much less likely to try to understand it, as many of us did when we came fresh to brain research, by analogy with some kind of 'input-throughput-output' transmitting system. Think for example of a human organization, say a sales organization. Out at the periphery there are any number of individual purchases being made; but the managing director, or his committee, don't want information about the individual purchases. They want information which will help them to make high-order selections in the hierarchy of conditional policies for the organization. So of course only a tiny fraction of the informational traffic which impinges on the periphery of the organization reaches the high-level committee. My point is that if you had thought of the whole organization as an information channel, as those of us who started from communication engineering and communication theory might be tempted to do, you would find it derisorily inefficient. No doubt you could make something of a joke by pointing out how 'wasteful of information' is the 'channel' from the periphery to the managing director of the organization. But if you said it seriously and tried to persuade the managing director that he ought to have a high capacity channel carrying all the information from the periphery to his office, the joke would be on you. For you would have missed the point of the whole organization.

In relation to the central nervous system, I think this is very important to keep in mind. Nothing is easier than to show that, viewed as an information channel, the pathways from the retina to neurons deep in the central nervous system carry only a fraction of the informational traffic impinging on the sense receptors. As a joke we could describe this as a 'terribly inefficient' system. But if we try to understand it as a system whose function is to update an organization, then of course its true efficiency would be measured by something quite inverse: namely, the effectiveness with which it *prevents* information that is not relevant to decision making from troubling the central system. Again, you could make a joke of this and say that one of the main functions of the sensory information processing in the brain is to throw away information. But once we've smiled, let's see the point. That is what it is designed to do. That is what it ought to do if it's to be biologically effective. Measurement of information flow can be quite inept as a guide to efficiency. If we keep this in mind it may throw a good deal of light on the reasons for and the probable functions of the various visual mechanisms that are going to be described this morning.

One other point. We are going to have some evidence from the field of visual psycho-physics. Indeed, all of us are likely to make reference to this. This, of course, means using your own sensory experience as a source of clues as to what is going on in your sensory system. There is a particular trap here that is worth noticing. So far I have avoided mentioning perception. I've spoken only of the sensory system as keeping up to date the organism's conditional readiness for action in the world. The link between the mechanical and the perceptual is very much a matter of controversy and speculation; yet if we don't think carefully about it we can easily misinterpret the physiological significance of psychological data. The trap is to assume that what we see in sensory experience must have some direct *analogue* in the physiological activity in our retina, or in our visual pathway. To show that this need not be the case, I would like us to consider (without implying that it is the only candidate) the possibility that what correlates directly with our conscious experience in perceiving (i.e., with what we actually perceive, as distinct from all the information that bombards our sensory surface) is an internal *matching response* of the organization to the sensory demands on it. In other words, I suggest that the *updating of the central organizing system to match selected features of the sensory input* meets some of the main requirements for a physical correlate of our conscious experience. If we agree that this is at least an open possibility, then we will have to keep in mind two quite different possible sources of optical illusions and the various other informative psychophysical effects that we'll be looking at. On the one hand the sensory channel itself is likely to be plastic, being made of biological material. If you feed it with an abnormal sensory diet you may expect changes in its transfer characteristics which may give rise to perceptual illusions. An example of this is the familiar complementary aftereffect of looking at patches of different colours. This is typical of many aftereffects which can be studied psychophysically and which throw some light on mechanisms on the input side. Notice, however, that if you took this as your only or your main guide, the first thing it might suggest to you is that you ought to find, perhaps in the retina, two colour coding channels: one of them for red-green and the other for blue-yellow. You wouldn't on the basis of this be led to suspect that there are in fact three colour pigments, and that colour information is initially represented by the balance of activity of cells sensitive to three different regions of the spectrum, rather than four or two. So, on the one hand, a psychophysical effect can usually suggest a good experimental question; but on the other hand you mustn't be surprised if the answer is quite different from what subjective experience might suggest, because of course there are many more levels of processing, as we'll be hearing; and it's probably due to a combination of signals at a later state that this particular polarity between blue and yellow, and red and green, is observed.

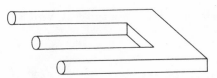

Fig. 2. A well-known optical illusion which seems likely to reflect a mismatch on the 'response' side of the organization of the brain.

By contrast, figure 2 shows a familiar perceptual anomaly which I think it is fair to say is unlikely to be due to any anomalous process at the retina or on the input side. Here one could say that the author has wickedly drawn the thing so that the internal mechanism for matching your readiness to the object gets conflicting clues; and in terms of the concept of perception I have been advocating, one would look on the response side of the organization of your brain, rather than on the input side, for the physiological correlates of the subjective anomaly.

In summary, note that from this operational standpoint we have no reason to expect any kind of 'television screen' somewhere in the brain. That's not the obvious function of the visual system, nor of any of the sensory systems. It is rather an *organizing* function. Secondly, we must beware of regarding all the subjective phenomena that we come across in vision as some sort of privileged way of witnessing the physiological events in our retinae. The illusions we'll be seeing, as Dr. JULESZ in particular will be emphasizing, may have their origin at many levels in the information processing system. The old question: Is what we see 'really' on the retina or 'really' in the external world? is a philosophical howler: a badly posed question. Seeing, I have been suggesting, is matching one's state of organization to the visually-mediated demands of the world. The only 'location' for what is seen is in that world. (Even when a visual experience is artificially induced by electrical brain stimulation, the *perceived* location of what is seen is out in front of the subject.) Your retina is one of the links in the chain of demand on your brain organization; and the *physical correlates* of what you see have, of course, their locations all along that chain. But I can't imagine any circumstance in which it would be sensible to say that *what we see* is located 'at' the retina, rather than 'at' the visual cortex, or halfway down the light path from our eyes to the object. Forgive me for having laboured what may seem obvious; but it is sometimes neglected in psychophysiological discussions, with confusing results.

Author's address: Prof. D. M. MacKay, Department of Communication, University of Keele, *Keele ST5 5BG* (England)

Proc. 3rd int. Conf. From Theoretical Physics to Biology, Versailles 1971, pp. 328–373
(Karger, Basel 1973)

The Logic of the Retina

F. Ratliff

The Rockefeller University, New York, N.Y.

Introduction

The retina, as its name implies, is a network. The functional units in it are not independent—they are interconnected and interact with one another; the neural activity generated by illumination of any one photoreceptor may influence, or be influenced by, the activity generated by many others. Consequently, many visual phenomena may depend as much upon the properties of the network of interconnected receptors and retinal neurons, acting as a whole, as they do upon the properties of the individual components, acting alone.

This vital and organic activity of the retina may be summarized briefly as follows: light absorbed by photoreceptors stimulates them to generate neural activity in the form of graded local excitatory and inhibitory influences. These opposed influences are then integrated, over both space and time, by the neural networks in the retina. By means of this simple calculus the retina abstracts information about significant features of the spatial and temporal pattern of light and shade on the receptor mosaic. Ultimately, these abstractions are expressed in a neural code signalled by a pattern of discrete identical nerve impulses that are transmitted along the optic nerve to the brain. This set of abstractions and transformations may quite properly be called 'the logic of the retina'.

Here the word retina is used as a broad generic term, for this analysis is frankly comparative. Although the highly developed eyes of vertebrates (such as our own), of molluscs (such as those of the squid), and of arthropods (such as the lateral eye of the horseshoe crab) are, of course, different in many respects, they nevertheless have many basic structural and functional features in common. The aim of this paper is first to elucidate a few

fundamental processes and general principles that are common to the
retinas of a wide variety of species at many different phylogenetic levels.
Later it is shown how complex special properties of particular retinal
networks, and of networks elsewhere in the nervous system, may arise from
simple variations of these fundamental processes. The approach is quantitative throughout, but always with the constraint that any mathematical
formulation be derived from, and tested by, direct experimental observation
on real neural networks.

Historical Background

The basic idea that the retina has a certain logic of its own based on
the interplay of opposed influences is not new—truly scientific work on
the subject began over 150 years ago. And, as will be shown later on,
practical aspects of the problem may be traced even farther back in history.
Indeed, the work of some artists and artisans in ancient times foreshadowed
some of the most recent scientific advances that have been made in this
area. But first, as an introduction to modern work on the logic of the retina,
let us consider briefly the contributions made during the 19th century by
four giants in the field: GOETHE, CHEVREUL, HERING, and MACH.

JOHANN WOLFGANG VON GOETHE, the famous German poet and author
(1749–1832) was also an amateur scientist—an amateur whose ideas were
not always well received by the scientists of his day. GOETHE's *Theory of
Colors* (1810), which seemed to contradict at every point the well-estab-
lished and widely accepted Newtonian theory of light, was severely criticized
from all sides [see MAGNUS, 1961]. Today most physicists regard it as little
more than a curious aberration in the history of science. But much of the
confusion about and criticism of GOETHE's work has resulted from a failure
to understand and to appreciate the distinction between the physical and
the psychophysiological aspects of color. What was not generally recognized
at the time—and not always even today—was that many of GOETHE's
conceptions about what he called 'physiological colors' were absolutely
correct. Most important was his view that the appearance of a color
depended not only on light impinging on a particular place on the retina
at any particular time—but also upon interactions within the visual system
resulting from the temporal succession and spatial distribution of different
kinds of illumination—that is to say, on the processes of successive contrast
and simultaneous contrast. Many basic principles of these phenomena he

demonstrated by impeccable experiments; most important and incontro-
vertible, perhaps, where those on colored shadows and similar contrast
phenomena. His general views are well expressed in this one summary
statement on simultaneous brightness contrast:

'A grey object on a black ground appears much brighter than the same
object on a white ground. If both comparisons are seen together the
spectator can hardly persuade himself that the two greys are identical. We
believe this again to be a proof of the great excitability of the retina, and
of the silent resistance which every vital principle is forced to exhibit when
any definite or immutable state is presented to it. Thus inspiration already
presupposes expiration; thus every systole its diastole. It is the universal
formula of life which manifests itself in this as in all other cases. When
darkness is presented to the eye it demands brightness, and *vice versa;* it
shows its vital energy, its fitness to receive the impression of the object,
precisely by spontaneously tending to an opposite state.'

Concerning color itself GOETHE wrote: 'If a coloured object impinges
on one part of the retina, the remaining portion at the same moment has
a tendency to produce the compensatory colour . . . . Thus yellow demands
purple; orange, blue; red, green; and *vice versa:* thus again all intermediate
gradations reciprocally evoke each other; the simpler colour demanding
the compound, and *vice versa.*'

No matter how far from the mark GOETHE may have been in his views
on the physical aspects of light, it is now evident that his concept of a
simultaneous reciprocal interaction of opposing or complementary influences among neighboring parts of the retina—derived mainly from his
studies of color contrast and brightness contrast—is basic to our understanding of the psychophysiology of the visual system.

MICHEL EUGÈNE CHEVREUL (1786–1889), the great French chemist
whose work on fats and organic analysis opened up two major fields of
research in biochemistry, also made important contributions to the
psychophysiology of vision. In 1824, CHEVREUL was called upon to direct
the dye plant of the famous Gobelin Tapestry Works in Paris. At the time
of his appointment many complaints were being made about the quality
of some of the pigments prepared at the plant. To preserve the good name
of the Gobelin, which had been founded in the 1500's and later made a
royal manufactory under Louis XIV in 1662, it was necessary for CHEVREUL
to discover and to eliminate the cause of these complaints. He found almost
at once that some of the complaints about gradual fading of certain light
colors were well founded and he undertook chemical research to discover

the causes of the impermanences. But he could find no chemical basis at all for persistent complaints that the weavers made about the lack of vigor in some other colors. He first theorized—and subsequently demonstrated by experiment—that the cause was not in the dying, but in the weaving: the appearance of a particular yarn in a woven tapestry was determined not only by the color and tone of the yarn itself, but also by the color and tone of other yarns woven with it.

CHEVREUL's rediscovery and analysis of these contrast phenomena was of great significance because the practical problems he faced led him to formulate general laws of simultaneous contrast, soundly based on experimental observations. His great work on 'The Principles of Harmony and Contrast of Colors and Their Application to the Arts', in which he published these observations and general laws in 1839, is still used by artists and artisans. Indeed, it is still so much in demand that MARTEL's English translation was reprinted just a few years ago [CHEVREUL, 1967].

CHEVREUL's main conclusions were: 'Whenever the eye sees two differently colored objects simultaneously, the analogous character of the sensation of the colors undergoes such a diminution that the difference existing between them is rendered proportionately more sensible in the simultaneous impression of these two colors on the retina.' Furthermore, his experiments seemed to show: 1) that the contrast effect radiates out from the line of juxtaposition of the two colored surfaces; 2) that it is reciprocal between the two surfaces juxtaposed, and 3) that the effect still exists when these two surfaces are at a distance from one another, only it is less evident than when they are juxtaposed.

The evident simultaneity and reciprocity of the contrast of contiguous colors and contiguous tones led CHEVREUL to attempt to represent the phenomena in quasi-mathematical terms, which would express all the facts succinctly. These expressions were rather like the set of simultaneous equations used today to describe complex systems of mutually interacting components—but they were not too successful, mainly because he had no psychophysical suitable methods for the proper measurement of the quantities involved. Nevertheless, many basic ideas—generally accepted today—were implicit in CHEVREUL's laws and equations.

EWALD HERING (1834–1918) studied medicine at Leipzig and was a practicing physician for a few years. Most of his scientific life, however, was devoted to the study of general physiology, including such diverse subjects as the microscopic structure of the liver and the interrelation between neural centers for the control of breathing and for vasomotor control. This latter work led to the correct explanation of periodic variations in certain blood-pressure records now known as Traube-Hering waves. But HERING's major interest was in the psychophysiology of vision [see HERING, 1964]. In this field he was a strong opponent of the then popular mentalistic explanations of sensory phenomena couched in such terms as 'judgment' or 'unconscious inferences'. Rather, he sought the laws of sensory behavior and from them hoped to deduce the underlying laws of physiology.

To HERING belongs much of the credit for recognizing that reciprocal interactions in the retina underly our normal everyday visual experience as well as the so-called 'illusions' and 'distortions' such as border contrast and color contrast. Concerning this HERING wrote:

'The idea that the phenomena of simultaneous contrast are not merely "optical illusions", but rather the expression of an important vital characteristic of the visual system, has been emphasized especially by Plateau and Mach, although these men developed these ideas in a way that is different from the one I shall try . . .'

'The most important consequences of reciprocal interactions are not at all those expressed in contrast phenomena, that is, in the alleged false seeing of the "real" colors of objects. On the contrary, it is precisely the so-called correct seeing of these colors that depends in its very essence on such reciprocal interactions, and it is much more important to investigate the latter in the situation where we are not at all aware of them rather than where they attract our attention as contrast phenomena. It is to reciprocal interactions in the somatic visual field that we owe, to a large extent, our visual acuity . . . as well as the possibility of recognizing objects by their colors . . . A closer familiarity with the consequences of this reciprocal interaction is essential for understanding the nature of our vision . . .'

HERING's extremely important hypothesis about the role of reciprocal interactions in normal color vision, although long supported by psychophysical evidence, has only recently been vindicated by direct electrophysiological experiments. The classical Young-Helmholtz trichromatic theory must be extended, as HERING insisted, to include excitatory and inhibitory interactions among various pairs of the three basic cone types with their three different sensitivities to the visible spectrum (maxima in the red, green, and blue).

The life and work of ERNST MACH (1838–1916), the well-known Austrian physicist-philosopher-psychologist, are characterized by the great diversity of his interests and the great extent of his influence on modern science and philosophy. In these days of supersonic planes and rockets,

MACH's name is practically a synonym for supersonic phenomena. Indeed, Mach numbers are almost household terms. But MACH made many other and much more important contributions to science and philosophy [see COHEN and SEEGER, 1970] for which he is less well known today. His experimental research in both physics and psychophysiology, his critical-historical analyses of the origin and development of scientific concepts, and his view of science and epistemology resulting from these various investigations brought him recognition in the fields of physics, psychophysiology, and philosophy; and they exerted a profound and lasting influence on subsequent developments in those areas. Most significant for this topic, however, was MACH's interest in the theory of knowledge, which frequently caused him to turn from the study of physics to the analysis of the senses. And early in his career he had a practical and more compelling reason—he could not afford the expensive equipment necessary for the experimental work that he planned to do in physics. As a result, much of his early work was devoted to the study of vision, which could be carried out with little or no optical equipment, other than his own eyes.

During one of these excursions into visual psychophysiology MACH discovered, more or less by accident, a visual phenomenon which now bears his name—the Mach bands. While experimenting with rotating discs used to produce various spatial distributions of light and shade, he found light and dark bands appearing where, according to physical calculations, none were expected. Formerly, such phenomena as these Mach bands had generally been attributed to 'unconscious inferences' or 'errors of judgment' on the part of the observer, or had been dismissed as 'mere optical illusions' unworthy of study. But for MACH these were not explanations, they were merely various ways of expressing the still unexplained facts. Through careful psychophysical experiments MACH sought to find an explanation in terms of the interdependence of neighboring points on the retina, an interdependence which he believed could be accounted for in terms of the functions of the known neural interconnections among those points. As MACH expressed it [see RATLIFF, 1965, p. 306]:

'Since every retinal point perceives itself so to speak, as above or below the average of its neighbors, there results a characteristic type of perception. Whatever is near the mean of the surroundings becomes effaced, whatever is above or below is disproportionately brought into prominence. One could say that the retina schematizes and caricatures. The teleological significance of this process is clear in itself. It is an analog of abstraction and of the formation of concepts.'

The general concept of a simultaneous reciprocal interaction of opposing or complementary influences among neighboring parts of the retina formulated more or less independently by GOETHE, CHEVREUL, HERING, and MACH, is basic to a proper understanding of the psychophysiology of the visual system. But their ideas were not widely accepted in their own time. Indeed, until the direct electrophysiological experiments on the whole optic nerve by ADRIAN and MATTHEWS [1928] and by HARTLINE [1932, 1938] on single optic nerve fibers showed otherwise—psychophysiologists generally regarded the retina as a mere passive and rather mechanical transducer, rather than the vital integrative organ which it is. To illustrate a few major aspects of the integrative action of the retina let us now confine our attention for a while to just one visual phenomenon: the Mach bands.

The Mach Bands

The Mach bands may be seen quite distinctly if one pays close attention to any half shadow such as that cast by an object in good sunlight. For example, stand with your back to the sun and look closely at the shadow of your own head and shoulders on a sidewalk. The objective illumination is high and uniform in full sunlight, more or less uniformly graded in the half shadow, and uniformly low in the full shadow. That is to say, objectively, there are no maxima and minima. Yet you will see a narrow dark band at the dark edge of the half shadow and a narrow bright band at the bright edge. The effect will be enhanced somewhat if you sway from side to side so as to produce a moving pattern of light and shade. (It is best to do this in some secluded spot, unless you want to gain a reputation as an eccentric.)

The apparent brightness is not isomorphic with the physical distribution. There are no such maxima and minima in the actual distribution of illumination in these shadows. Indeed, the subjective appearance and the objective distribution of light and shade are seldom, if ever, isomorphic. If one looks at any pattern of illumination in which a dark area is contrasted with a light area along a margin of intermediate contrast, as shown in figure 1, one always sees instead of the objective rectilinear distribution (represented by the dashed lines) an apparent darkening and lightening (represented by the solid curved lines). Everyone does not see them in exactly the same way as illustrated here. In fact, these very graphs show that there

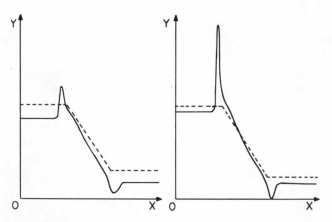

Fig. 1. The subjective appearance (solid lines) of a rectilinear objective distribution of illumination (dashed lines). X represents the angular width of the pattern of illumination (the gradient subtends an angle of about 40 min of arc at the eye). Y represents both the objective luminance (dashed line) and the apparent brightness (solid line), properly scaled. The apparent bright and dark bands (the maximum and minimum in each subjective distribution) are the so-called Mach bands. The two graphs represent results obtained from two skilled observers. Note that although the Mach bands are qualitatively the same for the two observers, there are large quantitative differences. Redrawn after FIORENTINI and RADICI [1957].

are substantial differences between the results obtained from two skilled observers in careful psychophysical experiments. Although the maxima and minima (the Mach bands) are qualitatively the same for the two observers, there are large quantitative differences.

The subjective appearance of the Mach bands is so striking that many people do not believe, at first, that the bright and dark bands do not have objective counterparts. Indeed, more than once they have been mistaken for objective phenomena. For example. shortly after x-rays were discovered, the Mach bands were mistaken by several investigators for maxima and minima of an x-ray diffraction pattern and the wavelength of x-rays incorrectly deduced from them [see RATLIFF, 1965, pp. 209—214]. As we shall see, what these investigators probably measured was the effect of neural networks in their own eyes and brains.

Since the Mach bands delineate contours that we expect to see, no one but the most careful observer (or someone who has reason to measure objectively the distribution of light) is apt to notice them or to realize that they are a caricature of the objective distribution of illumination. Some artists are careful observers of the world about them, and this is often reflected in their work. A fine example, relevant to this topic, is the painting by Signac shown in figure 2. Here he has represented Mach bands and related contrast effects in the various half shadows throughout the entire picture. Particularly striking is the shadow of the match box. The effects we see, of course, depend partly on what Signac has painted and partly on how the eye responds.

Signac, and other neo-impressionists, did not attempt to paint a picture that would look exactly like the subject of their painting. Indeed, no painter does; instead they abstract and emphasize certain striking aspects of what they see. In this painting where Signac saw contrast, he painted contrast—whether it was objectively present in the original scene or not. And when we view the painting our own eyes further accentuate the contrast that he painted. As a result, the contrast is doubly exaggerated and the painting appears to have much more contrast than would the original scene.

Fig. 2. Breakfast (Le petit déjeuner), 1886–1887, by Paul Signac. Oil on canvas, 89 × 115 cm. Collection Rijksmuseum Kröller-Müller, Otterlo, The Netherlands. Reproduced by permission. Photographed in the Neo-Impressionism Exhibition, February 9–April 7, 1968. The Solomon R. Guggenheim Museum, New York. For a colored illustration of this painting and a catalog of the exhibition, see ROBERT L. HERBERT [1968].

Such effects give the painting a certain vitality, for they depend much upon
the viewer's distance and position with respect to the painting and consequently change in interesting ways as one moves toward or away from it.
As Shakespeare's poet remarked when commenting that his artist friend's
painting was more than a mere imitation of nature:

'It tutors nature; artificial strife
Lives in these touches, livelier than life.'—'Timon of Athens', I, 1.

The Nerve Impulse

Let us turn now to the neurophysiological mechanisms that interpret
and add to this 'artificial strife'. The optic nerve forms a bottleneck through
which must pass all the information that the central nervous system receives
from the retina. Consequently, the activity observed in this nerve may be
expected to bear a close relation to the physical and chemical events which
occur at the site of its generation in the retina and to the subsequent visual
experience and behavior which it initiates. The most striking feature of the
activity in the optic nerve—and of other nerves as well—is that it appears
in the form of discrete impulses, each one very much like every other one.
As shown in figure 3, variations in an external stimulus are signalled, not
by changes in these impulses themselves, but by changes in their temporal
pattern. The individual impulses are independent of the strength of the
stimulus eliciting them; they depend upon the local state of the axon at
the time they are conducted. If a nerve fiber responds at all, it responds
with impulses of nearly uniform size and duration. It is the *intervals* between
impulses rather than the characteristics of the individual impulses themselves that distinguishes these two records of neural activity from one
another.

Excitation and Inhibition in the Eye of Limulus

An excellent preparation for the study of the activity of the optic nerve
and for analyzing some of the interactions of opposed influences alluded
to above is the lateral eye of the horseshoe crab, *Limulus*. *Limulus* is a
marine arthropod which is found only on one side of each of the major
landmasses—that is, along the east coast of America, from Maine to
Yucatan, and along the east coast of China, Indochina, and Japan. *Limulus*

Fig. 3. Oscillographic records of the electrical signs of impulses discharged in a single optic nerve fiber in response to strong illumination of the eye (upper record) and in response to weak illumination (lower record). Each record is 1 sec in duration. Note that the electrical signs of the impulses (i.e., the transient deflections of the traces) do not change in amplitude when the illumination is changed—only the intervals between impulses are altered. Records courtesy H. K. HARTLINE.

is quite large, for an arthropod—adults have carapaces 25 cm wide, or more. According to fossil evidence, *Limulus* has existed for some two to three hundred million years without changing much in its outward appearance.

The lateral eye of *Limulus* is compound (much like the more familiar eye of the bee). It is composed of about 1000 ommatidia (literally, little eyes), each of which—although consisting of many cells—appears to function as a receptor unit, somewhat analogous to the rods and cones in our own eyes. The eyes are enormous, relatively speaking—about 1 cm in length in the adult animal. The individual ommatidia are so large (about 0.25 mm center to center) that they can be seen with the naked eye. Figure 4 (upper half) shows a frontal view of one of these compound eyes. The dark spot near the center of the eye is a pseudo-pupil caused by the ommatidia in that area being oriented in the direction of the camera. Most of the light incident on them from the direction of the camera is absorbed; so little is reflected there that the ommatidia appear very dark. Such a pseudo-pupil moves, of course, as the observer moves with respect to the eye. In figure 4 (lower half) is shown a slightly enlarged section cut perpendicularly to the frontal view of the eye. The cornea has been stripped off and here one

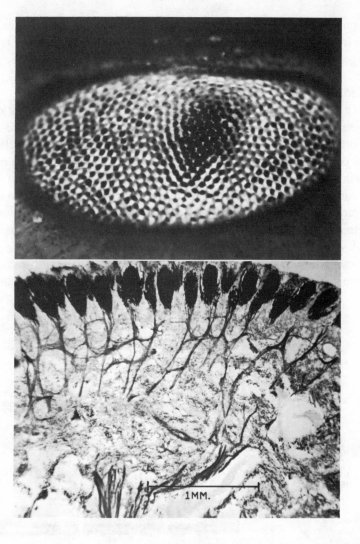

Fig. 4. The lateral eye of the horseshoe crab *Limulus.*—Upper half: corneal surface. The optical axes of the ommatidia diverge so that the visual fields of all of those in one eye together cover approximately a hemisphere.—Lower half: photomicrograph of a section of the eye cut perpendicularly to the plane of the figure above and shown here at a slightly higher magnification. Samuel's silver stain reveals the heavily pigmented sheaths of the ommatidia (receptor units) at the top of the section, the nerve fibers emerging from them and coming together to form the optic nerve at the bottom of the section, and a plexus, or network, of lateral interconnections among the nerve fibers immediately below the receptor units. (Figure prepared by WM. H. MILLER. Micrograph of section from HARTLINE, WAGNER and RATLIFF [1956].)

sees only the heavily pigmented ommatidia or receptor units. Nerve fibers
arise from these ommatidia in small bundles and come together to form
a larger bundle—the optic nerve—which proceeds to the optic lobe of the
brain. A plexus, or network, of nerve fibers interconnects all bundles
immediately behind the photoreceptors.

The eye of *Limulus* provides an especially favorable preparation for
the analysis of the functional properties of a neural network that arise from
the interaction of excitatory and inhibitory influences: it has a relatively
small population of interacting elements (1000 as compared to 100 million
or so in our own retinas); the three-dimensional network or plexus of neural
interconnections forms a true retina; the interaction mediated by this plexus
is almost purely inhibitory; and both the excitatory and inhibitory influences within the retina can be observed directly.

A small bundle of fibers may be separated, by microdissection, from
the main trunk of the optic nerve and placed on electrodes to record the
electric signs of the nerve impulses generated by light on the eye [HARTLINE

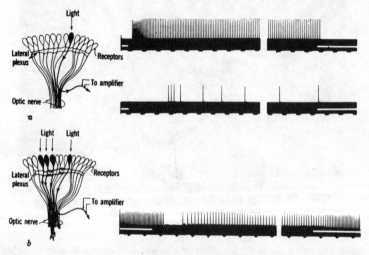

Fig. 5. Oscillograms of action potentials in single optic nerve fibers of *Limulus*. The
experimental arrangements are indicated in the schematic diagrams.—a) Response to steady
illumination of a single ommatidium. For the upper record the intensity of the stimulating
light was 10,000 times that used to obtain the lower record. The signal of exposure to light
blacks out the white line above the $\frac{1}{5}$ sec time marks. Each record interrupted for approximately 7 sec.—b) Inhibition of the activity of a steadily illuminated ommatidium produced
by illumination of a number of other ommatidia near it. The blackening of the white line
above the $\frac{1}{5}$ sec time marks signals the illumination of the neighboring ommatidia. Records
from HARTLINE, WAGNER and RATLIFF [1956].

and GRAHAM, 1932]. Figure 5 illustrates the activity recorded from one of these fibers in response to illumination of the eye. The upper pair of records shows responses typical of many simple sense organs. Weak stimulation (lower record of the pair) produces a low-frequency response, strong stimulation (upper record) produces a high-frequency response. (This resembles our own subjective experience: a strong light appears bright, a weak light appears dim.) Most important for this discussion of the logic of the retina, however, is the effect of illumination on neighboring receptors. This produces a similar discharge of impulses in their axons (not recorded) and a concomitant inhibitory effect (bottom record) on the discharge of impulses from the axon of the receptor under observation [HARTLINE, WAGNER and RATLIFF, 1956].

Steady State Equations

The inhibitory influences in the eye of *Limulus* are exerted mutually among the receptors so that each inhibits, and is inhibited by, its near neighbors. This reciprocal action is readily apparent in simultaneous records of the activity of two interacting elements. Many such sets of records, at many different levels of activity, show the quantitative aspects of this mutual inhibitory interaction (fig. 6). The decrease in frequency of impulses in one element, is linearly related to the increase in frequency of impulses in the other—once a threshold has been reached. This may be expressed by a pair of simultaneous linear equations [HARTLINE and RATLIFF, 1957]:

$$r_1 = \varepsilon_1 - k_{1,2}(r_2 - r_{1,2}^0) \qquad\qquad (1)$$
$$r_2 = \varepsilon_2 - k_{2,1}(r_1 - r_{2,1}^0)$$

The activity of a receptor unit, its response r, is to be measured by the frequency of discharge of impulses in its axon. This response is determined by the external stimulus ε, that excites the receptor, diminished by whatever inhibitory influences are exerted on that receptor as a result of the activity of the neighboring receptor units. The threshold frequency that must be exceeded by one receptor before it can exert inhibition on another is represented by r^0. It and the 'inhibitory coefficient' k are labeled in each equation to identify the direction of the action: $r_{1,2}^0$ is the frequency of receptor two at which it begins to inhibit receptor one, $r_{2,1}^0$ the reverse; $k_{1,2}$ is the coefficient of the inhibitory action of receptor two on receptor one; $k_{2,1}$ the reverse. (There are numerous restrictions on these equations

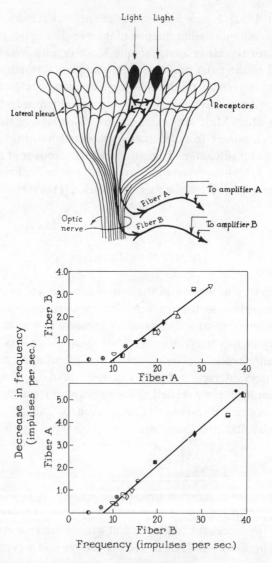

Fig. 6. Mutual inhibition of two receptor units in the lateral eye of *Limulus*. The discharge of impulses was recorded simultaneously from two optic nerve fibers as indicated in the diagram at the top. In each graph (bottom) the magnitude of the inhibitory action (decrease in frequency of discharge) of one of the receptor units is plotted on the ordinate as a function of the frequency of the concurrent discharge of the other on the abscissa. The different points were obtained by using various intensities of illumination of receptor units A and B in various combinations. The slope of the line gives the value of the inhibitory coefficient. The intercept with the x axis gives the value of the threshold. From HARTLINE and RATLIFF [1957].

which will not be detailed here; most important of these is that negative frequencies are not allowed.)

Other experiments have shown that the effects of the third, fourth, and *n*th neighbors, simply add arithmetically. This means that the interaction can be expressed in terms of a simple set of simultaneous equations [HARTLINE and RATLIFF, 1958]:

$$r \quad r_m = \varepsilon_m - \Sigma\, k_{m,n}(r_n - r^0_{m,n}). \tag{2}$$

Note that the inhibition depends upon the response of neighbors—not the stimulus to them. In general, near neighbors inhibit one another more strongly than do distant neighbors, that is, the strength of the inhibition, $k_{m,n}$, diminishes with increasing distance between m and n. This dependence upon distance introduces a topographic factor into the inhibitory interaction, which gives the inhibition its special significance in spatial vision. These distance effects have been determined [RATLIFF and HARTLINE, 1959; BARLOW, 1969] and the set of equations (2) may be represented approximately by the 'weighting function' shown in figure 7. This is to be interpreted as follows: a stimulus localized at the point X_0 produces excitatory influences (positive values above Y_0) which in turn produce inhibitory

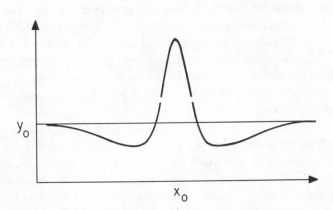

Fig. 7. A graphical representation of a field of excitation and inhibition. The excitatory component centered on X_0 is actually much larger than the sum of the two flanking inhibitory components. It is drawn on a reduced scale here, as indicated by the gaps in the curve. From RATLIFF [1971].

influences in neighboring areas (negative values below Y_0) which first rise to a maximum and then diminish with distance X, more or less as shown.

With no more than this brief sketch of the nature and extent of the inhibitory influence in mind, it is easy to see how an apparent maximum and minimum at or near the contour between two differently illuminated areas may be produced by inhibition. As MACH postulated a century ago, fields of inhibitory influences that extend for some distance laterally are all that is required. Consider, for example, a simple step in intensity of illumination. Because of the considerable extent of the inhibitory influence, a unit of the retinal network within the dimly illuminated area but near the boundary will be inhibited not only by its dimly illuminated neighbors, but also by some brightly illuminated ones. The total inhibition on it will, therefore, be greater (and its response less) than that of dimly illuminated neighbors farther from the boundary. Similarly, a unit within the brightly illuminated area, but near the boundary and, therefore, within the range of inhibitory influences from dimly illuminated neighbors, will have less inhibition on it (and a greater response) than elements far from the boundary that are completely surrounded by brightly illuminated neighbors. Because of these differential effects, maxima and minima (similar to the Mach bands) appear in the response of the network, even though there are no such maxima and minima in the distribution of the stimulus itself.

Such effects have been observed directly in the eye of *Limulus* [RATLIFF and HARTLINE, 1959]. Figure 8 shows the results of a similar experiment carried out recently by ROBERT B. BARLOW, Jr. Notice the similarity of the maxima and minima to the Mach bands (fig. 1). Theoretical calculations, using the equations (2) slightly modified to include some known non-linearities in the system, agree very well with the observed results. [For similar experiments on the vertebrate visual system see: BAUMGARTNER, 1961; JUNG and BAUMGARTNER, 1965; JUNG, 1967; GORDON, 1969; DeVALOIS and PEASE, 1971.]

Contrast phenomena are by no means restricted to the nervous system. Similar effects are to be found wherever fields of opposed influences are integrated. A good example which most everyone has seen, is the contrast effect produced by electrostatic fields in xerography (fig. 9). As is commonly observed, the xerographic process reproduces only the edges of extended uniform areas unless some special precautions are taken. Similar effects occur in photography, in television—indeed, wherever information is transmitted from one place to another by way of interdependent channels. Such effects are frequently put to good use to 'crispen' contours in photo-

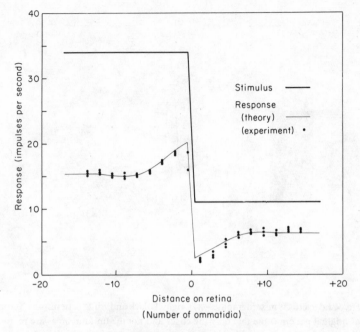

Fig. 8. Response of an optic nerve fiber in the compound eye of *Limulus* to a step pattern of illumination. Unpublished experiment by R. B. Barlow, Jr. Theoretical calculations by D. Quarles and R. B. Barlow, Jr. Compare with figure 1.

graphy, to restore degraded images transmitted by radio from spacecraft, etc. Not only are the effects similar—the causes are similar. In xerography the electrostatic field is positive where a dark point is imaged on the copy plate, but has a negative component which 'inhibits' darkening of the copy in neighboring areas [see Neugebauer, 1967]. Similarly, in photography the chemical byproduct of the development process at one point diffuses to neighboring points and 'inhibits' further development there; and similarly again, in television the electrons resulting from secondary emission at points on the signal plate struck by the scanning beam fall on neighboring points and 'inhibit' the signal from them. The essentials of all of these interactions are similar and all can be represented in a similar abstract mathematical form such as the set of simultaneous equations (2) or the weighting function in figure 7. The resemblance of the contrast effects in such diverse systems is not a trivial coincidence. It is, instead, an indication of a 'universal' that transcends certain particulars: contrast depends upon the *relations*

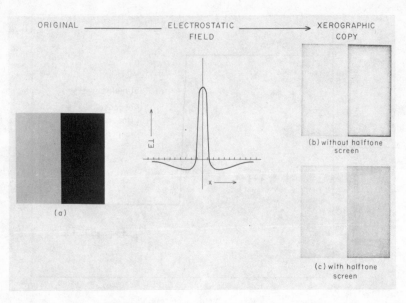

Fig. 9. Edge effects in xerography.—a) Original black and white pattern.—b) Xerographic copy of original pattern (note that only the edges and not the uniform areas are reproduced). —c) Xerographic copy made with a halftone screen placed over the original. Inset in center, the electrostatic field resulting from a dark point imaged on the xerographic copy plate. From RATLIFF [1971].

among interacting elements in a system—not upon the particular mechanisms that achieve those relations.

The Dynamics of Excitation and Inhibition

We have considered, thus far, only the spatial influences in the steady state. Let us turn now to an analysis of the dynamics.

Figure 10 shows the 'instantaneous rate' of two receptor units. One, represented by the filled circles, was illuminated steadily throughout the period shown. The other, along with some of its neighbors, had the steady illumination on it increased to a new steady level at $t = 0$, and then reduced to the original steady level at $t = 2$. Note the enormous initial transient at the onset of this increment and the undershoot at the cessation. Because of the lateral inhibition the steadily illuminated neighbor goes through similar but opposite and slightly delayed transients. As in the spatial effects in the steady state, the responses are greatest where change occurs.

Again this illustrates a general principle; such dynamic interaction

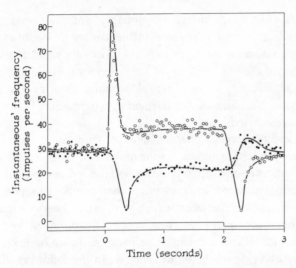

Fig. 10. Simultaneous excitatory and inhibitory transients in the responses of neighboring ommatidia in the lateral eye of *Limulus*. From RATLIFF [1961].

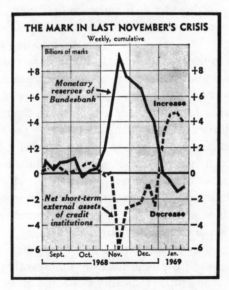

Fig. 11. Simultaneous 'excitatory' and 'inhibitory' transients in an economic system. From the *New York Times*, May 11, 1969.

occurs frequently in nature. For example, all of the main features of the dynamics of the reciprocal interaction in the retina of *Limulus* appear also in a graph (fig. 11) representing some of our economic difficulties during recent years. The resemblence is probably not altogether superficial. Opposed 'excitatory' and 'inhibitory' influences are no doubt at work here, too, but the economists have yet to find out exactly what they are. In any event, a scientific understanding of such temporal interactions—whether in a national economy or in a neural network, requires a knowledge of the underlying mechanisms—in particular, where and when the opposed influences are exerted.

In the eye of *Limulus* one can deduce, from input-output measurements, some of the characteristics of the underlying or intervening processes. Also, one can measure directly some of the processes that determine important time constants and coupling coefficients. This permits a high degree of interaction between experiment and theory: sometimes the mathematical theory points the way to a new experiment to determine a basic process, sometimes a crucial experiment reveals a basic process which reshapes the evolving mathematical theory. In the following discussion of the application of linear systems analysis to the dynamics of excitation and inhibition in the eye of *Limulus*, examples of both will be shown.

The precursor of the discrete impulses in optic nerve fibers of the eye of *Limulus* is the continuous and relatively smooth generator potential that is initiated by light. This can be recorded by inserting an electrode into a proper location in the ommatidium. In the most favorable locations one can see signs of the discrete impulses riding on the generator potential, as they are passively conducted to the recording site from their site of origin some distance away. Typical records are shown in figure 12.

The light is turned on and off as indicated by the step signal below each record. When the light is turned on the 'generator potential' increases—that is, the cell is depolarized; when it is turned off the potential decreases—that is, the cell becomes repolarized. The effect of lateral inhibition (or self-inhibition) on the generator potential is similar to turning off, or reducing, the light—that is, the generator potential decreases. The frequency of discharge of impulses varies accordingly. In the steady state, rate of discharge is proportional to the level of the generator potential. The generator potential itself is proportional to the logarithm of the intensity.

The eye is a strong filter. Indications of some of its filter characteristics—in particular, a high frequency cut-off—may be seen directly in these records. They are rather complicated, however, because the variations are step-like and cross threshold repeatedly at low frequencies. In order to apply methods of linear systems analysis it is necessary to work above threshold in the usual operating range of the eye where impulses are always being

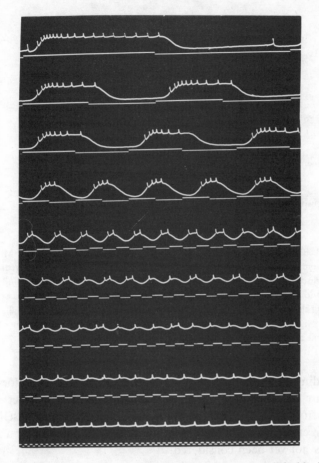

Fig. 12. Oscillograms of generator potentials and nerve impulses elicited by intermittent illumination of a receptor unit. Light and dark periods are signalled by upward and downward steps in line below each record. Duration of each flash in upper record was 0.5 sec. From MILLER, RATLIFF and HARTLINE [1961].

discharged. In brief, these methods consist of sinusoidally modulating an input about some mean suprathreshold level at various frequencies and measuring the amplitude and phase of the corresponding sinusoidal output.

Figure 13 shows the several components of the system that have been analyzed. [For a review see DODGE, SHAPLEY and KNIGHT, 1970.] Light produces a generator potential which in turn, through a voltage to frequency converter, generates a train of impulses. This generator may be produced naturally by light, or artificially—as indicated—by passing current through

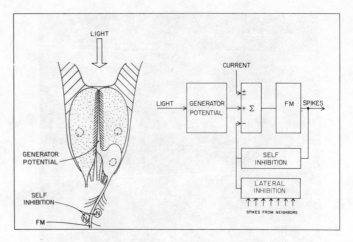

Fig. 13. Block diagram (right) of the *Limulus* photoreceptor unit (left). FM represents the impulse (spike) generating mechanism as a voltage (generator potential) to frequency (rate of impulse discharge) converter. Σ is the place in the cell where the excitatory and inhibitory influences add to drive the voltage-to-frequency converter. Current may be passed through the recording electrode to produce artificially a positive (excitatory) effect such as that normally produced by light, or a negative (inhibitory) effect such as that normally produced by nerve impulses (spikes). From DODGE, SHAPLEY and KNIGHT [1970].

the recording electrode. Inhibitory effects opposing the excitatory effects of light are of two kinds: 1) the lateral inhibition (an indirect negative feedback from activity of neighbors) already described, and 2) self-inhibition (a direct negative feedback from the activity of a particular unit itself) which has not yet been considered.

Let us go, step by step, through these several processes. Figure 14 shows direct graphical transcriptions of samples of raw data and the corresponding transfer functions (frequency responses) obtained from measurements on large sets of similar data. At the top left is the first component—a generator potential obtained in response to sinusoidally modulated light. (Nerve impulses have been abolished by the drug tetrodotoxin, which has no effect on the generator potential itself.) Middle left is the second component —instantaneous frequency or rate of spikes (impulses) resulting from a steady light (to set the receptor in the middle of its operating range) and superimposed sinusoidally modulated current passed through the electrode. (This contains the self inhibition.) At the bottom is the overall process from modulated light to the spike (impulse) frequency. At the right are shown the corresponding transfer functions. For the light to spike rate the data

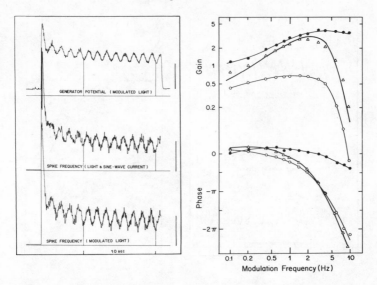

Fig. 14. Left: sample data records for the component processes in a single receptor unit (no lateral inhibition). The upper record shows modulation of the generator potential by modulated light, the middle record shows modulation of spike (impulse) rate by modulated current superimposed on steady light, and the lower record shows modulation of the spike (impulse) rate by modulated light. Smooth curves fitted to the 10 sec span of data between the vertical lines show output of data processing scheme.—Right: frequency responses for light to generator potential (○), current to spike (impulse) rate (●), and light to spike (impulse) rate (△). The smooth curve fitting the composite frequency response (light to spike rate, △) is computed from the empirical frequency responses of the two components (○ and ●). From DODGE, SHAPLEY and KNIGHT [1970].

points are from the actual experiment; the smooth curve is the theoretical prediction made by combining the two component processes.

Figure 15 shows the effect of lateral inhibition—which can best be controlled by exciting the axons of the neighbors electrically. Although such artificial stimulation produces so-called antidromic impulses—i.e., travelling toward the receptor instead of toward the brain, the inhibitory effects are the same as those produced by normal stimulation of these neighbors by light. The center record shows the resulting modulation of the inhibitory potential, and the upper record the modulation of the spike rate. Note that the mean rate is below the steady rate, a short segment of which is shown at the left of the record.

The characteristics of directly observed unitary lateral inhibitory potentials are shown in figure 16. These were first calculated from the frequency responses (see insert fig. 15) and, subsequently, directly recorded

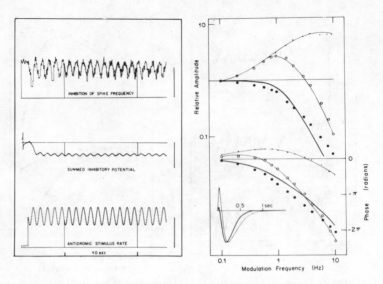

Fig. 15. Dynamics of lateral inhibition.—Left: sample data records. Modulated anti-dromic stimulus rate at bottom. Top record is modulated spike (impulse) rate produced by this time varying inhibition. Middle record is the varying potential produced by modulated antidromic inhibition.—Right: frequency response of the summed inhibitory potential (●), for inhibition of spike (impulse) frequency (○) and for self inhibition (·). Insert: wave form of the unit inhibitory potentials computed by taking the Fourier transform of the frequency response of the summed inhibitory potential (thin line) and of the frequency response of the spike (impulse) data (heavy line). From DODGE, SHAPLEY and KNIGHT [1970].

by means of intracellular electrodes. Note the initial depolarization preced-ing the inhibitory hyperpolarization. This introduces a substantial and important delay into the lateral inhibition. Similar measurements, however, show that the onset of the self-inhibition is almost instantaneous [STEVENS, 1964; PURPLE and DODGE, 1965].

Dynamic Equations

Let us return now to the set of equations (2). It is evident that self-inhibition—which was excluded from these steady state equations—must be taken into account in the dynamics. This may be done simply by inserting a separate term for self-inhibition as follows:

$$r_m = \varepsilon_m - Kr_m - \sum_{n \neq m} k_{mn} r_n \qquad (3)$$

Here r_m is the firing rate of the *mth* ommatidium, Kr_m is its self-inhibitory coefficient, ε_m the excitation of m (the generator potential) and k_{mn} the lateral inhibitory coefficient. For simplicity, the thresholds have been omitted.

With the steady state equations in this form they may now easily be extended to include the dynamics. Let $\varepsilon_m = G(f)I =$ the transfer function from light I to generator potential. Then:

$$r_m = \varepsilon_m - KT_s(f)r_m - T_l(f) \sum_{n \neq m} k_{mn} r_n \tag{4}$$

where T_s is the self inhibitory transfer function T_l the lateral inhibitory transfer function. (Recent experiments indicate that the phase shift is the same for all inhibitory terms, $k_{mn}r_n$, so it is not necessary to include T_l in the summation.) Given these equations, one should be able to predict (within the linear range) the response of the entire eye to any arbitrary stimulus form. Let us consider just one example: the 'tuning' and 'amplification' that results from the pronounced delay in the lateral inhibition caused by the peculiar biphasic shape of the inhibitory potential.

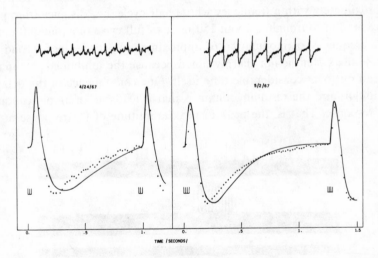

Fig. 16. Comparison of the unit of inhibitory potential predicted by Fourier analysis (line) with the averaged transient response observed directly (points). The insert above the date of the experiment shows several cycles of the raw data. The 'impulse' of excitation to produce the unit of inhibitory potential consisted of a burst of 3 antidromic stimuli spaced 10 msec apart (left graph) and 4 such stimuli (right graph). From KNIGHT, TOYODA and DODGE [1970].

Tuning and Amplification by Delayed Lateral Inhibition

As was shown above, the initial depolarization preceding the inhibitory hyperpolarization results in a very substantial delay to the maximum inhibitory effect. In the discharge of impulses this clearly reveals itself as shown in figure 17. In the preparation from which these records were obtained steady light was turned on one receptor to provide the steady discharge of impulses in the record shown at the top and then a neighboring receptor, along with a few of its neighbors, was illuminated with a very short flash. This short and very intense flash produced, after a short latent period, a fairly compact high frequency burst of impulses. From the onset of this burst to the first perceptible inhibitory effect on the steadily illuminated receptor, however, required a period of about 150 msec or so—as one would expect from the shape of the inhibitory potential shown in figure 16.

Such a delay literally tunes this neural network to particular temporal frequencies just as an electronic network is tuned by delays in an RC circuit. If the whole eye or a large area of it is illuminated with sinusoidally modulated light, then the greatest lateral inhibitory influences are produced about 150 msec after the greatest excitation. This, of course, coincides approximately with a frequency whose half-cycle, i.e., distance from peak of excitation to trough is about 150 msec (a full cycle of about 300 msec), or a frequency of modulation of approximately 3 cycles per second. Of course the system is not sharply tuned because the inhibitory potential is spread out over considerable time itself. One consequence of the delayed inhibition and the resultant tuning is that it leads to an amplification of the response. That is, the peak to peak amplitude of the response at the

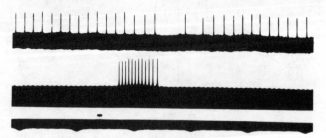

Fig. 17. Transient inhibition of the discharge of impulses from a steadily illuminated receptor unit (upper trace) by a burst of impulses discharged from a second receptor unit nearby (bottom trace) in response to a 0.01 sec flash of light (signalled by the black dot in the white band just above ½ sec time marks). From HARTLINE, RATLIFF and MILLER [1961].

best tuned frequency is greater with the inhibition than it is without. In a sense, the network is an AC amplifier at these particular frequencies. Thus, the retina abstracts and accentuates information about certain significant features of the stimulus at the expense of information about other presumably less significant features.

The results of an experiment demonstrating this tuning and amplification in the lateral eye of *Limulus* are shown in figure 18. The solid points were obtained by confining the light to just one receptor unit. In this way the effects of lateral inhibition are eliminated. The low frequency cut-off here results from self-inhibition and from some effects of adaptation to the slowly changing light intensity [DODGE, KNIGHT, and TOYODA, 1968]. If the spot size is increased (open circles), the low frequencies are further attenuated as would be expected by the introduction of strong lateral

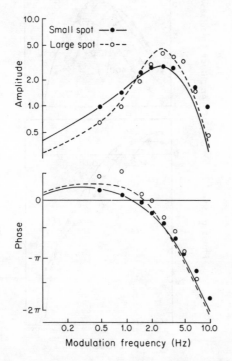

Fig. 18. Tuning and amplification by lateral inhibition. Theoretical (solid line) and observed (solid circles) light to impulse-rate frequency response of a single receptor unit qith no lateral inhibition (small spot), and theoretical (dashed line) and observed (open circles) frequency response for the same receptor unit with lateral inhibition (large spot). Data from RATLIFF *et al.* [1967] Theoretical predictions from KNIGHT, TOYODA and DODGE [1970].

inhibition. But, as the theory predicts (dashed line), the response is actually amplified at those intermediate frequencies to which the system is best tuned because of the inhibitory delays.

There is no question that these effects are due to the delayed lateral inhibition. It is easy to demonstrate this by introducing artificial delays. This can be done simply by delaying the stimulus to the neighbors that are exerting the inhibition on the test receptor. Figure 19 shows the results of such an experiment. As the delay is increased from 0 to 400 msec the frequency response, or 'tuning curve', becomes considerably amplified and shifts to lower and lower frequencies. As one would predict from the theory, second order maxima and minima also gradually appear.

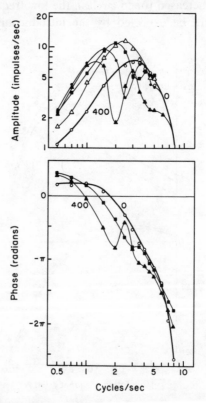

Fig. 19. Frequency response of a single receptor unit with artificial delays of lateral inhibition from neighbors. Delays of inhibitory modulation with respect to excitation were 0 msec (open circles), 100 msec (open triangles), 300 msec (filled circles), and 400 msec (filled triangles). Phase data are shown for 0, 200, and 400 msec only. From RATLIFF, KNIGHT and MILKMAN [1970].

Tuning and Amplification in the Spatial Domain

Since this is a linear system it is easy to transfer to the spatial domain the basic concepts that were outlined above for the temporal domain. Here, instead of sinusoidal variations in time, one simply considers sinusoidal variations in space—and, instead of the temporal distributions of excitation and inhibition, the spatial distribution. An idealized spatial transfer function for a visual system in which there is lateral inhibition is shown in figure 20. In the upper right is a curve representing the so-called line spread function for the dioptric apparatus of the eye. That is, the blurring and scattering of the excitation due to the various imperfections of the lens and other transparent media of the eye. As represented by the dashed line such a lens system transmits low spatial frequencies very well, but—of course —cannot resolve high spatial frequencies. As a result, the transmission curve falls off gradually. Second from the top is a curve representing a distribution of lateral inhibition. Just as in the examples we have seen in the temporal domain, such a lateral spread of inhibition will attenuate low

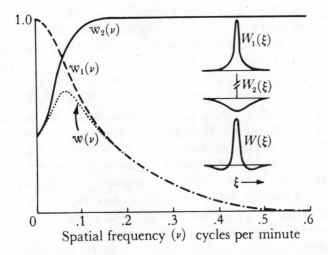

Spatial frequency (ν) cycles per minute

Fig. 20. An idealized composite spatial transfer function (dotted line) for the whole visual system. The composite function is the product of the transfer function (broken line) for the spread of excitation (including, for example, imperfections of the dioptric apparatus) and the transfer function (solid line) for the inhibitory network. The corresponding excitatory, inhibitory, and composite spread functions (or weighting functions) are shown at the right of the graph. From RATLIFF [1965].

spatial frequencies. This is represented by the solid curve rising from the left. The composite weighting function (central excitation flanked by inhibition) is shown at the bottom. This composite transfer function, obtained by combining the opposed excitation and inhibition, shows a pronounced maximum at intermediate spatial frequencies, as expected. Just as in the temporal examples considered above, the inhibition accounts for the low frequency cut-off and limitations in the excitatory mechanisms account for the high frequency cut-off.

The human visual system is much more complex than this and cannot be properly represented by a single transfer function. Nevertheless, the same basic concepts seem to apply, in part. This problem is treated in detail by Professor F. W. CAMPBELL elsewhere in these Proceedings and need not be discussed further here. Let us confine our attention to just one problem—a tuning and amplification in space, analogous to that described above for the temporal domain.

A recent investigation of the spatial distribution of inhibition in the *Limulus* eye [BARLOW, 1969] showed that inhibition is greatest not at points immediately adjacent to the inhibiting receptor units, but some distance from them. The exact shape of the distribution may seem unimportant; such effects as Mach bands and border contrast will be produced by practically any distribution of inhibition that diminishes with distance. Indeed, simple rectilinear approximations are frequently used in some models of neural networks. But just as in the temporal domain, the exact distribution of the inhibition is quite significant. In the left hand half of figure 21 are illustrated three physiologically realistic spatial distributions of inhibition, $k(x)$. At the top is an exponential distribution, in the middle a Gaussian distribution, and at the bottom a distribution of the type found in *Limulus* (approximated by a narrow Gaussian subtracted from a broad Gaussian). Below each is shown the corresponding spatial transfer function. Note that only the bottom distribution (with eccentric maxima) amplifies the response (i.e., yields values greater than 1.0). This occurs, of course, at or near spatial frequencies which correspond to the distance between the two maxima. The transfer functions of the rectilinear approximations to these three distributions yield almost exactly the same results at low frequencies but depart markedly from one another, and from the curvilinear distributions, at higher frequencies. Note especially that the rectilinear approximation to the Gaussian amplifies the response, while the Gaussian itself does not. It is evident that the exact characteristics of the spatial distribution of excitation and inhibition—which have been somewhat

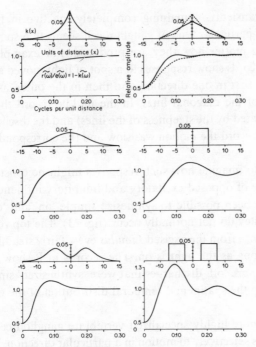

Fig. 21. Normalized transfer functions for (left) various curvilinear distributions of inhibitory fields, *k(x)*, and (right) for rectilinear approximations to them. See RATLIFF, KNIGHT and GRAHAM [1969].

neglected in the past—are just as important as are the temporal characteristics in reaching an understanding of the behavior of neural networks.

Excitation and Inhibition in the Vertebrate Retina

Typical records of nerve impulses obtained from an optic nerve fiber of the frog eye by Professor H. K. HARTLINE, about thirty years ago, are shown in figure 22. Vertebrate retinal ganglion cells, in general, are rather highly specialized. Unlike *Limulus* they do not usually give a simple maintained discharge to steady light. Although there are many different types of vertebrate retinal ganglion cells, the type of response shown here is of special interest because it illustrates how the vertebrate eye—like the *Limulus* eye—accentuates transients. The cell gives a vigorous discharge when the light is turned on—indicated by the black signal—and again, when it is turned off. But it goes a step farther than *Limulus* and responds

only to the transients—remaining completely inactive in between. Furthermore, this particular type of ganglion cell seems to respond to changes
in intensity no matter how they are produced. For example, the middle
and bottom records show responses to a spot of light moved across the eye.
The motion is first in one direction and then in the other, as indicated by
the direction of the diagonal lines. In the middle record the motion was
rapid (as indicated by the steepness of the lines) and the discharge vigorous.
In the bottom record the motion was slow and the corresponding discharge
was weak.

It is easy to imagine how such responses might be constructed by the
proper balance of opposed excitatory and inhibitory influences. In fact, in
Limulus it has been possible to synthesize simple 'on' and 'off' responses
even though they do not normally occur (fig. 23). The top record A is the
'normal' response from the excised *Limulus* eye—fairly steadily maintained
discharge so long as the light is on. The records in B show transient 'on'
and 'off', and pure 'off' discharges that were 'synthesized' simply by artificially adjusting the spatial and temporal distribution of the excitation and
inhibition.

An even more highly specialized vertebrate ganglion cell type is one
which responds selectively to motion in a particular direction. Records from

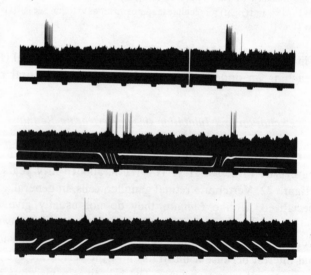

Fig. 22. The response of a vertebrate (frog) retinal ganglion cell to a spot of light turned
on and off (top record) and to a rapidly (middle record) and a slowly (bottom record) moving
spot of light. Time is marked in $\frac{1}{5}$ sec. From HARTLINE [1940].

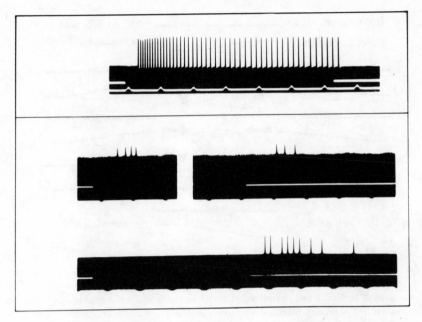

Fig. 23. Oscillograms of diverse 'types' of impulse discharge patterns in single fibers of *Limulus* optic nerve.—a) Typical sustained discharge of a single receptor unit in response to steady illumination.—b) Upper record: synthetic 'on-off' response produced by adjusting the magnitude and duration of the inhibition exerted on the same unit by its neighbors. Lower record: synthetic 'off' response. Time marked in $\frac{1}{5}$ sec. From RATLIFF and MUELLER [1957].

such ganglion cells in the retina of the rabbit are shown in figure 24. As indicated by the plus and minus signs in the center of the chart, this cell would respond to a stationary stimulus that was turned on and off. However, as in the frog retinal ganglion cell of figure 22, this ganglion cell also responded to motion. But here the response to motion was directionally sensitive. As indicated by the records of impulses corresponding to each of the arrows, a movement into the receptive field from one direction yielded a discharge of 79 impulses, whereas a movement of the same spot into the same receptive field from the opposite direction yielded only two impulses. Oblique motions produced greater or lesser responses depending on how far they were off of the axis of symmetry.

This fairly complex response may also be interpreted in terms of the spatial and temporal distribution of the opposed excitatory and inhibitory influences. Two major features are required, and experimental evidence supports this interpretation [BARLOW and LEVICK, 1965]. These two features

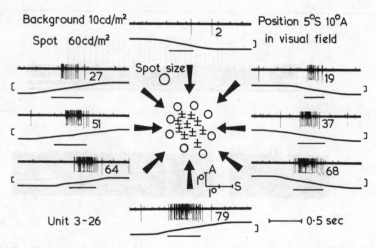

Fig. 24. Responses of a directionally sensitive ganglion cell (rabbit retina). The map of the receptive field is shown in the center. The following symbols are used: + for responses at 'on';—for responses at 'off'; ± for responses at 'on' and 'off'; 0 for no response. The orientation of horizontal and vertical meridians of the visual field is also shown: A = anterior; S = superior. Calibration marks, degrees of visual angle. Each record is the response to movement of a spot of light entirely across the receptive field in the direction of the adjacent arrow. The upper trace is a recording, with a tungsten microelectrode, from an axon in the layer of nerve fibers; the lower trace is a signal representing displacement of the spot across the visual field, vertical calibration bar 5°. The time for which the spot was within the receptive field is indicated by the horizontal bar in each record. The number of spikes is shown immediately after each response. All records read from left to right; increasing positivity in action-potential traces causes a downward deflection. From BARLOW, HILL and LEVICK [1964].

are an asymmetrical distribution of the inhibitory influence and a delay of the inhibitory influence in time. The sort of distribution required in space is simply the familiar weighting function for inhibition (fig. 7) with one half of the flanking inhibition removed. That is to say, a stimulus centered on any point produces excitation in the vicinity of that point but produces inhibition only to one side. The temporal distribution of excitation and inhibition in time is, of course, also asymmetrical. Any stimulus will produce excitation shortly after it is applied, and this excitation—in turn—will produce inhibitory influences which may persist until some time still later. Thus, any stimulus moving across the receptive field in the direction of the asymmetrical inhibitory field will produce a wave of inhibition exerted in that same direction. Furthermore, the inhibition will persist somewhat (because of the temporal asymmetry). Consequently, with motion in this direction the excitatory influences and inhibitory influences will tend to arrive at some common point in the network at about the same time and cancel one another out. Thus, little or no response is observed. A stimulus moving in the other direction, however, will produce excitation followed by a wave of inhibition. That is, inhibition always trails after excitation in both space and time. Consequently, there is no interference with the excitatory influence. It passes on through the network before the trailing inhibition arrives at the critical summing point. Of course, the actual retina is far more complicated than this simple scheme, only the essential features required to explain directional sensitivity have been outlined here. Actually, there is practically no limit to the complexity of the types of neural networks that might exist based on nothing more complicated than particular spatial and temporal distributions of excitation and inhibition.

In the vertebrate retina many very highly specialized systems are found; there are systems which respond best to small spots, there are systems which respond best to particular colors, and so on. However, a few words of caution are appropriate at this point, for many review articles and popular accounts would lead one to believe that retinal ganglion cells and various cells in the visual cortex of the brain are much more highly specialized than they really are. As a matter of fact, few, if any, respond selectively to just one particular type of stimulus. For example, cells that respond to motion (as in fig. 22 and 24) also respond to stationary spots turned on and off. Furthermore, what specialization does exist is by no means absolute. The motion sensitivity, of course, is velocity dependent. And, in addition, for any particular velocity there is more than one direction of motion that will yield the same response; the effect of any motion oblique to the principal axis can be closely matched by its symmetrical mate.

Indeed, wherever there is symmetry there is ambiguity (fig. 25). If one saw only the word CHOICE one would be hard put to tell which was the mirror image and which was the actual printed word. As soon as one sees some neighboring area where the same peculiar symmetry does not exist, however, it is immediately obvious which is which. No doubt neural networks operate in this very way. Local information may be ambiguous, but additional information from neighboring areas or from other cell types immediately resolves the ambiguity. This, of course, requires some comparison or integration later on in the system. The point to be emphasized here is that the information conducted in a single neuron is often ambiguous in some respect or another. For example, the cell of figure 24, although directionally sensitive, doesn't know right from left. Indeed, as we shall see shortly, this problem of ambiguity extends to entire networks, in some cases.

Fig. 25. Symmetry and ambiguity.

Fig. 26. Two-Squares-and-a-Happening. From LAND and McCANN [1971].

Contour and Contrast

An experiment by LAND and McCANN [1971], which they call their 'Two-Squares-and-a-Happening', is illustrated in figure 26. Notice that the rectangle on the right is much brighter than the one on the left and that each of the two appears more or less uniform over its entire extent. These are the 'Two Squares' of the experiment. Now, the 'Happening': all it consists of is to place a pencil or other long narrow object over the contour between the two rectangles. Immediately they change from appearing quite different to being nearly indistinguishable from one another.

Surprisingly, neither of these appearances is isomorphic with the objective distribution, which is represented by the solid line in figure 27. The appearance of these two squares, when the contour between them is not obscured, seems to depend critically upon that contour. It is almost as if the abrupt step alone determines the apparent brightness of each of the two fields. Indeed, if one drew a uniform line to the right from the top of that objective step and to the left from the bottom of that step—as indicated by the dashed line—it would represent fairly accurately the subjective brightness of the two rectangles, that is, almost uniformly bright on the right and almost uniformly dark on the left. Evidently this contour is an important source of the information that is utilized by the visual system.

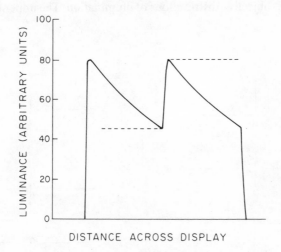

Fig. 27. Luminance as a function of distance in the 'Two-Squares-and-a-Happening' experiment in figure 26. Solid line: objective luminance. Dashed line: approximate subjective appearance. Redrawn after LAND and McCANN [1971].

Why should such a contour have any effect upon the apparent brightness of adjacent areas and—in particular—upon areas rather distant from it? Furthermore, why should such a contour cause the areas contiguous to it be almost uniformly bright on the one side and almost uniformly dark on the other side, as would objectively uniformly light and dark areas placed side by side? For two so different stimuli to appear similar, one can only assume that the dominant underlying neural events must also be similar even though they are elicited by different stimuli. The answer may reside, in part, in mechanisms already discussed in previous sections. As has been emphasized repeatedly throughout this paper, inhibitory networks tend to abstract information about changes in space and time. As was shown above (in fig. 8), when a simple step is processed by such a network, the output shows a pronounced maximum on the light side of the step and a pronounced minimum on the dark side. A similar result will also be obtained whenever any distribution of luminance containing an abrupt change is processed, no matter what the distribution may be in areas adjacent to the change which delineates the contour between them.

This similarity of responses to rather different stimuli is illustrated in the results obtained by a computer simulation of a neural network shown in figure 28. On the left, in the column labeled 'stimulus', are shown three very different objective distributions of illumination. The upper distribution

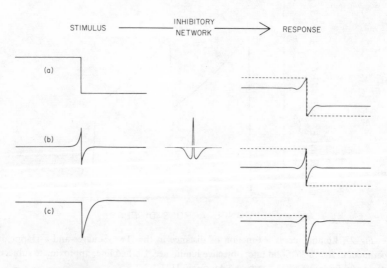

Fig. 28. Computer simulation of the responses (solid lines on the right) of the inhibitory neural network of *Limulus* (represented by the weighting function inset in the center) in response to three different stimulus patterns (solid lines on the left). To the human observer all three patterns appear very much the same—bright on the left and dim on the right with an abrupt step between. Note that this appearance is given simply by extending the maxima and minima of the neural response to the left and right (dashed lines). From RATLIFF [1971].

(a) is a simple abrupt step from a uniform high intensity to a uniform low intensity. The middle distribution (b) is uniform everywhere except for the spur of high intensity and the spur of low intensity at the contour between the right and left halves. Note that this resembles somewhat the distribution of illumination in LAND and MCCANN's 'Two-Squares-and-a-Happening' experiment. The third distribution (c) is uniform on the left and on the right, but has a single short downward spur at the contour between these two uniform areas. Here, however, the intensity of the uniform area of the right half of the distribution is slightly greater than the left half. To a human observer these very different distributions of illumination appear very much the same in certain respects. In particular, all appear to be brighter on the left side of the contour than on the right—as indicated by the dashed lines in the corresponding graphs in the column labeled 'response'.

It is not surprising that the objective step at the top appears subjectively as a step (with perhaps some slight contrast effects near the border which are not represented here). But it is surprising to find that the middle distribution, which objectively is not a simple step between two uniform areas, nevertheless appears so—as in the 'Two-Squares-and-a-Happening'. Incidentally, it might be more appropriate to call this the 'Craik-O'Brien Effect' because it was discovered by CRAIK in England about 30 years ago and, independently by O'BRIEN in America some years later. A similar illusion in the perception of loudness was reported earlier by RAWDON-SMITH and GRINDLEY [1935]. An even more remarkable effect, first discovered by O'BRIEN [1958], results from the distribution at the bottom of the figure. This stimulus which is objectively darker on the left side of the contour appears subjectively—as indicated by the dashed line—to be just the reverse: bright on the left and dark on the right. As noted above, it seems reasonable to assume that very different stimuli, such as the three illustrated here, can appear the same only if they produce similar neural responses at some point in the visual system.

As shown by the calculated responses (solid lines in the right hand column) the mere. processing of these different stimulus distributions by a simple *Limulus* type inhibitory network is sufficient to produce considerable similarity in the neural responses. All show a pronounced maximum on that side of the contour which subjectively appears bright and a pronounced minimum on that side of the contour which subjectively appears dark. If this one feature of the neural response in *Limulus* is also common to the corresponding neural responses in our own visual systems, it might be sufficient reason for the three objectively different stimuli to appear subjectively somewhat the same. But even so, much would remain to be explained. Why should the influence of the contour be exerted over the whole extent of an adjacent area, rather than only locally where the maxima and minima are in the neural response? And why should all three different stimuli look most like one particular one of them—namely the step at the upper left—rather than one of the others?

The answers to these questions are not known. It may be that the influence of local contour on extended neighboring areas in our visual systems is similar to some of the so-called methods of 'data compression' now used in communication engineering. In some such methods information about contours only is abstracted from a picture and transmitted; the uniform areas, consisting almost entirely of redundant information are restored later at the receiving end. Absolute levels are determined by the information in the amplitudes of the maxima and minima at the contours. This results in a considerable saving in the amount of information transmitted. But any such abstraction which discards information always carries with it the danger of ambiguity or misinterpretation. The brain has no way of knowing what information the eye has discarded. And stimuli that are quite different physically may appear to be the same visually merely because the retina abstracts and transmits certain features which they happen to have in common. Although this type of 'data compression' leads to some ambiguities and uncertainties, it is nevertheless likely to turn out to be an important general principle of nervous activity because of the great economies in both structure and function that it can achieve: fewer channels are required for transmission and less energy is required to transmit the compressed information.

Art and Science

Techniques developed by artists and artisans in the practice of the visual arts have many times anticipated important discoveries in the visual sciences. Indeed, these very same effects of contour and contrast that have only recently been discovered in science have been an 'artist's secret' in practical use for at least a thousand years. Many examples have been found in Oriental art as far back as the Sung Dynasty (960–1279) in China.

The moon shown in figure 29 is a fairly recent example of an artistic use of the effect of contour on contrast. It is in a Japanese ink painting by Keinen, about 1900. The moon is objectively only very slightly lighter

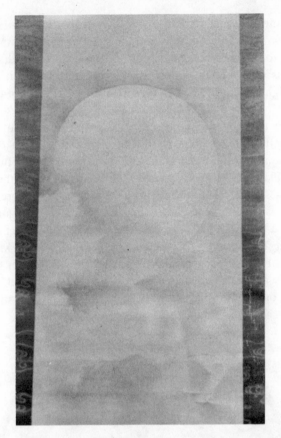

Fig. 29. Detail of ink painting 'Autumn Moon' by Keinen, about 1900. In the collection of the late Mr. AKIRA SHIMAZU, Nara, Japan. Photograph courtesy Dr. H. ASANUMA. From RATLIFF [1971].

than the background. The single deft stroke of the brush has left a gradient of ink similar to the downward spur of one of the distributions illustrated in figure 28. The apparent brightness difference between the moon and its surroundings is caused almost entirely by this hard edge. This, one can demonstrate immediately by alternately covering and uncovering the edge of the moon with a loop of heavy string. Notice the enormous difference in the apparent brightness of the moon's disk.

A still earlier example is a painting of the moon on a Korean vase of the Yi Dynasty [for a photograph, see RATLIFF, 1971]. In this case although the moon's disk appears fairly bright it is actually darker than the sky surrounding it. Measurements of the photograph with a brightness

spot meter under ordinary room lights showed that the luminance of the
moon was about 15 foot lamberts and the space, 1 moon diameter directly
below it, 20 foot lamberts. The contour effect is so strong that, as in one
of the effects described by O'BRIEN (the bottom curves in fig. 28), the
apparent brightness of the two areas is actually the reverse of the objective
luminance.

A similar effect was produced in still another way about a thousand
years ago in the famous Ting ware white porcelain and in the northern
Celadon ceramics of the Sung Dynasty. For example, the lotus arabesque
on the Ting plate in figure 30 appears slightly brighter than the surrounds.
In the Japanese ink painting and in the design on the Korean vase the
technique consists simply in controlling the gradation of pigment applied
directly with a brush. In the Ting ware the control of the effect is somewhat
more indirect. The design is first incised in the soft clay with a knife, making
the cut so that it has a sharp inner edge and a graded outer edge. When
the clay is dry and firm it is then covered with glaze and fired. The slightly
creamy cast of the transparent white glaze filling in the cut makes the

Fig. 30. Ting Yao saucer with incised lotus arabesque. Creamy white glaze over withe
porcelain, rim fitted with a cooper band. Sung Dynasty. From RATLIFF [1971].

necessary gradient from light to dark. The effect is much more subtle than
in the Japanese painting and the Korean vase. But subtlety and restraint
are usually characteristic of the Sung ceramics.

Conclusion

In recent years biology has become more and more analytic, more and
more confined to the investigations of the structure and behavior of single
cells and of the molecular and sub-molecular events within them. Indeed,
the application of methods and concepts developed in the physical and
chemical sciences to the study of living organisms has become so tho-
roughgoing that it is now difficult to tell where the physical sciences end
and the biological sciences begin. No one can deny the great productivity
and remarkable achievements of this analytic approach. But—even so—it
does not come to grips with one of the most fundamental problems facing
modern biological science: how unitary structures and elementary processes
are organized into the complex functional systems that make up organs
and organisms.

To continue in this analytic direction alone and to ignore the molar
or holistic approaches would be a most serious error—for it is probably
the elaborate organization of unitary structures and elementary processes
alone that distinguishes living beings from lifeless things. But this type of
error is not confined to biological science. Indeed, in every field and at
every level of human knowledge and endeavour the necessity for understanding the interdependence of the parts that are organized into wholes
asserts itself again and again. The consequences of our failures to do so
permeate every aspect of our existence and are near to overwhelming us.
Mere scientific advances are not at stake here—it is the survival of man
himself.

References

ADRIAN, E.D. and MATTHEWS, R.: The action of light on the eye. III. The interaction of retinal
    neurones. J. Physiol., Lond. *65:* 273–298 (1928).
BARLOW, H. B.; HILL, R. M. and LEVICK, W. R.: Retinal ganglion cells responding selectively
    to direction and speed of image motion in the rabbit, J. Physiol., Lond. *173:* 377–407,
    (1964).
BARLOW, H. B. and LEVICK, W. R.: The mechanism of directionally selective units in the
    rabbit's retina. J. Physiol., Lond. *178:* 477–504 (1965).

372   RATLIFF

BARLOW, R. B., jr.: Inhibitory fields in the *Limulus* lateral eye. J. gen. Physiol. *54:* 383–396 (1969).

BAUMGARTNER, G.: Der Informationswert der on-Zentrum- und off-Zentrum-Neurone des visuellen Systems beim Hell-Dunkel-Sehen und die informative Bedeutung von Aktivierung und Hemmung, in Neurophysiologie und Psychophysik des visuellen Systems. R. JUNG und H. KORNHUBER (Eds.) (Springer, Berlin 1961).

CHEVREUL, M. E.: The principles of harmony and contrast of colors and their application to the Arts (Reinhold, New York 1967). (Based on the first French edition of 1839.)

COHEN, R. S. and SEEGER, R. J. (Eds.): Ernst Mach: physicist and philosopher. Boston Studies in the Philosophy of Sciences, vol. VI (D. Reidel, Dordrecht, Holland, 1970).

CRAIK, K. J. W.: The nature of psychology, pp. 94–97 (Cambridge University Press, London 1966).

DEVALOIS, R. L. and PEASE, P. L.: Contours and contrast: responses of monkey lateral geniculate nucleus cells to luminance and color figures. Science *171:* 694–696 (1971).

DODGE, F. A.; KNIGHT, B. W. and TOYODA, J.: Voltage noise in *Limulus* visual cells. Science *160:* 88–90 (1968).

DODGE, F. A.; SHAPLEY, R. M. and KNIGHT, B. W.: Linear systems analysis of the *Limulus* retina. Behav. Sci. *15:* 24–36 (1970).

FIORENTINI, A. and RADICI, T.: Binocular measurements of brightness on a field presenting a luminance gradient. Atti Fond. Ronchi *12:* 453–461 1957).

GORDON, J.: Edge accentuation in the frog retina; thesis, Brown University, Providence, R. I. (1969).

HARTLINE, H. K.: The receptive fields of optic nerve fibers. Amer. J. Physiol. *130:* 690–699 (1940).

HARTLINE, H. K. and GRAHAM, C. H.: Nerve impulses from single receptors in the eye. J. cell. comp. Physiol. *1:* 277–295 (1932).

HARTLINE, H. K. and RATLIFF, F.: Inhibitory interaction of receptor units in the eye of *Limulus.* J. Gen. Physiol. *40:* 357–376 (1957).

HARTLINE, H. K. and RATLIFF, F.: Spatial summation of inhibitory influences in the eye of *Limulus,* and the mutual interaction of receptor units. J. Gen. Physiol. *41:* 1049–1066 (1958).

HARTLINE, H. K.; RATLIFF, F. and MILLER, W. H.: Inhibitory interaction in the retina and its significance in vision; in Nervous Inhibition, pp. 241–284 (Pergamon Press, New York 1961).

HARTLINE, H. K.; WAGNER, H. G. and RATLIFF, F.: Inhibition in the eye of *Limulus.* J. Gen. Physiol. *39:* 651–673 (1956).

HERBERT, R. L.: Neo-Impressionism (The Solomon R. Guggenheim Foundation, New York 1968).

HERING, E.: Outlines of a theory of the light sense (Harvard University Press, Boston, Mass. 1964).

JUNG, R.: Neurophysiologie des Konturensehens und Graphik; in BAMMER, H. G. Zukunft der Neurologie (Thieme, Stuttgart 1967).

JUNG, R. und BAUMGARTNER, G.: Neuronenphysiologie der visuellen und paravisuellen Rindenfelder, Proc. 8th Int. Congr. Neurology, Vienna 1965, vol. 3, pp. 47–75 1965).

KNIGHT, B. W.; TOYODA, J. and DODGE, F. A., Jr.: A quantitative description of the dynamics of excitation and inhibition in the eye of *Limulus.* J. Gen. Physiol. *56:* 421–437 (1970).

LAND, E. H. and MCCANN, J. J.: Lightness and retinex theory. J. Opt. Soc. Amer. *61:* 1–11 (1971).

MAGNUS, R.: Goethe as a Scientist (Collier Books, New York 1961). (Transl. by HEINZ NORDEN from: Goethe als Naturforscher, Leipzig 1906.)

MILLER, W. H.; RATLIFF, F. and HARTLINE, H. K.: How cells receive stimuli. Scientif. Americ. *205:* 222–238 (1961).

NEUGEBAUER, H. E. J.: Development method and modulation transfer function of xerography. Appl. Optics *6:* 943–945 (1967).

O'BRIEN, V.: Contour perception, illusion and reality. J. Opt. Soc. Amer. *48:* 112–119 (1958).

PURPLE, R. L. and DODGE, F. A.: Interaction of excitation and inhibition in the eccentric cell in the eye of *Limulus. Cold Spr. Harb. Symp.* 30: 529–537 (1965).

RATLIFF, F.: Inhibitory interaction and the detection and enhancement of contours; in ROSENBLITH, W. A. Sensory communication, pp. 183–203 (M.I.T. Press/Wiley, Cambridge, Mass./New York 1961).—Mach bands: quantitative studies on neural networks in the retina (Holden-Day, San Francisco 1965).—Contour and contrast. Proc. Amer. Phil. Soc. *115:* 150–163 (1971).

RATLIFF, F. and HARTLINE, H. K.: The responses of *Limulus* optic nerve fibers to patterns of illumination on the receptor mosaic. J. Gen. Physiol. *42:* 1241–1255 (1959).

RATLIFF, F.; KNIGHT, B. W. and GRAHAM, N.: On tuning and amplification by lateral inhibition. Proc. Nat. Acad. Sci., Wash. *62:* 733–740 (1969).

RATLIFF, F.; KNIGHT, B. W. and MILKMAN, N.: Superposition of excitatory and inhibitory influences in the retina of *Limulus:* the effect of delayed inhibition. Proc. Nat. Acad. Sci., Wash. *67:* 1558–1564 (1970).

RATLIFF, F.; KNIGHT, B. W.; TOYODA, J. and HARTLINE, H. K.: The enhancement of flicker by lateral inhibition. Science *158:* 392–393 (1967).

RATLIFF, F. and MUELLER, C. G.: Synthesis of 'on-off' and 'off' responses in a visual-neural system. Science *126:* 840–841 (1957).

RAWDON-SMITH, A. F. and GRINDLEY, G. C.: An illusion in the perception of loudness. Brit. J. Psychol. *26:* 191–198 (1935).

STEVENS, C. F.: A quantitative theory of neural interactions: theoretical and experimental investigations; thesis, The Rockefeller Institute, New York (1964).

Author's address: Dr. FLOYD RATLIFF, The Rockefeller University, *New York, NY* (USA)

Proc. 3rd int. Conf. From Theoretical Physics to Biology, Versailles 1971, pp. 374–384
(Karger, Basel 1973)

The Transmission of Spatial Information
Through the Visual System

F. W. CAMPBELL

The Physiological Laboratory, Cambridge

The theorem of FOURIER (1768–1830) enabled HELMHOLTZ (1821–1894) to make his contribution to our understanding of the physical nature of music and to the first serious attempt to unravel the physiology of hearing. We know from his biographer, KOENIGSBERGER [1965], that HELMHOLTZ 'had for years gone to bed and got up again with Fourier's series in his mind' and it is, therefore, surprising that he did not apply with equal success this technique to optics and to the physiology of the eye.

HELMHOLTZ did come very near to considering it, for KOENIGSBERGER reports that he thought that 'the eye has no harmony in the same sense as the ear; it has no music'. In making this comparison between the eye and the ear, HELMHOLTZ was contrasting colour vision with that of harmony, for the ear can analyze a compound sound into its frequency components while the eye cannot resolve compound colour into its component wavelengths. Half a century had to pass before the first step was taken by DUFFIEUX [1946] to apply Fourier theory to physical optics.

I hope to show you that music is certainly present in the visual sense, but it is in the domain of space and not colour.

We know from experience that as an object recedes from us it becomes increasingly difficult to perceive its details until, at a sufficiently great distance, the object itself disappears from view. Alice, in 'Through the Looking-Glass', once remarked 'I see nobody on the road' and the King replied in a fretful tone, 'I only wish I had such eyes, to be able to see nobody and at that distance too'.

Very many factors have been put forward to account for this everyday experience. For example, the limits of visual acuity may be restricted by the optical properties of the eye, or by the dimensions of the foveal mosaic of cones, or by the rate at which light quanta are captured by individual photoreceptors [ROSE, 1970] or even by limitations within the visual nervous system itself. In order to solve this problem in an analytical manner it is necessary to measure the transmission of spatial signals through each part of the system. To do this it is essential to choose an input signal which produces a fairly simple output signal. We have chosen to use the simplest spatial signal of all—a grating whose luminance varies sinusoidally along one axis [see CAMPBELL, 1968, for illustrations]. This is the equivalent of a pure tone in the auditory system.

We are going to consider the following elements to be in cascade, and investigate the transmission properties of each section in turn: — Object → Image → Cone mosaic → Ganglion cell → Geniculate fibre → Visual cortex cell → Psychophysics.

This approach for investigating the transmission of information through a system is really borrowed from electrical engineering where the attenuation and amplification of temporal signals is important. HOPKINS [1962], LINFOOT [1964] and O'NEILL [1963] have already successfully developed Fourier theory to a point where many optical systems can be precisely described in this manner.

ARNULF and DUPUY [1960] and WESTHEIMER [1960] were the first to realize the power of this approach in visual optics and they used a technique invented earlier by LE GRAND [1937]. They passed coherent beams of light into the eye and set up a sinusoidal interference grating on the retina using the principle of Thomas Young [YOUNG, 1800]. This ingenious technique by-passes the defects that might arise from aberrations in the optical components of the eye, permitting them to establish the resolving power of the retina, coupled to the brain, in isolation from the dioptrics. Unfortunately, the success of this experiment depends upon having a monochromatic source with a high degree of coherence and also high luminance.

Such a source became available when the neon-helium laser was developed. Using it, CAMPBELL and GREEN [1965] were able to show that the fundal-image formed by a well-focussed eye, with a normal-sized pupil, was surprisingly good and that most of the loss of contrast in the perception of fine gratings was due to the properties of the retina and/or brain. The transmission properties of the dioptrics have also been measured directly using an objective method [CAMPBELL and GUBISCH, 1966] and the results obtained by these two fundamentally different methods agree well.

The effects of change of pupil size have been investigated, as well as those of focus [GREEN and CAMPBELL, 1965]. GREEN [1968] has demonstrated the effects of off-axis aberrations and CAMPBELL and GUBISCH [1967] have shown the effect of the chromatic aberration present in the eye on the contrast sensitivity function.

Thus, the nervous system itself is responsible for most of the loss of perceived contrast.

These approaches have so far been useful in finding out the nature of the fundal-image [GUBISCH, 1967]; but can these be used to further our understanding of how spatial signals are transmitted and transformed by the nervous system?

The first convenient level at which spatial signals can be detected is at the ganglion cells that transmit the signals from the retina to the geniculate body. ENROTH-CUGELL and ROBSON [1966], using sinusoidal gratings generated on the face of an oscilloscope, have studied the response of these cells in the cat. In this animal, the direction of movement of a grating (or other stimulus) is unimportant as the cells respond equally to movement in all directions. They found one class of cells that responded in a linear manner to their stimuli. A finding, which may be of great significance, was that each cell responded only over a limited range of spatial frequency.

The fibres from the geniculate body terminate in the striate visual cortex and their activity can readily be monitored by microelectrodes [HUBEL, 1957]. COOPER and ROBSON [1968] and CAMPBELL, COOPER and ENROTH-CUGELL [1969] find that at this level the geniculate units again respond only to limited bands of spatial frequency. Like the ganglion cells, the geniculate units respond to movement in all directions.

However, the cells in the striate visual cortex behave quite differently, as has been most elegantly shown by HUBEL and WIESEL [1959, 1962, 1965] in the cat and also in the monkey [HUBEL and WIESEL, 1968]. Here the cells are very sensitive to the orientation of an edge or bar when it is moved across the receptive field. Using grating patterns CAMPBELL et al. [1968] have measured quantitatively this selectivity to orientation in the cat. These orientationally selective cells will, of course, also respond to a grating providing it is moved close to the optimum orientation. These cortical cells also respond only to limited bands of spatial frequencies, each cell responding to a different range in the spectrum of spatial frequencies [COOPER and ROBSON, 1968; CAMPBELL, COOPER and ENROTH-CUGELL, 1969]. There is some preliminary evidence [CAMPBELL et al., 1969] that neurons in the cortex of the monkey are similarly organized.

The neurophysiological findings in the cat and monkey suggest that two important properties of an image have been coded. Firstly, the information about orientation of an edge, bar or grating is extracted and secondly, the spatial frequency content of the image is also extracted. This organization is strikingly similar to that found in the auditory system where units are found that respond to tone bursts over a limited range of pitch [KIANG, 1966] rather like the visual cortex cells that respond over a limited range of spatial frequency. Again, in the auditory cortex are found cells which only respond either to a rise or fall of pitch [WHITFIELD and EVANS, 1965]; indeed, because of their resemblance to the visual cortical cells, they have been called 'directional cells' [WHITFIELD, 1967].

Is there any evidence that the visual system of man is similarly organized? If we can demonstrate that it is, it will greatly strengthen the argument that these neurophysiological findings are directly relevant properties of the mechanism by which we perceive, and possibly recognize, objects: and also that the neurophysiological findings are not incidental artifacts irrelevant to our understanding of the visual mechanism.

CAMPBELL and KULIKOWSKI [1966] attempted to show that man has channels sensitive to the orientation of a grating by measuring psychophysically the threshold of a test grating in the presence of a high-contrast masking grating. They changed the orientation of the masking grating relative to the test grating and found that maximum masking occurred when the two gratings had the same orientation and that no masking occurred when the masking grating was at right angles to the test grating. For intermediate angles, the masking effect decreased exponentially; the masking effect was reduced by half when the angle between the gratings was only 12 to 15°. Man seems to have a higher orientational selectivity than the cat [CAMPBELL et al., 1968] and may resemble more closely the monkey [HUBEL and WIESEL, 1968] where the selectivity was found to be higher than that of the cat. Thus, as far as orientation is concerned, there is a striking agreement between the neurophysiology and the psychophysics.

Can we show that man also has channels selectively sensitive to spatial frequency? In a preliminary note CAMPBELL and ROBSON [1964] suggested that Fourier theory might be applied to the psychophysics of spatial vision. In their main paper, CAMPBELL and ROBSON [1968] measured the contrast threshold for a number of gratings each with a different waveform. They found that the threshold over a wide range of spatial frequency is determined by the amplitude of the fundamental Fourier component in the grating and that the higher harmonics do not contribute to the threshold, providing they are below their own threshold. In like manner, a square-wave grating can be distinguished from a sinusoidal-wave grating when the contrast is sufficiently high for the third harmonic to be detected in the presence of the fundamental. Their findings led them to suggest that there must be a number of channels in the human visual system each tuned to different spatial frequencies. The effective band-width of each channel is probably not greater than about one octave of spatial frequency.

In 1966, CAMPBELL and KULIKOWSKI used the technique of masking a low contrast test grating with a high contrast grating of the same orientation and spatial frequency. By studying the effect of changing the orientation between the gratings they found the orientational properties of the visual system. In undertaking these experiments they found that it was necessary to have the two gratings of the same spatial frequency, or the masking effect was much less. GILINSKY [1967, 1968] developed this masking method further by adapting for some time to a grating of a given frequency and then subsequently inspecting the test grating which was at a different orientation. Using her adaptation method, PANTLE and SEKULER [1968] studied the influence of gratings with the same orientation but with different spatial frequencies. They summarize their findings as: 'These conclusions are similar to those reached by Campbell and Robson (1968) from a Fourier analysis of the visibility of gratings of different spatial frequencies and waveforms.'

Unfortunately, PANTLE and SEKULER [1968] used square wave gratings to produce their adaptation effects; this complicated their data since the higher Fourier components in their gratings also influenced the results. By using sinusoidal gratings covering a wide range of spatial frequency CAMPBELL and BLAKEMORE [1969] were able to show the true band-width of each channel and also to demonstrate the fact that there were multiple channels. In their experiments, the subject adapted to a given frequency, and the loss in contrast sensitivity was measured at—and on either side of—that frequency (open circles, and right hand and upper scale of fig. 1). The 'band-width' of the adaptation effect is about ± one octave. Also on this figure is plotted the response of a single neurone from the cortex of the cat (closed circles, and left-hand and lower scale). The adaptation technique of BLAKEMORE and CAMPBELL shows that there are channels tuned to spatial frequencies ranging from 3 c/deg up to the upper-limit of resolution at about 48 c/deg. It is not yet clear what happens at frequencies less than 3 c/deg.

CAMPBELL and MAFFEI [1970] and MAFFEI and CAMPBELL [1970] have

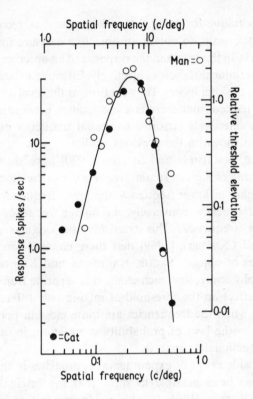

Fig. 1. The closed circles (●) are the responses (spikes/sec) from an orientational cortical neuron from a cat. The sinusoidal grating stimulus had a contrast of 0.5 and was drifting in the preferred direction so that one bar of the grating passed the receptive field each second. (Results supplied by J.G. ROBSON and G.F. COOPER, the Physiological Laboratory, Cambridge.) — The open circles (○) represent the relative threshold elevations in man, produced by adapting to a sinusoidal grating of contrast 0.7 and spatial frequency 7 c/deg [BLAKEMORE and CAMPBELL, 1969].

To compare the response of this particular cat neuron with the psychophysical adaptation, the data has been super-imposed to obtain the best fit by eye and the appropriate scales attached. It must be emphasized that there is a range of cortical neurons, each responding maximally to a different spatial frequency [CAMPBELL, COOPER and ENROTH-CUGELL, 1969 and CAMPBELL et al., 1969]. Likewise, in the human, adaptation to different spatial frequencies reveals a comparable range of tuned channels [BLAKEMORE and CAMPBELL, 1969].

The curve through the results is,

$$S = (e^{-f^2} - e^{-(2f)^2})^3$$

where S = contrast sensitivity and f = spatial frequency.

developed a technique for using the evoked potential, recorded from the
visual area of the scalp, to obtain an objective measure for the existence
of these channels. In this manner, the response of an observer is not required
for they can determine thresholds objectively. Effectively they have removed
the *psycho* from phychophysics. They confirm, at the level where the evoked
potential is generated, that there is a mechanism generating an electrical
signal which is selectively sensitive to spatial frequency and orientation,
just as has been shown in the previous studies.

BLAKEMORE, NACHMIAS and SUTTON [1970] have shown that if one
adapts to a grating of a given spatial frequency and then looks at a grating
of a slightly higher or lower frequency, the lower frequency appears to be
of even lower frequency; conversely, the higher frequency appears to be
of even *higher* frequency. This confirms the original suggestion by
BLAKEMORE and CAMPBELL [1969] that these channels may be used for
coding the sizes of objects. SACHS, NACHMIAS and ROBSON [1971] have
shown, psychophysically, that each channel is separate from its neighbour
by testing the effect on the threshold of mixing two different spatial frequencies. Even when the frequencies are quite close in period, they only
add, according to the laws of probability summation; in other words the
channels are functionally separate.

Using this wide range of experimental approaches in the cat, monkey
and man, it has been possible to show that the original paradigm of
CAMPBELL and ROBSON [1964, 1968] is productive in that it has suggested
a number of novel experiments, each of which confirmed their original
suggestion: 'Thus a picture emerges of functionally separate mechanisms
in the visual nervous system each responding maximally at some particular
spatial frequency, and hardly at all, at spatial frequencies differing by a
factor of two. The frequency selectivity of these mechanisms must be
determined by integrative processes in the nervous system and they appear
to a first approximation at least, to operate linearly' [CAMPBELL and
ROBSON, 1968].

Why has the experimental application of elementary Fourier theory
worked so well so far? Its strict application demands that the system being
studied is linear; that is, that the principle of superposition holds. We have
used it only because it is the 'queen' of all description and also because
it is easy to explain the results and their implications to a wide public. It
does not follow that it is the best one, or even ultimately the correct one.
Since the visual system has neurons tuned to each orientation it seems
sensible to use a one-dimensioned function for studying these early stages

of signal transmission. It is true that the fact that a threshold exists, means that there is indeed a nonlinearity. However, if we confine our attention to threshold measurements, this nonlinearity is easy to handle. The visual system is grossly nonlinear when the dynamics of dark and light adaptation are involved. We have always avoided this complication by restricting our studies to the performance of the eye at one mean light level. Likewise, the visual system is nonlinear if it is overloaded with very high contrast levels. We have confined ourselves to levels of contrast less than 0.7, which covers almost all of the range used in normal vision, providing that one neither looks directly at light sources, nor subjects the eye to glare.

It would be disappointing, were this internal consistency for the prediction of thresholds for different types of objects restricted to repetitive patterns, like gratings, and not to other more interesting and realistic objects that surround us in daily life. CAMPBELL, CARPENTER and LEVINSON [1969] have attempted to predict the thresholds of thin lines and bars from the contrast sensitivity function. They find that it is possible to do so, again using theory (Fourier).

I may not have convinced you that the visual system really does have harmony, but you may now agree that it will be in the domain of contrast and not colour where we may appreciate the music of which HELMHOLTZ might well have dreamed. Today it would be difficult to imagine our concepts of sound, music and audition if Fourier had not given us this approach. In audition it has brought some order out of chaos, although much remains to be understood. The history of the application of Fourier's series in vision lies before us.

References

ARNULF, A. et DUPUY, O.: La transmission des contrastes par le système optique de l'œil et les seuils des contrastes rétiniens. C.R. Acad. Sci., Paris 250: 2757–2759 (1960).

BLAKEMORE, C. and CAMPBELL, F. W.: On the existence in the human visual system of neurons selectively sensitive to the orientation and size of retinal image. J. Physiol. Lond. 203: 237—260 (1969).

BLAKEMORE, C.; NACHMIAS, J. and SUTTON, P.: The perceived spatial frequency shift: evidence for frequency-selective neurones in the human brain. J. Physiol., Lond. 210: 727–750 (1970).

CAMPBELL, F. W.: Proc. IEEE 56: 1009 (1968).

CAMPBELL, F. W.; CARPENTER, R. H. S. and LEVINSON, J. Z.: Visibility of aperiodic patterns compared with that of sinusoidal gratings. J. Physiol., Lond. 204: 283–298 (1969).

CAMPBELL, F.W.; CLELAND, B.G.; COOPER, G.F. and ENROTH-CUGELL, C.: The angular selectivity of visual cortical cells to moving gratings. J. Physiol., Lond. *198:* 237–250 (1968).

CAMPBELL, F.W.; COOPER, G.F. and ENROTH-CUGELL, CHRISTINA: The spatial selectivity of the visual cells of the cat. J. Physiol., Lond. *203:* 223—235 (1969).

CAMPBELL, F. W.; COOPER, G. F.; ROBSON, J. G. and SACHS, M. B.: The spatial selectivity of visual cells of the cat and the squirrel monkey. J. Physiol., Lond. *204:* 120–121 (1969).

CAMPBELL, F.W. and GREEN, D.G.: Optical and retinal factors affecting visual resolution. J. Physiol., Lond. *181:* 576–593 (1965).

CAMPBELL, F. W. and GUBISCH, R. W.: The effect of chromatic aberration on visual acuity. J. Physiol., Lond. *192:* 345–359 (1967).

CAMPBELL, F.W. and KULIKOWSKI, J.J.: Orientational selectivity of the human visual system. J. Physiol., Lond. *187:* 437–445 (1966).

CAMPBELL, F.W. and MAFFEI, L.: Electrophysiological evidence for the existence of orientation and size detectors in the human visual system. J. Physiol., Lond. *207:* 635–652 (1970).

CAMPBELL, F. W. and ROBSON, J. G.: Application of Fourier analysis to the modulation response of the eye. J. opt. Soc. Amer. *54:* 581 (1964). — Application of Fourier analysis to the visibility of gratings. J. Physiol., Lond. *197:* 551–566 (1968).

COOPER, G.F. and ROBSON, J.G.: (1968). Successive transformations of spatial information in the visual system. Conf. on Pattern Recognition, N.P.L. Inst. of Electrical Engineers, London 1968.

DUFFIEUX, P. M.: L'intégrale de Fourier et ses applications à l'optique (Privately printed. Besançon 1964).

ENROTH-CUGELL, C. and ROBSON, J.G.: The contrast sensitivity of retinal ganglion cells of the cat. J. Physiol., Lond. *187:* 517–552 (1966).

GILINSKY, A.S.: Masking of contour-detectors in the human visual system. Psychon. Sci. *8:* 395–396 (1967). — Orientation-specific effects of patterns of adapting light on visual acuity. J. opt. Soc. Amer. *58:* 13–18 (1968).

GUBISCH, R. W.: Optical performance of the human eye. J. opt. Soc. Amer. *57:* 407–415 (1967).

HOPKINS, H. H.: 21st Thomas Young Oration. The application of frequency response techniques in optics. Proc. phys. Soc. Lond. *79:* 889–919 (1962).

HUBEL, D. H.: Tungsten microelectrode for recording from single units. Science *125:* 549–550 (1957).

HUBEL, D. H. and WIESEL, T. N.: Receptive fields of single neurones in the cat's striate cortex. J. Physiol., Lond. *148:* 574–591 (1959). — Receptive fields, binocular interaction and functional architecture in the cat's visual cortex. J. Physiol., Lond. *160:* 106–154 (1962). Receptive fields and functional architecture in two nonstriate visual areas (18 & 19) of the cat. J. Neurophysiol. *28:* 229–289 (1965). — Receptive fields and functional architecture of monkey striate cortex. J. Physiol., Lond. *195:* 215–243 (1968).

KIANG, N.Y-S.: Discharge patterns of single fibres in the cat's auditory nerve (M.I.T. Press, Cambridge, Mass. 1966).

KOENIGSBERGER, L.: Hermann von Helmholtz (Dover Publications, New York 1965).

LE GRAND, Y.: La formation des images rétiniennes. Sur un mode de vision éliminant les défauts optiques de l'oeil. 2e Réun. de l'Institut d'Optique, Paris 1937.

LINFOOT, E. H.: Fourier methods in optical image evaluation (The Focal Press, London/New York 1964).

MAFFEI, L. and CAMPBELL, F.W.: Neurophysiological localization of the vertical and horizontal visual coordinates in man. Science *167:* 386–387 (1970).

O'NEILL, E. L.: Introduction to statistical optics (Addison-Wesley, Reading, Mass./Palo Alto, Calif./London 1963).

PANTLE, A. and SEKULER, R.: Size-detecting mechanisms in human vision. Science *162:* 1146–1148 (1968).

ROSE, A.: Quantum limitations to vision at low light levels. Image Tech. *12:* 13–31 (1970).

SACHS, M.B.; NACHMIAS, J. and ROBSON, J.G.: Spatialfrequency channels in human vision. J. opt. Soc. Amer. *61:* 1176–1186 (1971).

WESTHEIMER, G.: Modulation thresholds for sinusoidal light distributions on the retina. J. Physiol., Lond. *152:* 67–74 (1960).

WHITFIELD, I. C.: The auditory pathway (Arnold, London 1967).

WHITFIELD, I.C. and EVANS, E.F.: Responses of auditory cortical neurons to stimuli of changing frequency. J. Neurophysiol. *28:* 655–672 (1965).

YOUNG, T.: Outlines of experiments and enquiries respecting sound and light. Phil. Trans., p. 106–150 (1800).

Author's address: Prof. F.W. CAMPBELL, The Physiological Laboratory, *Cambridge* (England)

Discussion

H.C. LONGUET HIGGINS: Just a very brief question, addressed to Dr. CAMPBELL whose talk I found marvelously entertaining. When you showed us the Fourier transform of your window, that of course was a real function, it wasn't, strictly speaking a Fourier transform, because it omitted phase-information. Now, of course, to recover the original picture you need to supply the phase-information, and the recording of phase-information is one of the difficult questions raised by any theory. Which postulates that, Fourier transformation or anything like it, is done by the visual system. I wonder if you'd like to comment about that particular problem.

F. W. CAMPBELL: Yes, you've homed in, straightaway on the '$ 64,000 question' of phase, and in the auditory system, phase is very unimportant under most conditions, but in the spatial domain, of course, it is important. Now the awkward point, which we haven't straightened out at all is, need we postulate that the image is, in the end, completely re-formed? If it is, then we must retain phase information completely. Even if it is not re-formed in its complete sense, then we must have some partial phase-information. There are a number of possible experiments one could do to establish this and we haven't had the opportunity to do so. I think, therefore, that the answer is simply that we don't know, at this stage.

A. ROSE: I would like to ask Dr. CAMPBELL whether he knows of any evidence whether the eye can resolve beyond the diffraction limit, as might be expected in principle, from an optical system?

F.W. CAMPBELL: That is an awkward one. The classical case is Vernier visual acuity where you have two lines butting each other and you try to adjust them until they are lined up. The resolution then is of the order of a few seconds of arc. But it is not, strictly speaking, the same thing as resolution. You have got to have the lines fairly long—about 30 min long in fact. So there certainly seems to be some integration along length, and in the same way you can detect a very thin line of half a second or so, but that is because your contrast sensitivity is very high. The actual dip in the line spread function is of the order of 1%. So it is a rather semantic argument, what you mean by resolution here.

A. ROSE: There is reason to believe that one can go beyond the diffraction limit in a two-dimensional pattern. The Vernier acuity I think is, as you say, a special case. However, if you have a large enough signal-noise ratio it is, in principle possible to exceed the diffraction limit for a two-dimensional pattern.

F. W. CAMPBELL: Yes, that would be true. The trouble is you might get things like spurious resolution; you would not necessarily see what was actually there but you might be able to say there was a difference in the input.

B. JULESZ: I would like to comment on LONGUET-HIGGINS' question, which is a very important one. The analogies between audition and vision, which Prof. CAMPBELL and his co-workers pioneered in these last few years, are really astonishing even with regard to phase. It is really not true that the auditory system is not sensitive to phase. The story for both modalities is almost identical. As long as each of the frequency-tuned channels are below threshold, it is immaterial what the phase relationship is between signals in the various channels. The very moment two or more channels reach threshold, you can, of course, see phase dependence with the eye, and you can also perceive phase in audition. After all, an auditory click and auditory white noise have the same amplitude spectrum, but since the phase is different in the two cases, they sound very differently. What was so epoch-making in these recent visual experiments is that for 100 years the idea lay dormant that our vision is so similar to audition.

E. HAKEN: When I saw the step-function and the 'overshoot', in the perception I was reminded of the Gibbs phenomena in TD Fourier analysis. Then I thought I must drop this idea because you eliminate the mechanism of inhibition. But then, in the second talk, the Fourier analysis in space came up again, so I wonder if there is anything one can say about this 'overshoot' problem on the one hand, and Gibbs phenomenon and Fourier transform on the other hand?

E. HAKEN: I see: then, you would say that the inhibition could also be interpreted as a truncated-representation of the delta function?

F. W. CAMPBELL: I think that if I do the transform of one of these channels into the space domain, it might become a bit clearer. What we seem to have is an excitatory center here, surrounded by an inhibitory region. There is also another region of disinhibition needed to explain this steep attenuation. It's rather small, unfortunately, it's only about 8% of the peak and it requires averaging techniques to show it up, neurophysiologically.

Proc. 3rd int. Conf. From Theoretical Physics to Biology, Versailles 1971, pp. 385–387
(Karger, Basel 1973)

A Synopsis of Cyclopean Perception

B. JULESZ

Bell Telephone Laboratories, Incorporated Murray Hill, N.J.

With random-dot stereograms (i.e., special stimuli that were developed by the author a decade ago) one can 'repeat' most of the classical experiments in perception from optical illusions to figural aftereffects. Such a random-dot stereogram is shown in figure 1. Since these stimuli appear as random aggregates of dots when cast on the anatomical retinae, and portray the desired information first at a site where the two monocular pathways combine to form a percept of depth, it is thus possible to skip operationally several synaptic levels. The central site where the information is portrayed is called the 'retina of the cyclops'. When, say, a classical optical illusion is presented by cyclopean techniques and the amount of the perceived illusion is identical to the classical case, one can assume that the processes responsible for the illusion must reside after the 'cyclopean retina of steropsis'. Thus, it is possuble to locate the site of many perceptual processes as to being central or peripheral.

Such a process localization is demonstrated by experiments shown in figures 1 and 2. Figure 2 portrays the well-known Müller-Lyer illusion, portrayed by the usual brightness gradients. When figure 1 is stereoscopically fused (by crossing the eyes, or viewing the stereogram through prisms), the same Müller-Lyer illusory figures appear in vivid depth portrayed by a textured area hovering over a textured background. In this case, cyclopean Müller-Lyer illusion is perceived identically to its classical counterpart. Thus, this illusion must be central.

The technique of random-dot stereograms can be generalized. Random-dot cinematograms, for instance, can present the desired information on the cyclopean retina of movement perception. This is the essence of the investigations that are being demonstrated in a book by the author, entitled:

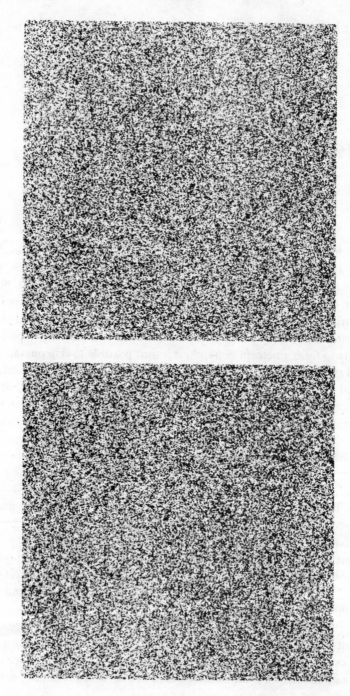

Fig. 1. Random-dot stereogram.

Fig. 2. Müller-Lyer illusion.

'Foundations of Cyclopean Perception' [University of Chicago Press, 1971]. See also "Binocular Depth Perception in Man—A Cooperative Model of Stereopsis" reprinted in this book.

Author's address: Dr. BELA JULESZ, Bell Telephone Laboratories Inc., *Murray Hill, NJ 07974* (USA)

Round Table Session

Proc. 3rd int. Conf. From Theoretical Physics to Biology, Versailles 1971. pp. 388–403 (Karger, Basel 1973)

Binocular Depth Perception in Man

A Cooperative Model of Stereopsis[1]

B JULESZ[2]

Murray Hill, N.J.

Under ordinary conditions the world appears very similar to us whether we view it with both eyes or one eye alone. Except for a slight increase in plasticity—owing to stereopsis (stereoscopic depth perception)—the contours and textures of objects seem unchanged. Because of this similarity between the monocular and binocular views it was commonly believed that understanding binocular vision must await the understanding of monocular vision. What is more, since the binocular percept exhibits an attribute—three-dimensionality—that is missing in the left and right vertical projections, it was widely accepted that seeing with two eyes is more complex than seeing with one eye. For example SHERRINGTON [1906], the leading physiologist of his time, concluded 'that during binocular regard . . . each uniocular mechanism develops independently a sensual image of considerable completeness. The singleness of binocular perception results from union of these elaborated uniocular sensations'. This opinion prevailed (and became even more exaggerated by the Gestaltist school) until the invention of random-dot stereograms in 1959.

Such a random-dot stereogram is shown in figure 1 [JULESZ, 1960 a, b]. When viewed with one eye alone the left and right arrays appear as a random assemblage of dots devoid of all form and familiarity cues However, when stereoscopically fused a diamond shaped area is perceived hovering above the background in vivid depth. Where monocularly only randomness was perceived, in the binocular view forms and contours can

1   Reproduced with the kind permission of Dr. C. J. GRUSSER and the Springer Publishing Co, Berlin, from: 'Pattern Recognition in Biological and Technical System', 1971.
2   I thank University of Chicago Press for permitting me to use figures 1, 8 and 9 which were published in *Foundations of Cyclopean Perception* by B. JULESZ.

Fig. 1. Random-dot stereogram. When stereoscopically fused a diamond appears in vivid depth above the randomly textured surround. (Courtesy of University of Chicago Press, 'Foundations of Cyclopean Perception' by B. JULESZ.)

be experienced. What is important, the left and right arrays separately do not contain the diamond form. It is only the *relation* between the two arrays that contains this information. In figure 1 both the background and the diamond are covered with the same textures, except that the diamond is horizontally shifted in one array in the nasal direction as if it were a solid sheet. The use of a digital computer and a high resolution plotting device assures that no monocularly perceivable gaps or other cues will spoil this perfect camouflage. Of course, instead of a diamond any desired form can be portrayed such that this form does not exist on the anatomical retinae and is first portrayed at a stage where the two monocular views are combined.

This demonstration thus showed that monocular form recognition is not necessary for stereopsis. A later demonstration by JULESZ [1966, 1967] showed that molecular forms clearly seen by one eye can be scrambled in the binocular view. This is shown in figure 2 where the bilateral symmetry along the horizontal axis is apparent in the left array, yet in the binocular view it cannot be perceived. Thus, whenever binocular fusion occurs it dominates monocular vision.

As a result of these findings it is now clear that the Sherringtonean view is incorrect, what is more, paradoxically its opposite is true. Binocular vision (when restricted to binocular localization) is simpler than monocular vision. While in monocular vision the enigmatic processes of form recognition must operate in order to separate complex objects from their

Fig. 2. Random-dot stereogram, in which the monocularly apparent bilateral symmetry (across horizontal axis) in the left view is scrambled in the stereoscopic view.

backgrounds, in binocular vision objects can be separated without having to recognize them first. This simplification of problems brought about by random-dot stereograms has been realized by the neurophysiologists as emphasized by BISHOP [1969] and as attested by the recent emphasis on finding the physiological basis of binocular depth perception [BARLOW, BLAKEMORE and PETTIGREW, 1967; PETTIGREW, NIKARA and BISHOP, 1968; HUBEL and WIESEL, 1970].

The implications of random-dot stereograms for binocular vision are

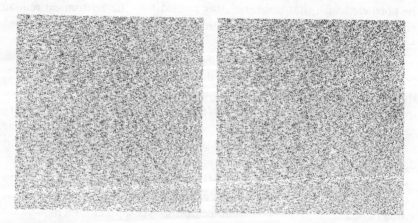

Fig. 3. Random-dot stereogram which, when monocularly viewed, appears uniformly random, yet when stereoscopically fused, portrays a spiral surface in depth.

manifold. First of all, although WHEATSTONE [1838]—by inventing the stereoscope—showed that binocular disparity is a basic depth cue, quite recently many researchers (for instance GIBSON [1950] and ITTELSON [1952]) regarded disparity of minor importance. They assumed that the many other monocular depth cues from retinal gradient of texture to movement parallax are more powerful. However, as figure 3 demonstrates, in the absence of any monocular depth or familiarity cue a surface of great complexity can be perceived using binocular disparity as a single cue. Were not binocular disparity such a powerful depth cue, how could we understand why the highest animal forms sacrificed panoramic vision and developed stereopsis instead. In order to obtain stereopsis, the two eyes' views had to overlap (thus reducing the panoramic visual field by half) and complex processes of head movement and eye-coordination had to evolve. Obviously, stereopsis had to offer substantial advantages over monocular movement parallax and the other depth cues.

Many researchers believe that stereopsis enables predators to judge the distance of their prey while motionless, while animals that rely on monocular movement parallax must be in motion (and thus less easy to hide). However, this argument is not very powerful, since stereopsis gives only the sense of *relative depth*. In contributing to *absolute depth* judgments, stereopsis is just one of the many depth cues. There must be some other reasons why stereopsis evolved and the previous demonstrations can give a possible answer. Since time immemorial animals developed camouflage and easily blended with the background. It is possible to hide successfully under monocular vision and the predator has to possess complex form recognition to see, say, a moth on a tree when both are covered with similar textures. However, as figures 1 and 3 demonstrate even under *ideal* monocular camouflage, the hidden objects jump out in depth when stereoscopically viewed. What is more, this object separation does not necessitate any familiarity with the stimulus, and therefore stereopsis could evolve at a relatively early stage in the hierarchical processing chain of vision.

The previous lecture summarized some of the recent findings on the neurophysiological basis of stereopsis. These findings were obtained in the cat and monkey. BOUGH [1970] demonstrated stereopsis in the macaque monkey by using random-dot stereograms. With this development much of the psychological findings on human stereopsis can now be related to physiological evidence. One of the advantages of random-dot stereograms is that they provide an objective and unfakeable test for stereopsis. We are not asking the subject whether he perceives depth *per se*, but simply what does he see. If an animal is first taught to discriminate between a cross or a circle, and then these figures are portrayed by random-dot stereograms, then transfer of discrimination to these stimuli establishes stereopsis.

This unfakeable aspect of random-dot stereograms is very important. Besides providing an objective test for stereopsis in humans, random-dot stereograms can help to test some enigmatic abilities which previously belonged to the realm of anecdotal accounts. Let us briefly review such a feat that took place during my sabbatical stay at the Psychology Dept. of M.I.T. STROMEYER and PSOTKA [1970] presented one image of a random-dot stereogram to one eye of a 23-year-old woman, who possessed eidetic imagery (photographic memory). She would build up the eidetic image of the array by scanning it for a few minutes. Then a day later the other image would be presented to her other eye. She was able to fuse her eidetically stored image with the physically presented one and correctly reported a square in depth and accurately pointed to its corners. This study alone would deserve several pages, but here it should suffice that stereopsis and random-dot stereograms established the existence of a detailed texture memory for arrays 1000×1000 in size lasting for days (at least for one eidetiker). Whether the rest of us have the same detailed memory as an eidetiker only without being able to recall it remains to be seen.

In recent years neurophysiologists found many units in the visual cortex that respond to stimulation of both eyes. These binocular units fire for edges

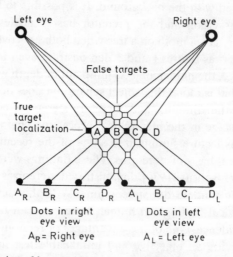

Fig. 4. Illustration of how ambiguous localization arises in binocular vision.

of the same orientation that cast on the two retinae [HUBEL and WIESEL, 1962]. However, in Area 17 of the cat [BARLOW *et al.*, 1967] and in Area 18 of the monkey [HUBEL and WIESEL, 1970] there are units that also require a very specific binocular disparity for optimal firing. Both facilitation and inhibition can be obtained in very narrow disparity ranges [BISHOP, 1969]. In spite of these very important physiological findings, I think they represent only the first steps in the understanding of binocular localization.

In order to understand the basic problem of binocular localization one must realize that the visual system is confronted with ambiguities. This becomes obvious from figure 4. Here four objects (having the same ordinates) cast their projections on the left and right retinae. If these objects are similar (having the same brightness, color, shape, and orientation) it is ambiguous which projected object in one view belongs to which projection in the other eye's view. In case of four objects, already 16 localizations are possible out of which only four are correct while 12 are 'phantom' targets. With increased number of objects the probability of false localization quickly becomes unity.

This problem of how the visual system selects the real solution from millions of false possible localizations is particularly severe for random-dot stereograms. Hundreds or thousands of identical dots lie on the same horizontal line in the left and right arrays. There are millions of ways the dots could be localized. Already by chance alone, half of the black and white dots are in alignment at zero disparity. Similarly, at one unit disparity (in the nasal or temporalward direction), again owing to chance, half of the stimulus points are in registration. Why does the visual system ignore these solutions at small disparity values and searches for the diamond at a large disparity value? The answer is simple: the stereopsis mechanism searches for the *densest* possible solution. Instead of stopping at a 50% localization it searches for a disparity shift such that near 100% of the stimulus point in an area should become aligned. If by various horizontal shifts each area of the left and right arrays can be brought in near perfect alignment the visual system regards this the solution of the binocular localization.

Present neurophysiology has found only the first stage of this mechanism, the *local* disparity extractors. However, the next required stage, which evaluates all the outputs of these local units and selects that *global* solution which has the densest cluster of firing, units with the same disparity, has not been found yet. In 1962 I tried out such a simple model of stereopsis using computer simulation [JULESZ, 1962]. In this model (called AUTO-

MAP-1) the left and right arrays were horizontally shifted with respect to each other, and after each increasing shift the difference between the two arrays was taken. Then these difference fields (with increasing shifts) were stacked above each other. In some of these difference fields, clusters of adjacent dots with minimum values were formed. These clusters corresponded to the cross-sections of the original objects. I do not want to go into the details and refinements of this model, since it could not cope with some other basic phenomena. For instance, a 15% expansion of one image of figure 1 still retains stereopsis [JULESZ, 1960a, b]. Such an invariance under expansion-dilation is not *inherently* incorporated in the model. Furthermore, some recently discovered phenomena by FENDER and JULESZ [1967] shed light on a basic aspect of stereopsis that cannot be explained by any simple model. Before a new model of stereopsis will be described, we have to summarize a few new psychological findings.

The first such finding was obtained under binocular retinal stabilization [FENDER and JULESZ, 1967]. Before this work, it was assumed that Panum's fusional limit was a rigid 6 min arc disparity value in the fovea, above which the fused image would break apart and appear as double. Indeed, under binocular retinal stabilization (using the usual contact lenses and mirrors but for both eyes) the subject cannot fuse the images, which move together with the retinae, until they are actually shifted on the retinae within 6 min arc alignment. Then they suddenly coalesce. However, after fusion the images can be slowly pulled apart by large amounts in the horizontal direction without losing fusion. The limit of pulling depends on the stimulus. For classical targets (a vertical line) the limit is about 40 min arc, whereas for random-dot stereograms it is in excess of 120 min arc. After this limit is reached the fused images suddenly break apart and have to be brought back again within 6 min arc to fuse again. For random-dot stereograms this hysteresis phenomenon of fusion is shown in figure 5.

The importance of this finding is that stereopsis is clearly a *cooperative* phenomenon. Cooperative phenomena in physics exhibit hysteresis (a lag between cause and effect—a primitive type of memory) and the amount of this hysteresis depends on the number of elements that participate. The finding that a simple line fails to stimulate the entire stereopsis mechanism while complex stereograms do, has many consequences. Moreover, any advanced model of stereopsis must also account for whether two points are 'corresponding points' or not according to their past history.

Another fact that a model of stereopsis has to explain is the long 'learning' required for some stereograms that portray complex surface.

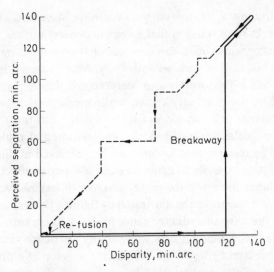

Fig. 5. Hysteresis phenomenon of binocular fusion during binocular retinal stabilization [FENDER and JULESZ, 1967]. Abscissa: physical pulling; ordinate: perceived separation.

While figure 1 can be perceived in depth almost immediately, figure 3 initially requires several seconds. It often takes a minute or more till the percept stabilizes and reaches its final extent. But once such a complex stereogram has been fused at a second trial it can often be fused very rapidly.

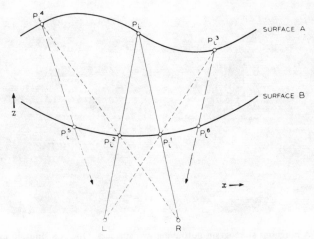

Fig. 6. Algorithm for generating ambiguous stereograms portraying two surfaces [JULESZ and JOHNSON, 1968].

A third finding is of interest too. We have seen how the stereopsis mechanism seeks for a solution that gives the densest surface. What would happen if stereograms could be generated that contained more than one dense surface? In case of such an ambiguity, what factors might influence the final percept? That ambiguous stereoscopic depth percepts exist is demonstrated by the classically known 'wallpaper-effect'. When one views a periodic structure (e.g., an old radiator, or bathroom tiles) it is possible to fuse the images at several disparity values (integral multiples of the periodicity). The result is that a plane can be perceived at different depth levels. Would it be possible to generalize the wallpaper-effect and instead of multiple planes have two (or more) surfaces of *any shape*.

This problem was solved as illustrated by figure 6 [JULESZ and JOHNSON, 1968a, b]. In the case of ordinary random-dot stereograms, the surfaces can be speckled with random dots without any restriction; for ambiguous stereograms this is not true. When we select the color of a dot on surface 'A' as P_A, this selection constrains P_B on surface 'B' to take the same color. However, since P_B can be viewed by the other eye as well, the line of sight through P_B will intersect surface 'A' in P_A which has to be identical to P_B and in turn to P_A. This reasoning must continue and a single dot will force its color an a set of points. Theoretically, this set has infinite points, but with finite resolution, the set has finite numbers of elements. The result is a 'holistic' organization where each element is coupled to

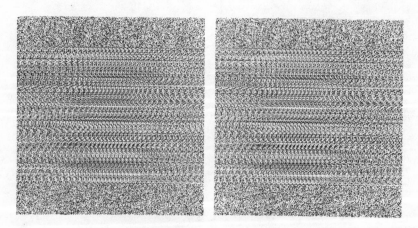

Fig. 7. Ambiguous stereogram portraying two surfaces. One is a sinusoid cylinder, the other a sinusoid cylinder with less amplitude and half wavelength. A top and bottom margin of the unambiguous surfaces aids in perceptual reversal [JULESZ and JOHNSON, 1970].

the rest. Surface 'A' can be the front surface of an object, while surface 'B' can be the back (hidden) surface. With such a stereogram one can view the surface from the front or from behind at will. Instead of physically moving around the object (as is the case with holograms) the viewer can sit still and let his mind do the wandering.

Such an ambiguous stereogram is shown in figure 7. The stereogram contains two surfaces: 'A' a sinusoidal cylinder, and 'B' another sinusoidal cylinder with lesser amplitude and half period. In order to help fusion, the top and bottom margins of the two surfaces are unambiguously portrayed. If one looks up, one organization prevails; if one looks down, the entire area (except the top margin) changes into the other surface. Without the unambiguous margins it is difficult to switch organizations. Whichever of the organizations has been obtained (either by chance or some natural bias) will tend to prevail, and in order to change to the other surface one has to destroy the state in which he is in by a quick convergence movement or some other means. In figure 7 the two organizations are relatively easy to change. One can notice that after prolonged viewing the prevailing organization weakens and for some time both surfaces can be seen one behind the other. But then the old percept wanes and the new one becomes dominant. A satisfactory model must also explain these observations.

After these preliminary findings, I shall review the model. The model was first formulated in a book *Foundations of Cyclopean Perception* just published [JULESZ, 1971]. It is spring-coupled, magnetic-dipole model. For cybernetics the model has two useful properties. Since stereopsis does not require *semantics* (familiarity cues) the model can be much simpler than models used in cognition. At the same time, since it is a cooperative model it has many interesting and unexpected states that are usually missing in models of sensory psychology.

This model of stereopsis uses only such simple elements as magnets and springs. A model should not only be isomorphic to the real phenomenon (or an important aspect of it) but should also be better matched to human thought processes than the real phenomenon. Mechanical models are closer to human intuition than any other kind, and though coupled oscillators with frequency shifts may have provided a more modern version, I prefer the mechanical-magnetic model. I also wish to refrain from neurophysiological models, because we do not possess adequate physiological evidence for a global model of binocular depth perception.

The model is shown in figure 8. A two-dimensional array of magnetic dipoles is mounted in ball-joint bearings such that the dipoles can rotate

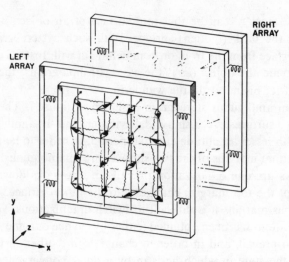

Fig. 8. A spring-coupled, magnetic-dipole model of stereopsis. (Courtesy of University of Chicago Press, 'Foundations of Cyclopean Perception' by B. JULESZ.)

out of the plane of the array. A black or white dot cast on the retina would correspond to a south (S) or north (N) pole turned towards the reader. One such oriented dipole array corresponds to the cortical state that results from stimulating one retina. The other array corresponds to the cortical state elicited by stimulating the other retina. The two arrays should be imagined in close vicinity sliding over each other. If the patterns on the retina are identical and the two arrays overlap then the S and N poles of corresponding dipoles *interlock*. Identical arrays of interlocked dipoles in one dimension are shown in figure 9a.

The essence of the model, however, is the spring-coupling of each dipole to its adjacent neighbors. This coupling of all dipoles through their neighbors is essential for hysteresis. (Such coupled dipoles are used in Ising's model of ferromagnetism and were used in brain modeling in some other connotation [CRAGG and TEMPERLEY, 1954].) The importance of this coupling becomes apparent when we try to explain the fusion of figure 1. Before the arrays are fused, by chance 50% of the dipoles become interlocked. Because of the spring-coupling the other 50% of dipoles are forced

Fig. 9. Explanation of how the spring-coupled, magnetic-dipole model works for various stimulus conditions. (Courtesy of University of Chicago Press 'Foundations of Cyclopean Perception' by B. JULESZ.)

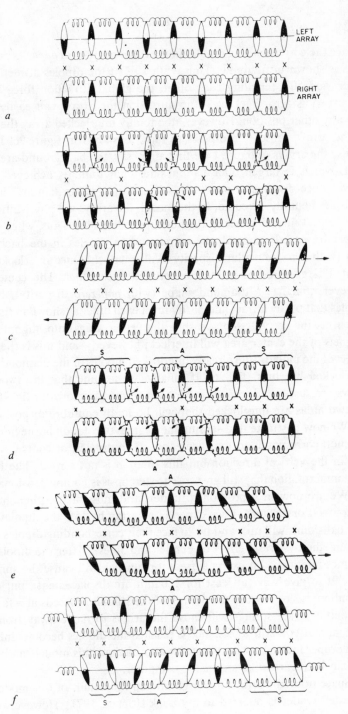

to face the corresponding dipoles in the other array with *opposite* polarities. This is shown in figure 9b. Thus, with 50% of the dipoles attracting each other, and 50% repelling each other, the global attraction force between the arrays is zero. Because of no attraction the arrays can easily be slid into any other position. However, in a totally interlocked array the pulling of the arrays causes the dipoles to turn, as shown in figure 9c. In order to fuse the arrays of figure 1 we first align, say, the background areas. This shift, which is larger than Panum's fusional limit, is achieved by the convergence-divergence movements of the eye; within Panum's limit the shifts are neurally executed. In the model, the arrays are physically moved with respect to each other in the horizontal direction. Thus, when the two arrays are exactly aligned, the corresponding dipoles in the background area (with zero binocular disparity) will interlock. Once interlocked they exert a large force of attraction between each other. The center area, however, that has a diamond shape (with *non-zero* disparity), contains dipoles that partly attract and partly repel each other as shown in figure 9d. When, as the second step, we shift the center areas into alignment, the magnets of the center area will interlock; however,—and this is the crucial aspect of the model—the already interlocked dipoles of the diamond *remain* interlocked but turn (fig. 9e). If, finally, we assume that the two dipole arrays are suspended in a spring-loaded frame (as shown in fig. 9f), after the two areas are interlocked they will be in an equilibrium position.

We now postulate that sensing the degrees of rotation in the horizontal direction corresponds to local stereopsis. The dipoles, of course, can also turn in the vertical direction but this rotation is not sensed. The limit of horizontal rotation that still gives rise to stereopsis is Panum's fusional limit.

We are now in a position to understand most of the phenomena of stereopsis. For instance, if we expand one array by 15% the dipoles across their ball-joints will turn and try to face their corresponding dipoles. It also explains the hysteresis-effect of Fender and Julesz. After the dipole array becomes interlocked it exerts a great force which will cause the springs in figure 9f to give way (at least for 120 min arc displacement), but only if the pulling is slow. When the pulling of the stimuli proceeds at a fast rate, the inertia of the springs will cause the arrays to move away from each other, and after a critical distance is reached the dipoles become unlocked. With proper masses, spring constants and frictions this model can be built and can be regarded as an analog computer.

Space does not permit me to go into all the details of this model. The interested reader is referred to my book [Julesz, 1971]. However, let me note how the model can clarify the problem of perceptual learning of fusion. When the stereograms are simple, such as figure 1, there are only two steps required for fusion: first aligning the background, and then trying a nasal or temporal shift in order to find the other depth plane. If by chance, the center area is fused first, a quick cortical shift will fuse the background as well. The only ambiguity is due to the dichotomy of performing a nasal or temporal shift. However, for stereograms that contain complex surfaces (having many depth planes) this dichotomous decision has to be made for each depth plane. In case of a wrong decision a false localization can be obtained, and often one has to backtrack several decisions and start it from scratch. Nevertheless, if this decision-tree (of nasal-temporal forks) is learned, refusion becomes greatly facilitated.

Finally, the model explains the problem of ambiguities, too, If a solution is reached, that is, all dipoles became interlocked, it is impossible to obtain another global solution until the interlocked state is destroyed by some drastic way. The only way to surely find another solution is to unlock the local interlocked dipoles. This can be achieved by using electromagnets with short time constants rather than permanent magnets. After the local elements become unlocked, the global fusional process has another chance to search for a different global solution. This explains why in the central nervous system adaptation to stimuli occurs at very early stages! Were adaptation a highly central phenomenon we would never be able to 'shake' an obtained global solution. On a higher level, we would never be able to enjoy a pun or a good joke which is based on ambiguous meanings. I am confident that binocular depth perception involves more than localizing objects in depth.

This model of stereopsis could be generalized to cope with some other perceptual processes, perhaps even with cognitive processes. Yet, let me finish my talk with another thought. Random-dot stereograms and stereopsis permit the portrayal of almost any form, such that the early processing stages are operationally 'skipped' and the global information is presented in Area 18 or even higher in the visual cortex [HUBEL and WIESEL, 1970]. Since the stimulus information is not even physically presented on the anatomical retinae, but only at some central site where the two monocular views become combined, one can use random-dot stereograms to separate central processes from peripheral ones. In the last 2 years, I started to 'repeat' the known phenomena of visual perception by random-dot stereograms and similar techniques. From optical illusions to aftereffects I tested the location of the processes. For instance, if figure 1 instead of a diamond portrays the well-known Müller-Lyer illusory figures, and, as is the case, the illusion is not diminished, one knows that the illusory process takes place after the site of global stereopsis. Thus, we can trace the information flow of the visual system without opening the black-box. I devoted an entire book to this 'psychoanatomical' investigation [JULESZ, 1971]. A synopsis of these psychoanatomical investigations, called cyclopean perception, is briefly given elsewhere in this book. Here it should suffice, that stereopsis and some modern stimulus techniques permit us to go much beyond the study of binocular depth perception into the study of perceptual processes in general. Thus, stereopsis is not only helping us to localize objects in space but permits us to localize processes inside our brains.

References

BARLOW, H. B.; BLAKEMORE, C. and PETTIGREW, J. D.: The neural mechanism of binocular depth discrimination. J. Physiol., Lond. *193:* 327–342 (1967).

BISHOP, P. O.: Neurophysiology of binocular single vision and stereopsis; in JUNG, R. Handb. of sensory physiology, vol. 7. (Springer, Berlin/Heidelberg/New York: 1969 [in press]).

BOUGH, E. W.: Stereoscopic vision in the macaque monkey: a behavioural demonstration. Nature, Lond. *225:* 42–44 (1970).

CRAGG, B. G. and TEMPERLEY, H. N. V.: The organization of neurons; a cooperative analogy. Electroenceph. clin. Neurophysiol. *6:* 85–92 (1954).

FENDER, D. H. and JULESZ, B.: Extension of Panum's fusional area in binocularly stabilized vision. J. opt. Soc. Amer. *57:* 819–830 (1967).

GIBSON, J. J.: The perception of the visual world (Houghton Mifflin, Boston 1950).

HUBEL, D. H. and WIESEL, T. N.: Receptive fields, binocular interaction and functional architecture in the cat's visual cortex. J. Physiol., Lond. *160:* 106–154 (1962).—Stereoscopic vision in macaque monkey. Nature, Lond. *225:* 41–42 (1970).

ITTELSON, W. H.: The Ames demonstrations in perception (Princeton Univ. Press, Princeton, N.J. 1952).

JULESZ, B.: Binocular depth perception of computer generated patterns. Bell Syst. Techn. J. *39:* 1125–1162 (1960a).—Binocular depth perception and pattern recognition; in CHERRY, C. Information theory. 4th London Symp. 1960 (Butterworth, London 1961b). —Towards the automation of binocular depth perception (AUTOMAP-1); in POPPLEWELL, C. M. Proc. IFIPS Congress, Munich 1962 (North-Holland Publ. Comp., Amsterdam 1963).—Binocular disappearance of monocular symmetry. Science *153:* 657–658 (1966).—Suppression of monocular symmetry during binocular fusion without rivalry. Bell Syst. Techn. J. *46:* 1203–1221 (1967).—Foundations of cyclopean perception. To be published by (University of Chicago Press [to be published] 1971).

JULESZ, B. and JOHNSON, S. C.: Stereograms portraying ambiguously perceivable surfaces. Proc. nat. Acad. Sci., Wash. *61:* 437–441 (1968a).—Mental holography: stereograms portraying ambiguously perceivable surfaces. Bell Syst. Techn. J. *49:* 2075–2083 (1968b).

PETTIGREW, J. D.; NIKARA, T. and BISHOP, P. O.: Binocular interaction on single units in cat striate cortex: simultaneous stimulation by single moving slit with receptive fields in correspondence. Exp. Brain Res. *6:* 391–410 (1968).

SHERRINGTON, C. S.: Integrative action of the nervous system (Yale Univ. Press, New Haven, Conn. 1906).

STROMEYER, C. F. and PSOTKA, J.: The detailed texture of eidetic images. Nature, Lond. *225:* 346–349 (1970).

WHEATSTONE, C.: On some remarkable, and hitherto unobserved, phenomena of binocular vision. Philos. Trans. roy. Soc. Lond. *128:* 371–394 (1838). Reprinted (pp. 371–377, 386–387) in: DEMBER, W. Visual perception: the nineteenth century (Wiley, New York 1964).

Author's address: Dr. BELA JULESZ, Bell Telephone Laboratories Inc., *Murray Hill, NJ 07974* (USA)

Proc. 3rd int. Conf. From Theoretical Physics to Biology, Versailles 1971, pp. 404–410
(Karger, Basel 1973)

Some Tools which Link Theoretical
Physics to Biology

B. KNIGHT

I was asked to make a few comments on some of the formal aspects of the theoretical part of our work on the *Limulus* eye, mostly for the benefit of the theoretical physicists present. I will do this with emphasis on some methods drawn 'from theoretical physics to biology', and will end by mentioning some of our current theoretical problems, on which we would welcome any professional assistance.

You recall that the *Limulus* eye puts the visual signal through several processing steps. There is the conversion of total intracellular voltage into the firing rate of nerve impulses. There is the conversion of firing rate back into voltage — both the processes of lateral inhibition and of self-inhibition. And there is the mixing of excitation, lateral inhibition, and self inhibition to form the total intracellular voltage. Laboratory measurements show that all of these several processes are *linear* to reasonably good approximation. We know something else about these processes: their laws of transduction, their input-output relations, remain unchanged with the progress of time. They all may be treated as time invariant linear processes.

A brief inspection of the first slide (see RATLIFF fig. 4, p.339, this volume) reminds us that the construction of the eye is also invariant under translations in the horizontal direction, to reasonable approximation. The whole system may be treated as linear and translationally invariant in both time and space.

Now each process within the eye (and also the eye as a whole) may be regarded as a linear operator which acts upon a visual signal which is extended over both space and time. Because of the invariance properties of this process operator, it commutes with operators of translation in space and time, and shares the same eigenfunctions. These eigenfunctions, the eigenfunctions of translation, are the complex-exponential (sinusoidal) running waves, and an arbitrary pattern in space and time may be expressed as a superposition of them.

This kind of situation, of course, is of common occurrence in theoretical physics. Maxwell's equations, for instance, are a familiar example. They have the linearity and the translational invariance properties. They define, implicitly, a linear operator (or Green's function) which acts upon specified charge and current density distributions, to yield the resulting electromagnetic field.

We may proceed to solve the problem of the *Limulus* visual response just as we would the electromagnetic problem. Because an arbitrary input may be Fourier-analyzed into complex-exponential running waves, it will suffice to find the response to such a wave, with arbitrary frequency and wavelength. Because this wave-form is an eigenfunction of all the operators involved, we may confidently assume that the response is given by the same complex exponential running wave, multiplied by a yet unknown complex coefficient (and we may assume the same about all intermediate variables). This assumption turns the equations of the system into simple algebraic relations among the coefficients. (See equation 4 of RATLIFF [p. 353, this volume], where for our case the job is partly done—in time but not in space.) We solve for the coefficient of the response wave. This will be in general a complex number which depends on frequency and wavelength, and is familiarly called the propagator (by physicists, or the transfer function by engineers).

Our procedure for the *Limulus* eye differs at one point from the electromagnetic example, in which the component pieces of the transfer function could be obtained simply by performing differentiations (specified by Maxwell's equations) upon complex exponential waves. In our case we determine the pieces by direct measurement in the laboratory. It is practical to make these measurements because all the transduction processes, excepting lateral inhibition, are local in space and hence independent of wave number, and depend only on frequency. The way the experiments were done has been discussed by RATLIFF, who also gives the measured amplitude and phase of the transduction coefficients from light to excitatory voltage and from voltage to firing rate in his figure 14 (p. 351, this volume).

In the case of lateral inhibition, plausible physiology suggests that *that* propagotor should be a product, of a function of frequency only and a function of wave-number only. An experiment has verified that this is so. The function of frequency has been measured at zero wave-number, and is discussed by RATLIFF (fig. 15, p. 352, this volume) and the function of wave-number has been measured at zero frequency by ROBERT BARLOW.

These measurements enable us to make predictions about how the *Limulus* eye will respond to a wide variety of visual stimuli. The predictions of the theory have been accurate so far. (See KNIGHT 1973a for a review.)

The same methods that we have applied to the *Limulus* eye, others are applying to more elaborate neural structures, such as the vertebrate retina and the olfactory bulb of the brain. Their tasks are more difficult and their results are less complete, but their early successes and their continuing progress are encouraging.

I will finish by mentioning a few unsolved problems which should present 'a challenge for theorists'.

1. Numerous situations arise in which we may directly determine the *amplitude* of a sensory transfer function, but not its *phase*. For example, the amplitude of response for the human visual system may be determined by making flicker-threshold measurements at various frequencies, but the corresponding absolute phase information is unattainable by direct measurement, since every human sense modality may introduce phase-shifts. Another example comes from relating the transfer function to certain fluctuation measurements, and will be discussed further presently.

Now the similar problem of retrieving the imaginary part of a transfer function from full knowledge of its real part, has a well-defined solution: if the system is stable as well as causal, we have no singularities in the upper half-plane and may use the Kramers-Kronig causality relations. But if we try to retrieve the phase from the amplitude in the same way (by taking a complex logarithm) the situation is far less clear. Surely a transfer function with a *zero* in the upper half-plane will yield an erroneous result; but what is the *physical* criterion (comparable to causality and stability in the previous example)? To one who has no deep insight into this matter it is a bit of a miracle that the Kramers-Kronig algorithm retrieves the phase properly in most cases. The transfer function for lateral inhibition in *Limulus* is an unusual but important case where it fails. What are the practical criteria that tell us whether we may proceed? Beyond that, what are the conditions, when we compose together transfer functions (as we do in the equations for the *Limulus* retina), under which the phase-retrieval property is inherited, or lost, or regained?

2. The voltage in response to light, in the *Limulus* visual cell, shows a certain amount of stochastic noise, as seen in figure 1a [from DODGE, KNIGHT and TOYODA, 1968]. On the basis of plausible physiology, we have

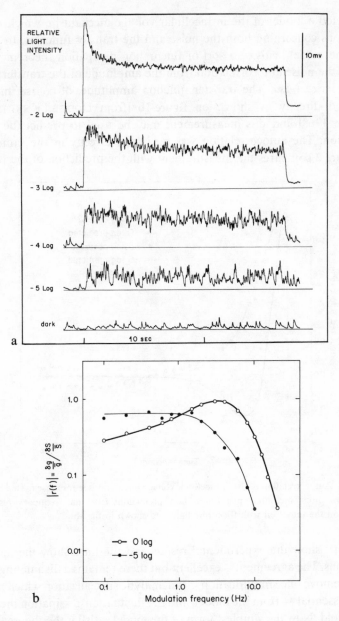

Fig. 1 a) Records of the voltage within a *Limulus* visual cell in response, to light of various intensities. *b)* Amplitude of the frequency response, measured at two light intensities The bright light (0 log) was 10^5 times as intense as the dim light (-5 log). (From DODGE, KNIGHT, and TOYODA 1968.)

constructed a model of the intracellular voltage mechanism, which makes predictions concerning both the noise and the transfer function. In particular, the model leads to a sort of fluctuation-dissipation theorem which predicts the noise autocorrelation from the amplitude of the transfer function, and *vice versa*. The transfer function amplitude, of course, may be measured directly as shown on figure 1b [from DODGE, KNIGHT and TOYODA, 1968] and this measurement may be used to predict the autocorrelation. The autocorrelation was measured directly in the same cell, and figure 2 compares the measurement with the prediction of the model.

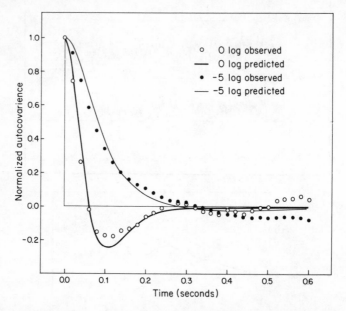

Fig. 2. Autocorrelation of the intracellular voltage fluctuations in response to steady light. Points show direct measurements. Lines show predictions from the frequency response, measured on the same cell with flickering light and shown in fig. 1b.

The points show the experimental results and the lines show the model's predictions. The agreement is excellent but there remains a disquieting fact: we can remove one simplification — an analytic specialization which seems quite inessential — from the model, and the fluctuation-dissipation theorem fails to hold. Now the simplest form of the model satisfies the theorem and so does the real-life *Limulus* cell. What special feature do they have in common? What criterion separates models that satisfy the theorem from those that do not? The answer would tell us something about visual neurophysiology. (The model is discussed in more detail by KNIGHT [1972 and 1973b].)

3. Currently, there is interest in the performance of multiple parallel nerve fibers. Often in the laboratory, when we are recording from a cluster of nerve fibers, we wish to deduce how many fibers are represented in the signal we receive, and what is the steady state firing rate of each individual. This leads to an unusual problem in analysis. In its simplest form, the problem is the same as if we had a room full of clocks, all of whose ticks sounded the same, and had to deduce the number of clocks and the frequency of each one's tick, just by listening to the pooled ticking. The power spectrum of the room noise is the sum of the power spectra of the clocks, and this leads to a way to solve the problem. If you try the natural things you will discover you are using Mellin transforms, and a formal solution is easily found. It is in terms of a contour integral, with poles at the zeroes of the Riemann Zeta function. As a practical algorithm the solution is unsatisfactory for other reasons as well. But if you express the Zeta function in terms of an infinite product over primes, a remarkable method of solution emerges (in terms of repeated prime dilations of the frequency scale) which is elegant and rapidly convergent. The trouble with this is that the problem that we *really* want to solve involves component impulse trains which show substantial irregularities in their interpulse intervals, as we are dealing with nerves, not clocks. Already in its idealized form the problem has led us to the borders of analytic number theory. We have gone beyond our depth, and need professional advice.

4. If a population of independent but similar sensory neurons respond to a common periodic stimulus, they will preferentially choose a certain part of the cycle in which to fire. The extreme case is if they all settle down to firing at one particular point in the cycle; this condition is called phase locking, and its presence or absence will profoundly affect the way in which the system processes information. Now if we fire a sensory neuron at a specified point in the stimulus cycle, its next firing in general will be at a different point in the cycle. Thus, the neuron furnishes a mapping from points in the cycle to new points. We can conceive of this as the neuron mapping the circumference of the circle onto itself. If the circumference of the circle has fixed points under this mapping then the neuron will be subject to phase locking by the stimulus.

A simplified model of this group of mappings is the group of fractional-linear (or projective) transformations which map the unit circle of the complex plane onto itself. These transformations already come in two types:

those that have fixed points on the circumference of the unit circle and those that don't. The group is isomorphic to the Lie group SL_2. In fact, if we inspect the big group (of all continuous mappings of the circle's circumference) near the identity, it is not hard to show that SL_2 is contained in the group an infinite number of times. The kinematics of phase-locking is remarkably rich.

As yet we are unable to answer simple questions like: will some iterate of the mapping have fixed points almost always? (Which is to ask: will a population of ideal neurons fall into an eventually repeating lock-step, in response to almost every periodic stimulus?) But tools—and probably powerful tools—are at hand. They come from topology and group theory, and are far more natural to theoretical physicists than to biologists.

So there are four problems—and I could name many more like them if time allowed—whose answers would be of great benefit, and lie much nearer the habitual domain of the theoretical physicist than that of the neurophysiologist. They are dignified problems, in that they are simply stated and their challenge stems from depth rather than from complexity.

I believe that it is by furnishing the methodology to overcome problems such as these that theoretical physics has the most to offer to biology.

References

DODGE, F. A., KNIGHT, B. W. and TOYODA, J.: Voltage noise in *Limulus* visual cells. Science *160:* 88 (1968).

KNIGHT, B. W.: Some point processes in motor and sensory neurophysiology. In LEWIS, P. A. W.: Stochastic point processes: statistical analysis, theory, and applications. Wiley, N.Y. (1972). (In this reference beware of 4 errors in figures 2, 8, 22 or their legends.)

KNIGHT, B. W.: The horseshoe crab eye: A little nervous system whose dynamics are solvable, in: Lectures on mathematics in the life sciences. 5. Some mathematical questions in biology. J. B. COWAN, eds., Am. Math. Soc., Providence, R. I. (1973a).

KNIGHT, B. W.: A stochastic problem in visual neurophysiology, in American Mathematical Society Symposium on Stochastic Differential Equations, J. KELLER and H. P. MCKEAN, eds., Am. Math. Soc., Providence, R. I. (1973b).

RATLIFF, F.: The logic of the retina.

Author's address: B. W. Knight, jr., The Rockefeller University, *New York, NY 10021* (USA)

Proc. 3rd int. Conf. From Theoretical Physics to Biology, Versailles 1971. pp. 411–432
(Karger, Basel 1973)

Panel Discussion

President: M. KAC, New York, N.Y. (USA)
Participants: J. D. COWAN, Chicago, Ill. (USA); A. LICHNEROWICZ, Paris (France);
H. C. LONGUET-HIGGINS, Edinburgh (Scotland); L. MAFFEI, Pisa (Italy); D. MACKAY,
Keele (England); I. PRIGOGINE, Oxford (England); L. ROSENFELD, Copenhagen (Den-
mark); H. SPEKREIJSE, Amsterdam (The Netherlands); S. ULAM, Boulder, Colo. (USA)

Excitatory and Inhibitory Interactions in Neural Tissue Models[1]

J. D. COWAN Department of Theoretical Biology, University of Chicago, Chicago, Ill.

Introduction

The problem of how the brain works is one of the ultimate problems of science. Almost
every aspect of biology, certainly of organismic biology, has its place in the so-called neural
sciences, from molecular biology and biochemistry to psychobiology. In addition, large areas
of computer technology and communications engineering are also of relevance. Even theoretical physics has found a role, although a minor one, in such studies. The entire subject
however, is still largely pre-theoretical. There is an acute lack of understanding of most of
the basic problems. This is clearly a result of the immense complexity of the actual phenomena
involved, and the consequent difficulty in interpreting in simple and concrete terms, what
is observed. There is always the fundamental epistemological problem: the brain is a physical
organ, but is also the mind. The role of symbolism in the functioning of the mind is subtle,
the manner in which the brain is organized for symbolic information-processing, complex.
For the past 25 years or more, many scientists and engineers have tried to use models of
the way the brain functions, derived from the study of computers, and from information
theory. MARR [1, 2] has recently provided models for the cerebellum and cerebral neocortex
which incorporate a whole range of ideas derived from the technology of random-access
memory systems, and from the theory of a certain class of modifiable computers called
Perceptrons [3]. The models are consistent with many neuro-anatomical details, and suggest
interesting experiments as to which are the modifiable synapses in the brain WINOGRAD
and I [4], some years ago, developed a theory for the design of reliable automata made from
unreliable components—which has certain features in common with the theory of Perceptrons.
The outcome of such ideas is a class of theoretical machines, any one of which is very sensitive
to external changes, but is remarkably insensitive to internal perturbations. The latter property
is shared by a wide variety of redundant systems, the most familiar of which is the hologram
[5], the neural analogies of which have been investigated, most notably by LONGUET-HIGGINS
and his collaborators [6].

1   Work supported in part by Alfred P. Sloan Foundation and the Otho S.A. Sprague
Memorial Institute.

A Neural Field Theory

I will now outline very briefly a different kind of theory, developed not for the inter-pretation of brain tissue as a specialized computer, but instead for the analysis of that large-scale activity which results from the localized interaction of millions of nerve cells. If one thinks about the nature of large-scale signal processing in the brain, one is led away from a purely local description of the individual activities of nerve cells toward a more economical description of the overall activity in terms of averages, both spatially and tempo-rally. I do not think one can hope for an understanding of pattern processing for example, by way of network analysis, or even by probabilistic considerations concerning local interac-tions. A more promising approach, in my opinion, is by way of a kind of non-linear field theory developed recently by WILSON and myself [7]. In this theory, neural tissue is seen as a highly redundant net, the number of degrees of freedom of which is so restricted that it can be analyzed as if it were a continuous medium. There are two source quantities, $E = $ the proportion of excitation cells; and, $I = $ the proportion of inhibitory cells in the tissue, activated per unit time. The net excitation produced in the tissue is given by the field quantity

$$\psi = E \otimes \xi_1 - I \otimes \xi_2,$$

where ξ is a connection function and $\otimes$ the convolution operator. Such a field produces new sources in a complicated fashion. The field excitation produces a depolarization or hyperpolarization of the cell membranes of the millions of neurons constituting the medium. In the field-theoretic description this results in a mean integrated excitation $\psi \otimes h$, where h is the Green's function of the nerve cell membrane. Such an excitation gives rise to a superthreshold response, i.e., the firing of a nerve impulse, by a proportion of cells in the medium that are sensitive to the excitation, and whose thresholds for impulse activation are less than the mean integrated excitation. At any given instant t, the new sources are determined by the relations:

$$E(r, t) = \begin{bmatrix} \text{proportion of sensitive} \\ \text{excitatory cells at} \\ r, \text{ t-}\tau \end{bmatrix} \times \begin{bmatrix} \text{proportion of cells} \\ \text{reaching threshold at} \\ r, \text{ t-}\tau \end{bmatrix}$$

$$I(r, t) = \begin{bmatrix} \text{proportion of sensitive} \\ \text{inhibitory cells at} \\ r, \text{ t-}\tau \end{bmatrix} \times \begin{bmatrix} \text{proportion of cells} \\ \text{reaching threshold at} \\ r, \text{ t-}\tau \end{bmatrix}$$

where τ is an operating delay that takes account of various retardations inherent in the transformation of source activities into integrated excitation, and back again.

Sensitive cells are those that are not refractory, i.e., have not fired an impulse within the time-interval t-τ-r to t-τ, where r is the refractory period or dead-time of the cell. Thus the proportion of sensitive excitatory cells at r, t-τ is given by

$$1 - \int_{t-\tau-r}^{t-\tau} E(r, t')dt'$$

with a similar expression for inhibitory cells. The second term in the expression for E(r, t) is taken to be some sigmoidal function of the mean integrated excitation $\psi \otimes h$. The follow-ing equations result:

$$E(r, t) = [1 - \int_{t-\tau-r}^{t-\tau} E(r, t')dt'] \times \mathscr{S}_e [E \otimes \xi_1 \otimes h - I \otimes \xi_2 \otimes h + P(r, t-\tau)]$$

$$I(r, t) = [1 - \int_{t-\tau-r}^{t-\tau} I(r, t')dt'] \times \mathscr{S}_i [E \otimes \xi_3 \otimes h - I \otimes \xi_4 \otimes h + Q(r, t-\tau)]$$

where $_e$ and $_i$ are the appropriate sigmoidal population response functions, and where P(r, t) and Q(r, t) are external excitations from other sources.

It will be seen that these equations are quite complicated, and as such, are hard to deal with. It is assumed however, that only slowly varying solutions (relative to r^{-1}) are of physiological interest, so that the equations can be time coarse-grained. This results in the somewhat simpler equations:

$$\tau \frac{\partial \bar{E}}{\partial t} = -\bar{E} + (1-r\bar{E}) \times \mathscr{S}_e [\alpha_1(\bar{E} \otimes \xi_1) - \alpha_2(\bar{I} \otimes \xi_2) + \bar{P}(r, t)]$$

$$\tau \frac{\partial \bar{I}}{\partial t} + -\bar{I} + (1-r\bar{I}) \times \mathscr{S}_i [\alpha_3(\bar{E} \otimes \xi_3) - \alpha_4(\bar{I} \otimes \xi_4) + \bar{Q}(r, t)]$$

The properties of these equations are of considerable interest, and have immediate relevance for psychophysics and physiology. I will briefly outline some of the simpler forms of behavior that can occur.

Examples

Consider first the spatially uniform case, i.e., consider a slab of neural tissue in which the cells are so richly interconnected that each one synapses with all the others. Figure 1

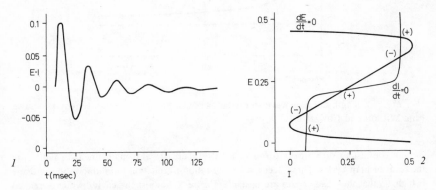

Fig. 1. Damped oscillatory behavior of E–I in response to a brief stimulus. [Redrawn from WILSON and COWAN, 7.]

Fig. 2. Phase-plane plot of stationary solutions of field equations for spatially uniform case. [Redrawn from WILSON and COWAN, 7.]

shows the response of such a slab to a brief stimulating pulse P(t). The net activity E(t) − I(t) is plotted as a function of time. It will be seen that this response is in the form of a rapidly damped oscillation, with a period of approximately 25 msec. Such damped oscillations are characteristic of time-averaged responses of neural tissue to pulsatile stimulation. Any neural tissue model must exhibit such responses as a kind of minimal constraint. Of much more immediate interest, however, are the maintained oscillatory and steady responses of the tissue. Figure 2 shows a plot of the stationary solutions of the field equations for the spatially uniform case. It will be seen that in this particular case there are five stationary states, of which three are stable and two unstable. The locus of points for which E is constant is shown for P(t) = o, i.e., for the case in which there is no external stimulus to the excitatory cells. The effect of such a stimulus is to translate the locus relative to the E-axis, and therefore to change the number of stationary states in the system, and also their stability properties. What this means is that the tissue can be *switched-on* from a low level of activity to a higher level, and back again, and also that hysteretic effects will occur. In addition there is a strength-duration curve for the switching-on from low to high activity. Figure 3 shows such a curve, a plot of the minimum stimulus intensity as a function of the minimum duration of application of such a stimulus required to switch-on the tissue. It will be seen that there is a *rheobase*, i.e., a stimulus, P = 0.375 that will switch-on the tissue only if applied for an infinite duration. Figure 4 shows the excitatory activity plotted as a function of the stimulus intensity P(t). The possibilities for switching and hysteresis are obvious.

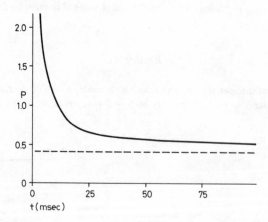

Fig. 3. Strength-duration wave for excitation of tissue from low to high activity. [Redrawn from WILSON and COWAN, 7.]

In contrast to such steady forms of activity, there are also maintained oscillations in the form of limit cycles. These occur when the ratio of excitation to inhibition in the tissue is high, so that the steady states are unstable. Figure 5 shows such activity, which is related to the predator-prey oscillations of population dynamics, and corresponds in the neural tissue model, to bursts of activity in the individual cells comprising the tissue. Consider now, the spatially non-uniform case of a tissue in which the probability of neural interconnection falls off exponentially with the distance between cells. Space constants can therefore be used to

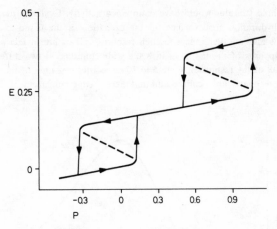

Fig. 4. Steady state values of E as a function of P (Q = 0), showing two hysteresis loops. [Redrawn from WILSON and COWAN, 7.]

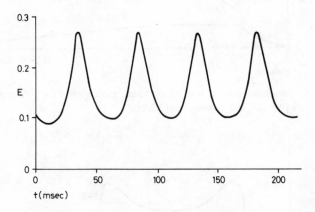

Fig. 5. E(t) for limit cycle activity. [Redrawn from WILSON and COWAN, 7.]

specify the richness of neural connectivity. As an example consider a tissue in which the excitatory cells exert a powerful, but short-range influence on neighboring cells, and the inhibitory cells a less powerful but longer-ranged influence. Thus the tissue exhibits recurrent lateral inhibition of the type discussed by RATLIFF [this volume] for the Horseshoe-Crab eye, and also a short-range, recurrent lateral excitation. In such a non-uniform tissue one would naturally expect to find responses rather more complicated than those seen in a uniform tissue. Figure 6 shows the responses of the tissues to a narrow, and to a broader stimulus both applied for the same duration. It will be seen that only the response to the broader stimulus switches the tissue on. Indeed the tissue is driven up to an unstable peak of activity from which it decays, back to the resting level. Such a transient response is a reflection of the refractory properties of the tissue: too strong a stimulus will cause it to switch-off, whereas a weaker but supra-threshold stimulus would have maintained activity. Evidently there exists a complicated strength-duration-areal surface for the excitation of the tissue to supra-threshold activity. There is also another aspect to such responses. They are in fact *localized*. Figure 7 shows the responses of the tissue to an 800 μ wide stimulus, at two different intensities, taken at the peak of the responses, some 5 to 10 msec after cessation of the stimulus. It will be seen that the responses are localized and that there is edge-enhancement. These properties

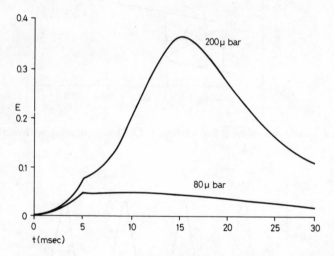

Fig. 6. Maximum response to two different bars, each applied for 5 msec plotted as a function of time [WILSON and COWAN, unpublished].

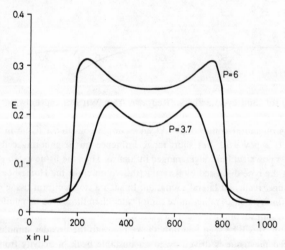

Fig. 7. Maximum response to two different bars each applied for 5 msec, plotted as a function of position [WILSON and COWAN, unpublished].

are mainly a result of the recurrent lateral inhibition in the tissue. The combination of such inhibition, with the recurrent lateral excitation, leads to many interesting spatial and temporal responses, the properties of which seem to be related to contemporary psychophysics and physiology, but we have not the time to discuss these here.

However, there is one last property that I want to discuss, which is of considerable interest outside the context of neural theory. This concerns the responses of a spatially non-uniform tissue composed only of excitatory cells. The properties of such a tissue were first investigated, some 25 years ago, by WIENER, PITTS, and SELFRIDGE [8] in connection with cardiac fibrillation, and later by BEURLE [9] in the neutral context, and recently from a mathematical viewpoint by GELFAND and KRINSKII [10] in the Soviet Union. Such a tissue model has also been discussed by GOODWIN and COHEN [11] in their theory of the control of embryonic differentiation; a related chemical analogue, the Zhabotinsky-reaction has been discussed by WINFREE [12] and by PRIGOGINE *et al.* The main difference between the responses of such a tissue, and one containing inhibitory cells, is that there are no stable localized responses. Instead, one sees only propagating waves of a highly non-linear character. Figure 8 shows such a system. The wave-fronts satisfy an eikonal equation, but are succeeded by refractory troughs which effectively prevent any interference of wave trains, or reflections at boundaries. Colliding wave trains generate refractory boundaries, and if there are several wave-trains coexistent in such a tissue, the fastest one will gradually propagate throughout the tissue, capturing the other wave-trains, and synchronising the activity. As soon as inhibition is introduced into such a tissue, such effects are diminished. There is, in fact, a critical level of inhibition, for a given strenght of excitation, which results in only localized responses, that reflect stimulus contours and other attributes. It is clearly the existence of *both* excitation and inhibition that is important for neural processing, and it may be important in other biological contexts.

Fig. 8. Wave propagation in a tissue composed only of excitatory cells. Dark areas contain sensitive cells, light areas refractory cells. Wave fronts propagating into sensitive regions consists of active cells. [Redrawn from R. L. BEURLE: J.I.E.E. *5:* Feb. 1959].

References

1 MARR, D.: J. Physiol., Lond. *202:* 437 (1969).
2 MARR, D.: Proc. roy. Soc. Lond. B *176:* 161 (1970).

3 ROSENBLATT, F.: Principles of neurodynamics (Spartan Books, Washington 1962).
4 WINOGRAD, S. and COWAN, J. D.: Reliable computation in the presence of noise (M.I.T. Press, Cambridge, Mass. 1963).
5 GABOR, D.: Proc. roy. Soc. Lond. A *197:* 454 (1949).
6 WILLSHAW, D. T.; BUNEMAN, O. P. and LONGUET-HIGGINS, H. C.: Nature, Lond. *225:* 178 (1970).
7 WILSON, H. R. and COWAN, J. D.: Biophysic. J. *12:* 1 (1972).
8 WIENER, N.; ROSENBLUETH, A. and PITTS, W.: Arch. Inst. Cardiol. Mexico *16:* 105 (1946); SELFRIDGE, O. G.: Arch. Inst. Cardiol. Mexico *18:* 177 (1948).
9 BEURLE, R. L.: Phil. Trans. roy. Soc. Lond. B *240:* 669 (1956).
10 GELFAND, I. and TSETLIN, M.: Dokl. Ak. Nauk. *131:* 1242 (1960); KRINSKII, V. I.: Biophysics *11:* 676 (1966).
11 GOODWIN, B. C. and COHEN, M. H.: J. theor. Biol. *25:* 49 (1969).
12 WINFREE, A. T.: in PYE, E. and CHANCE, B. Biochemical oscillators (Academic Press, New York 1971).

Author's address: Dr. J. D. COWAN, Chairman of the Committee on Mathematical Biology, University of Chicago, 959 East 57th Street, *Chicago, IL 60637* (USA)

A. C. LONGUET-HIGGINS: We seem to have been discussing the brain and the senses, about thought and perception. I think that the brain does in fact provide quite a good text on which to discuss mathematical models in biology. May I begin with one very general observation: it's rather interesting, sociologically, I think, that when a physicist discovers that within his own subject, one set of concepts cannot be explicated in terms of another—and I'm thinking particularly of statistical concepts which can't be explicated in terms of dynamical ones— he learns to live with this. But when he finds concepts in somebody else's subject which he can't immediately translate into physical concepts, this worries him. I don't think it should, because after all, for discussions of different types of phenomena, we shall need and do need and, shouldn't be ashamed of using—new concepts. In fact, the progress of any subject depends very largely upon the forging of new *concepts.* That's my first comment, but it's necessary in order to justify what one might call a proper scientific interest in some of the most complex phenomena in biology. One of these, which interests me very much at the moment, is language. Obviously, if we want to say something sensible about language we are not going to succeed in doing it in terms of concepts such as mass and charge and time and even linearity and nonlinearity. We are going to require those *concepts* which have been forged for the purpose of discussing language. What sort of a thing would a theory of language be? Language of course is a biological phenomenon. We emit sounds; I'm emitting sounds at the moment and they enter your ears and something happens in consequence. You may do something or you may say something or you may simply make a mental note. In making a model of this phenomenon we are in fact not at quite such a loss as one might think. Because we have in fact invented systems which we understand fairly well, called computing systems. We address these computing systems in languages which have been designed for interpretation by the system and then when we feed in a program in the language, something happens inside the computer and something may come out if we wish it to; something may get done if the computer is hitched up to some automation process or we may simply store information.

Now if you think about this process you realize that the concepts which are needed for constructing a theory of languages are not at all like those of classical mathematics or even classical physics. What we need are the concepts of *operational logic*. Now, it is not without significance that the people who design computers are at great pains to try to avoid any kind of linear phenomena. The essential ingredient of a computer is a bistable switch and nothing could be more non linear than a bistable switch. Again, if one thinks of the digital character of the input of the computer, nothing could be more nonlinear than that. While I very much admire the beautiful work which has been done on demonstrating how linear the visual system of *Limulus* is, that can never be enough for us if we wish to understand vision and perception. Because in vision and perception, the purpose of looking at something is to decide what it is. We can't make a decision that it is 30 percent you and 70 percent somebody else; that is not a decision. When a decision is required, this is essentially and basically a nonlinear phenomenon. Therefore it is not surprising to discover that in the nervous system we have got nonlinear elements and the nonlinearity is an integral part of the reason why the thing functions at all. If one is thinking about a neuron, for example, everybody knows that a neuron will only fire if its excitation exceeds a certain threshold. If one wants to design a classifying network one must have nonlinear elements in it. Because otherwise its output will simply be an equivocation between the different outputs that might correspond to pure inputs of one sort or another. I haven't got very many more moments, but I would just like to say I believe that thinking about computational logic and thinking in terms of algorisms, which of course are familiar to anybody who has ever written a program in Fortran, is a very fruitful and useful way of thinking about biological phenomena in general and in particular about perceptual and mental processes. There is a new subject, which as everybody knows is called artificial intelligence. This is a more interesting subject than it might seem. If it were merely the attempt to make machines which would walk around and do menial tasks, well that would be a convenience, but of no particular philosophical significance. In doing work in artificial intelligence, one is really expressing oneself in a new medium. One's models are no longer systems of equations or hardware models; they are computer programs. The computer program in fact enables one to express oneself in a precise manner, about phenomena which are extremely complicated, interlocking, mutually dependent and highly conditional upon this or that. A theory of language, or indeed, a theory of basal metabolism might well be couched in this sort of language. Of course a theory couched in the form of a computer program has the additional advantage that you can talk to the theory. If it's a theory about language, you can even put in questions in English and get out answers in English. In order to do this, which is not, I may say, a trivial matter, one has to really search one's mind about matters of syntax, semantics and so forth and one really discovers what the philosophical problems are. And I should say the scientific problems too, although I wouldn't to make a clear distinction between the two. What I'm saying is, that if we are thinking of the brain as a text for the use of mathematical models in science or in biology, I think it might be argued that in studying the brain we might do well to adopt a new medium of mathematical expression, namely the computer program. The more fully we understand logic and the way in which computer programs work, the better chance we have of understanding the phenomena which take place inside our own heads.

M. KAC: Speaking of languages we are also going to switch language and it gives me pleasure to say in French, donner la parole à M. LICHNEROWICZ qui va parler en français.

A. LICHNEROWICZ: Je pense que l'ordre a été bien choisi involontairement, car je suis tout à fait d'accord avec ce qu'a dit avant moi le Pr. LONGUET-HIGGINS. J'avais pris deux, trois notes avant et je puis ne pas les modifier. Je pense que la chose la plus importante est en effet d'avouer que les mathématiciens, ou les gens 'faisant du calculateur', n'ont peut-être pas encore assez travaillé pour nous fournir, au niveau voulu, les modèles voulus. Le mot modèle est un mot ambigu; on peut le prendre à ras de sol ou très haut. Je vais le prendre un peu plus haut qu'avant.

Il est certain que nos ordinateurs fonctionnant en calcul séquentiel ne sont pas adaptés, sauf utilisation en simulation, à vous fournir des modèles logiquement exploitables; au contraire je considère que le travail sur le calcul parallèle (qui doit fournir les prochains ordinateurs) est probablement un des points de rapprochement avec le fonctionnement du cerveau. Ce calcul parallèle est actuellement appliqué, en France, à Novosibirsk, aux Etats-Unis, en Angleterre, à certains types de problèmes qui sont les nôtres. L'étude des grands systèmes, l'identification des systèmes, commencent à être traités en calcul parallèle. Un autre problème est l'étude d'un réseau de très petits ordinateurs avec la parallélisation. Et on constate déjà que, on gagne par rapport aux algorithmes que l'on connaît, un facteur égal au carré du nombre de ces petits ordinateurs. Ceci pourquoi? Nous parlions des groupes tout à l'heure. Nous ne savons pas actuellement, pour l'architecture par exemple, représenter correctement le groupe des déplacements de manière opérationnelle car la mémoire demandée est trop grande. Or, la reconnaissance des formes telle qu'elle est faite par le cerveau fonctionne fort bien. Un troisième élément, qui est abordable en calcul parallèle et qui est lié au système nerveux, consiste dans l'étude des corrélations spatiotemporelles. Je crois que ce sont trois problèmes pouvant fournir des idées fondamentales pour des modèles utiles aux physiologistes. A un autre niveau d'intégration, figurent des choses assez simples que j'avais baptisées une *combinatoire topologique*. Il faudrait reconnaître, d'une manière plus précise qu'on ne l'a fait jusqu'à présent, les éléments de cette combinatoire et la notion de voisinage topologique correspondant qui permet l'identification. Je crois que c'est tout ce que je voulais dire, en me bornant au fond à nuancer.

Je ne suis pas d'accord sur un seul point avec le Pr. LONGUET-HIGGINS, sur ce qu'il a dit de la non-linéarité dans le domaine en question; car dans ce domaine, je ne comprends pas bien sa pensée. Si l'on parle de transformée de Fourier, tout est toujours linéaire. D'autre part, au bout d'un ordinateur, symboliquement il arrive toujours disons le chiffre 1 ou le chiffre 0, et votre décision sera oui si c'est 1 et non si c'est 0. L'ordinatuer n'est pas 'linéaire ou non'.

M. KAC: Merci bien. Now we will give the microphone to Prof. ULAM who will continue in English.

S. ULAM: I should like to direct my remarks to the central theme of our conference, the importance and use of ideas of physics and mathematics to biology. Before I do, however, I would like to tell you a short story that may illustrate the relationship which seems to exist between mathematicians and physicists. Now, everybody knows that mathematicians can be helpful—even to physicists. The story is as follows: One day a group of travellers came to town. They were all weary and tired, but they managed to get into a hotel and since they had to leave the next day and had some laundry to be done and the hotel could not do it, they went out on the town, looking around for a place to have their clothes washed. Many shops were already closed, but there was one shop with lights inside and on top of the marquee there were several signs, one of which said 'Laundry Taken In Here'. Our travellers went in and asked to have their laundry done. The man at the desk said, however, 'No. We don't do laundry here'. 'How is that', asked one of the visitors—'You have a sign out there'. 'Ah', said the man at the desk, 'we make signs'.

Listening to some of the interesting discussions during the last few days, I thought that one can iterate the parable and come to the conclusion that the relationships between physicists and biologists is perhaps similar to that of mathematicians to physicists. Seriously, though, it is not necessary to point out how decisively the ideas and the methods of physics are now of interest and use in the research in natural sciences, in biology first of all.

I am not speaking here of the well known use of mathematical statistics or analysis in problems which require it (the obvious low-brow applications, so to say), but in more conceptual ways. In fact, I think it is very much also the other way around. The problems and the general schemata of biology suggest wonderful new problems of mathematical character which are of novel types to mathematicians. These problems in combinatorial analysis especially—I am saying this in answer to Professor LONGUET-HIGGINS's question of which parts of mathematics will one day be useful in biology. In this sense I believe that there are applications, at least a stimulus for new imagination in mathematics itself, coming from biological set-ups. Somehow the logical situations and arrangements seem to mirror the ones in the non-organic world. Professor SOBOLEV mentioned some of the problems which certain mathematicians like himself have been working on, and which have direct meaning in biological sciences. I myself, together with several friends, have been studying the mathematics of patterns and processes, stimulated directly by what biology presents, in a very special set of cases, but suggesting possibly some new general theories. So, for example, there is the problem of how to compare, quantitatively, organs or organisms by introducing a notion of 'morphological' distances. This 'distance' is of a different type studied in problems of mathematics so far. To give an example, two sequences of 0's and 1's representing a code for some protein, should not be compared, position by position, to define a distance between them, but clearly one is allowed to shift the sequences right or left, delete or add a few symbols and compute the distance as a minimum module. Similarly, even about phenotypes and not merely codons.

The general questions of modeling in biological problems has to be discussed in great generality. Some models of life-like schemata are worth studying even if they are not too realistic, or even if they do not correspond too closely with what we find in actual life. One could imagine that some other types of life are theoretically possible, or perhaps exist on some other stars. By speculating about the logical schemata of the processes which we call life and in great generality, we might, with some luck, find answers to what happens in reality here on earth too.

The many discussions on the development of molecules, including especially the one by Professor EIGEN, seem to me to bring out a feature which is perhaps not sufficiently stressed. These molecules play a sort of chemical game or a number of games among themselves. It is these games that ultimately lead to out *ex post* definitions of *'sense'* or *'purpose'* and not the other way around. It appears to me sometimes that our sense of what we call *meaning* evolves from what these molecules established and not the other way round I thought of this very much during the talk by Professor LONGUET-HIGGINS. The evolution of the activities of molecules and the classes of molecules which we call organisms seems perhaps to have developed partly by chance and led ultimately to the inner logic of their own, and presented to us processes for which we unnecessarily look sometimes teleologically. The evolutionary games played between the species are hard to define. It is clear enough when a 'loss' occurs, but the real sense of 'win' in games played between branching chains each developing to infinity, is less clear or definable mathematically in presently used terms. Some of you have perhaps heard of or seen in the *Scientific American,* an account of games defined by very simple recursive rules leading to fascinating situations and problems, some even undecidable. I am speaking here of Conways' game and some of the recursively developing patterns studied by myself and collaborators. (An account of these can be seen in a book edited by A. Burks: Cellular automata [Illinois Press, 1970].)

Sequential Analysis of Luminance and Contrast Processing

H. Spekreijse

Laboratory for Medical Physics, University of Amsterdam, Amsterdam

Linear systems exist only in the mind; all real physical systems are nonlinear! By definition results of linear system analysis apply to all linear systems in all working conditions. In contrast, the analysis of a nonlinear system may be of little help in understanding either the behavior of the system in different circumstances or the behavior of other nonlinear systems. Why then use linear system analysis when dealing with physical systems? The answer is simple. Nonlinear systems can frequently be investigated in such a way that their behavior is approx. linear, e.g. by using 'small' signals. It turns out, however, that especially biological systems, cannot always usefully be described by the methods of linear system analysis, even when 'small' signals are used. The reason is that these systems contain essential nonlinearities. Their analysis may present a fresh problem to be solved on its own terms with a 'custom designed' approach. A tailored approach may pay off since in some cases more information about the system, such as the organization of successive stages, can be extracted from a non-linear than from a linear system.

The ganglion cells in the goldfish retina—which is an easy preparation for physicists to deal with—form a good example of an essential nonlinear system. At the output of this last stage of retinal signal processing a highly distorted response can be found. If the goldfish retina is stimulated with a sinewave modulated light—the signal which proved to be so useful in linear system analysis—only once during each stimulus cycle a burst of action potentials can be observed (fig. 1). In order to improve the signal to noise ratio, the shaped spike responses are period-averaged. An example of such an averaged response is depicted in the mid-column of figure 2. For half the stimulus period, the density distribution of the nerve impulses has roughly the same shape as the sinusoidal stimulus, but during the other half of the stimulus cycle, no spike discharge can be observed at all. Furthermore the amplitude of this average response is approx. proportional to the strength, i.e., modulation depth of the sinewave stimulus (fig. 2 C–E). These data suggest that to a first approximation, a half-wave linear rectifier can model the distortions observed in the discharge pattern of ganglion cells to 'small' stimulation. Although even at a modulation depth as small as 5%, the distortion is obvious in the response; at a still lower modulation depth (2%), an almost pure sinusoidal response is found (fig. 2F). At this modulation depth, the noise level becomes important, indicating that an essential nonlinear behavior can be masked by internal noise. If the strength of the noise exceeds the stimulus amplitude, the input-output relation may suggest that the

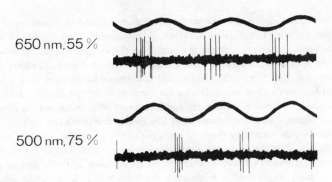

650 nm, 55 %

500 nm, 75 %

Fig. 1. Ganglion cell responses to sinewave modulated light of 3 Hz. The degree of modulation is expressed as a dimensionless parameter, $m = A/L_o$ in which A is the amplitude of the sinewave and L_o is the mean intensity of the stimulus signal. The parameter m varies between 0 (0%) for unmodulated and 1 (100%) for fully modulated light. The mean intensity of the stimulus passed through Ealing-TFP interference filters is approximately $6\ \mu W/cm^2$ for the red (650 nm) and $20\ \mu W/cm^2$ for the green (500 nm) stimulus.

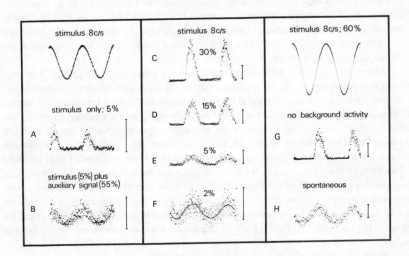

Fig. 2. The responses demonstrate the linearizing effect of an auxiliary signal (first column), internal noise (second column) and spontaneous spike discharge (third column). The circular light spot focused on the retina has a dia of 1.2 mm (first column), 6 mm (second column), and 2.5 mm (third column). The calibration bars are 20 spikes/bin. The bin duration is 625 μsec. In figures A, C, E and G the lowest points represent the zero count level. This holds approximately also for figures B and F, whereas for figure H, the lowest address points represent a firing rate of the order of 4 spikes/sec. The average intensity of the stimulus passed through a Wratten 29 filter is approximately $1\ \mu W/cm^2$. All responses are from red-'off' units.

system is linear, although it contains nonlinear or even essential nonlinear elements. The nonlinear characteristics in the system are seemingly modified by noise, resulting in a weaker nonlinear behavior and even sometimes in a linear one. For the ganglion cell the effectiveness of such noise is demonstrated in the right column of figure 2. Figure 2G shows the usual rectified response from a red-'off' ganglion cell to a sinusoidal stimulus with a frequency of 8 Hz and 60% modulation depth. During the experiment the spontaneous discharge rate was observed to be increasing. When it reached a frequency of 30 spikes/sec, an almost linear response was obtained even for a stimulus of 60% modulation depth (fig. 2H).

This masking or linearizing property holds not only for noise but also for almost any periodic signal, such as sinusoids, squarewaves, triangles etc. This follows directly from the data in the first column of figure 2, where the effect is shown of the adding of a 0.5 Hz triangular auxiliary signal with a modulation depth of 55% to a sinusoidal stimulus of frequency 8 Hz and a modulation depth of 5%. In contrast with the half-wave rectified response which results when the stimulus is a pure sine wave (fig. 2A), an almost pure sinusoidal response is found after addition of such a triangular auxiliary signal (fig. 2B). Therefore, if you get after period averaging a linear output from a biological system, do not be too certain that the system is indeed a linear one! It might be that internal noise or noise in the stimulus itself has masked the nonlinearities in the system studied.

The masking property of noise or of auxiliary signals can also be used to determine the functional sequence of linear and nonlinear transformations in a system which contains a nonlinearity. In addition the linearizing phenomenon can be employed to determine the characteristics of the various individual elements. This possibility contrasts with the case of linear systems where the sequence of transformations can never be detected from the response of a system as a whole. An application of this method can be found for the ganglion cell responses in the goldfish retina.

The results in figure 3 demonstrate the linearizing effect for low and high frequency sinusoidal auxiliary signals. As the amplitude ratio of the sinusoidal stimulus (5% modulation; 8 Hz) and the sinusoidal auxiliary signal (50% modulation depth) is kept constant, the amount of linearizing reflects the attenuation of the auxiliary signal in the elements preceding the half-wave linear rectifier. The data show low and high frequency cut-off, since for both low and high frequency auxiliary signals the amount of linearizing is reduced. For frequencies between 5 and 11 Hz the auxiliary signal is most effective. From this type of data it can be deduced that in the goldfish retina the elements preceding the rectifier have a low frequency cut-off which is probably of the first order, a pass band between 5 and 11 Hz and higher order high frequency cut-off in the high frequency region. These amplitude characteristics, derived from the linearizing effectiveness of auxiliary signals, resemble to a large extent the dynamic characteristics that FLOYD RATLIFF and BRUCE KNIGHT have shown for the *Limulus* eye.

The above type of sequential analysis can also be used to determine whether the spatial interaction in the goldfish retina occurs preceding or after the rectifying stage. In order to study the spatial interaction the retina is stimulated with a checkerboard pattern. The pattern contains two sets of elements, each of them can be modulated independently (fig. 4). Sinewave modulation of exclusively one set of squares gives a response that according to the data in figure 1 may be compared with a half-wave rectified sinusoid. Counterphase modulation of the other set of squares at the same mean intensity and modulation depth as the first set results, of course, in the same half-wave rectified sinusoidal response shifted 180° in phase. Simultaneous counterphase stimulation of both sets of checkerboard squares can give one

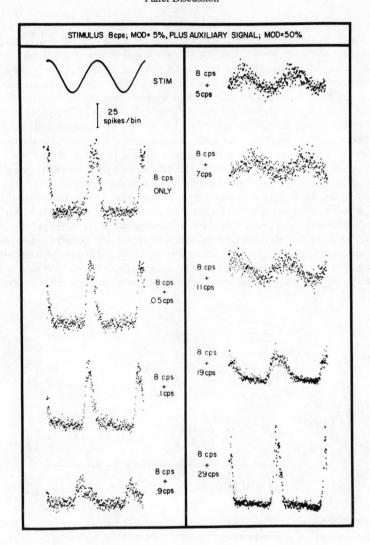

Fig. 3. Linearizing effect of sinusoidal auxiliary signals as a function of their frequency. The responses are from a pure red-'off' unit. The bottom points indicate for the distorted responses the zero spike count level. This may not be the case for the linearized responses. The average intensity of the stimulus passed through a Wratten 29 filter is approximately $1\,\mu W/cm^2$. The dia of the circular light spot focused on the retina is 2.5 mm. The number of summations is 600 and the bin duration is 625 μsec.

of the following two responses (fig. 4): either a full-wave rectified sinusoid or no spike discharge at all. This can be understood as follows: if the spatial summation takes place *after* the essential nonlinear element, then the response would consist of a full-wave rectified sinusoid with the same amplitude as the original half-wave rectified response (fig. 4a). On the other hand, if spatial summation occurs at a site *preceding* the nonlinear element, then the two sets of input sinusoids counterbalance each other completely, and no response results (fig. 4b). Figure 4c gives the actual data for a green-'off' centerprocess. As is evident from this figure, no spike discharge is observed when the two sets of squares are modulated simultaneously in counterphase. This holds, of course, only if the strengths of the responses to each set of squares are identical. Since neither the type of pattern nor the sizes of the pattern elements proved to be of importance, the data demonstrate directly that in the goldfish retina spatial summation takes place preceding the essential nonlinear element.

For setting the dynamic characteristics, the goldfish retina employs this simple algebraic summation within a receptive field. This follows directly from the data in figure 5. This figure shows that the shape of the amplitude characteristics depends on the energy per unit of time (power) falling within a receptive field rather than on the intensity of the stimulus spot. The same amplitude characteristic can be obtained by increasing the stimulated retina area by a factor of 32 and diminishing the intensity by the same factor or *vice versa* (curves 2 and 3). This means that, firstly, all the output signals of the early retinal processes are summed,

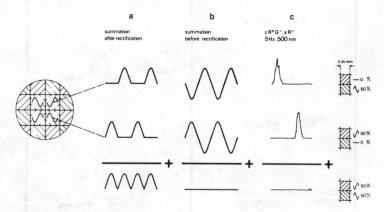

Fig. 4. Schematic representation (columns a and b) of retinal modes of summation between counterphase signals generated by the two sets of counterphase modulated squares of a checkerboard pattern. Column c gives the actual data for a green-'off' center process. The top of the figure is the average spike response to modulation (50%) of one set of squares (0.35 mm width). The other set of squares is not modulated at all but has a steady intensity, identical to the mean intensity of the modulated set. Reversing the condition and modulating the other set of squares with the same depth, but in counterphase, results, of course in the same rectified sinusoid but shifted 180° in phase (mid figure). Simultaneous counterphase modulation of both sets of squares does not result in a spike discharge at all (bottom figure). The mean intensity of the stimulus light passing through an Ealing FTP interference filter is approximately 4 μW/cm². The number of summations is 100; the modulation frequency is 5 Hz. The patterned spot focused on the retina has a diameter of 2 mm.

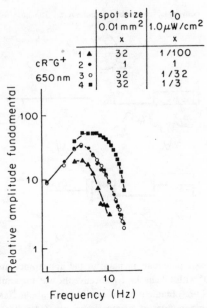

Fig. 5. Amplitude characteristics of the central red process of a spontaneous ganglion cell are depicted as a function of the energy per unit of time (power) of a Gaussian noise modulated ($\sigma = 40\%$) stimulus light. For spot diameters up to about 1 mm, a constant energy per unit of time, E, gives identical (curves 2 and 3) amplitude characteristics ($E = A \times I_o$, where A is the stimulus area and I_o is the mean stimulus intensity after passing through a 650 nm broadband interference filter). For a fixed diameter of the stimulus spot the sensitivity and shape of the amplitude characteristics are determined by intensity (curves 1, 3 and 4).

next the system is tuned—the term FLOYD RATLIFF used for the *Limulus* eye—according to the absorbed light power.

In the two types of experiments described above, spatially structured stimulus fields are used. This could give the impression that I am discussing spatial frequency processing. This, however, is not the case. All responses described can be understood simply by taking the luminance changes into account. On the other hand, there is overwhelming evidence, particularly from the Cambridge group, that the visual system processes spatial frequency or contrast in a separate way from luminance. Although evidence for such a processing is lacking in the goldfish retina, in man—visual evoked occipital potentials can be recorded that are a function of spatial contrast changes only. In accordance with psychophysical observations, the evoked responses to luminance and contrast seem rather independent. This follows from the data in figure 6 where the EP's to checkerboards of constant absolute contrast are shown. The contrast is produced in the following ways:

1. The luminance of one set of squares is kept constant and the luminance of the other set is increased by 10%. This results in a 10% contrasty pattern.

2. The luminance of one set of checks is kept constant, that of the other set is decreased with 10%. As in condition 1, again a 10% contrasty pattern is obtained. The subject cannot

428   Panel Discussion

2.1 c/s; both eyes; checkerboard pattern 15′; stimulus field 3°

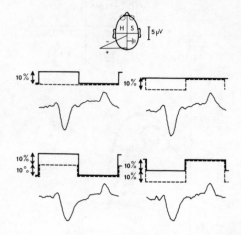

Fig. 6. Contrast-evoked potentials in man, to the appearance and disappearance of a checkerboard pattern of 3° with 15′ checks. Irrespective the way the luminance changes, highly similar contrast responses are obtained. Mean luminance is about 2000 asb. The number of responses averaged with a computer of average transients (CAT 4000 A) is 200.

2.1 c/s; one eye; checkerboard 40′; stimulus field 3°

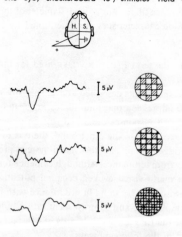

Fig. 7. Suppressing effect of a steady high contrast outlining of the borders of the checks in an appearing-disappearing checkerboard. Almost no suppression is found if the lines cross the centers of the checks. At intervals of 500 msec the pattern is presented for 200 msec. Mean luminance is about 5000 asb. The number of responses averaged is 200.

distinguish condition 1 from 2. Also the responses to both the appearance and disappearance of the checkerboard are identical.

3. Half of the checks is increased 20% in luminance, the other set 30%.

4. As in 3, but now the checkerboard is produced by a simultaneous lowering of the luminances of both sets of checks. One set is reduced 20% and the other set 30% in luminance.

Although the responses obtained in the last 2 conditions may differ slightly and although they are neither completely identical to the responses obtained in conditions 1 and 2, yet they support the general conclusion that these EP's are mainly produced by changes in spatial contrast, independent of the way this contrast is reached. Therefore, the data indicate that in the human visual system, contrast and luminance are processed to a large extent independently of each other.

The last point I want to stress is that contrast processing is difficult to explain with extensive retinal receptive fields. This becomes particularly obvious from the data obtained with appearing-disappearing checkerboards with outlined borders (fig. 7). The data show that the contrast EP is depressed by steady high contrasty outlining of the pattern. This suppression of the response occurs only if the lines are placed on the borders of the checks. Not much suppression is obtained if the lines cross the centers of the checks. This type of data seems to me very difficult to explain with receptive field models. Furthermore these data seem to conflict with the spatial frequency selective channel model, that Dr. CAMPBELL proposed today.

Author's address: Dr. H. SPEKREIJSE, Laboratorium voor Medische Fysica, Universiteit van Amsterdam, *Amsterdam* (The Netherlands)

Retino-Geniculate Communications and the Analysis of Contrast

L. MAFFEI

Laboratorio di Neurofisiologia del C.N.R., Pisa

The receptive fields of the cells of the lateral geniculate body (LGB) are organized in two antagonistic concentric regions like the receptive fields of the retinal ganglion cells. Even the size of the LGB and retinal receptive fields is, on the average, the same. Despite this similarity, the individual LGB receptive fields do not result from the projection of single retinal fibers. Indeed, there is anatomical evidence of a multiple fiber convergence from the retina upon individual LGB neurons. Moreover, from a functional point of view, the surround of LGB receptive fields has been found to be more powerful in antagonizing the center response than the surround of the ganglion cells [1].

The organization of the receptive field in antagonistic regions is usually interpreted as a mechanism for the analysis of contrast, i.e., of spatial luminance differences in the retinal image [3]. The current idea is that the analysis of contrast is performed in series, first in the retina and subsequently at geniculate and cortical level. In this line of thought, the stronger antagonistic effect of the surround observed at the geniculate level is interpreted as a mechanism for the reinforcement of the image contrast, already elaborated at a retinal level.

This hypothesis, of a first analysis of contrast being performed at retinal level (although written in any student textbook) has no experimental basis. The information elaborated at retinal level can be handed on to the geniculate only if we assume either a one-(retina)-to-one (geniculate) connection or a connection scheme in which the ganglion cells, projecting on to one geniculate cell, had their receptive fields concentrically superimposed. As to the one-to-one connection scheme, we have already mentioned that there is anatomical evidence to the contrary. As to the second hypothesis, there is already in the literature some indirect evidence that the surround of the geniculate receptive field could result from the projection of retinal units different from those projecting on to its center.

Now, it is logical that in the event that the center and the surround of LGB receptive fields were indeed made of different sets of retinal receptive fields, the elaboration performed in the retina by a single center-surround unit would get lost.

The experiments [2] were performed on cats. The single activity of retinal or geniculate neurons was recorded.

The results can be summarized as follows:

1) the response to illumination of the surround of the LGB cells does not disappear after dark adaptation as it does at retinal ganglion cell level;

2) the temporal properties of the LGB surround responses are different from those of the retinal antagonistic surround and very similar to those of retinal centers.

The conclusion is drawn that centers of retinal receptive fields build up the surround of the geniculate receptive fields. For instance, in an on-center geniculate cell the center of

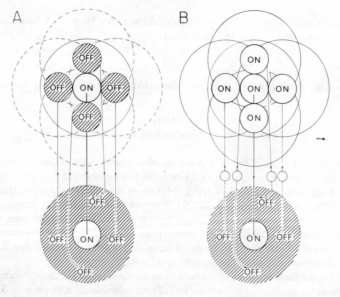

Fig. 1. Possible schemes of retino-geniculate connections. In each scheme, an on-center LGB receptive field is represented at the bottom and some retinal receptive fields projecting on it, at top. For the sake of simplicity, only one on-center retinal neuron in both schemes projects to the center of the LGB field and four off-center (A) or on-center (B), retinal neurons project to the surround. The small circles conventionally represent the centers and the large circles, the surrounds of the receptive fields. The peripheral on-center retinal fibers in B are connected to the geniculate surround via an interneuron.

its receptive field could be made up of on-center retinal units and the surround either of off-center retinal units or also of on-center units connected to the geniculate cell via an inhibitory interneuron (fig. 1). If so, the surrounds of the two sets of retinal fibers projecting on the center and the surround of an LGB receptive field, respectively, are widely superimposed and, since they have antagonistic effects, practically cancel each other. Therefore, the hypothesis that the organization of retinal receptive fields in two antagonistic regions is a mechanism for the analysis of contrast has to be reconsidered.

References

1 HUBEL, D. H. and WIESEL, T. N.: Integrative action in the cat's lateral geniculate body. J. Physiol., Lond. *155:* 385–398 (1961).
2 MAFFEI, L. and FIORENTINI, A.: Retino-geniculate convergence and the analysis of contrast. Submitted for publication.
3 RATLIFF, F.: Mach bands: quantitative studies on neural networks in the retina (Holden-Day, San Francisco 1965).

Author's address: Dr. L. MAFFEI, Laboratorio di Neurofisiologia del C.N.R., Via S. Zeno 49/A, *Pisa* (Italy)

L. ROSENFELD: In Prof. McKAY's introduction this morning, he drew a picture of sensory-motor schemes, and insisted that the limitation in reception is useful for the functioning of the muscular part. This is very much to the point, but what I wish to suggest is that, one can also look at this phenomenon from the other end. If you think how these organs have developed in the course of evolution, it would appear that the degree of sensitivity of the receptors at a certain stage, limits the possibilities of behaviour. These are two complementary aspects of the way in which you can look at a sensory-motor scheme. Another point in your picture was the intermediary role of concepts as a link between sensory reception and muscular action. Now, is the intervention of rational thinking not really more indirect? It may be rather direct for us, when we are exercising our abstract mode of thinking, but experience shows that for most adults in daily life, and also for any human being in the early stages of his mental development, rational thinking plays a very minor part, or even no part at all, in the sensory-motor chain. I would think that there is a double-code to consider. The primary one would consist of sensory-motor schemes with which you can operate directly; in fact, as is shown by experiences with children, you can construct at this level an operative logic which is very powerful, although not as well-defined as our conceptual logic. Besides that, you have the code system consisting of words which express abstract concepts, in correspondence with the sensory-motor schemes, and which in our rational thinking is even substituted for the direct consideration of those schemes.

D. McKAY: Yes, I agree, and indeed, am glad of the opportunity to say that in my brief sketch I was only mentioning samples of the function we must except to find at various levels in the hierarchy. In response to your points about development, I would say that the limits of the world of an organism at a given stage of development are set by the limits of the internal response repertoire which the organism has developed. The two indeed are interrelated, and I think that this is true of human beings too. I think the limits of our conceptual world are set by the repertoire we have developed.

I. PRIGOGINE: The first of my remarks is more or less an introduction to the second: I should like to offer a specific example of the application of biological concepts to the development of mathematical problems. Consider the problem of stability, which was studied by physicists essentially in connection with hydrodynamical problems. Now a problem which has probably not been mentioned in this conference, and which provides a striking application of mathematical methods, is the association of slime molds. The slime molds live as individual monocells during some period of their life cycle, and then agglomerate to a multicellular body. As has been shown in a paper by KELLER and SEGEL, and also by some work from Professor COHEN's group in Chicago, this phenomenon can understood as a problem of instability. This gives rise to many new and interesting mathematical problems. I think that it is completely right to connect the problem of language to the type of problems which appear in stability theory. Indeed, perhaps the main point is the evolution of the concept of meaning. Prof. EIGEN has emphasized this in his work on the evolution of biological macromolecules. Here, one deals with the problem of the evolution of concepts, in order to reach the stage at which you can speak about the language. There is, however, an essential point to take into account: we are not sure that natural languages are similar to artificial languages. The main difference may be in the problem of stability. Natural languages are characterized by a great degree of stability, they appear only after a slow evolution.

In this sense evolution of natural languages is a problem very much similar to the type of problems which EIGEN discussed.

H. C. LONGUET-HIGGINS: I wasn't suggesting that one could not use in the higher level problems, some of the lower level concepts. What I was trying to suggest was that higher level problems demand concepts that are not required in the discussion of lower level phenomena. For example, the concept of syntax, or of semantics, is simply inappropriate if one is discussing a chemical reaction producing local inhomogeneity. As one ascends the range of sophistication, so one needs more complicated concepts. Here is an example: one of the characterists of biology is the universal occurrence of structures of different kinds. The concept of structure and of a complex organized structure is one which is characteristically biological and we don't really find it among the concepts of the physical sciences.

M. KAC: May I say before we conclude this session, that this was an experiment in trying to find a way in which such meetings can best serve the Institut de la Vie purpose. Panel discussions have been proposed, so that the data will now be fed into a computer and analyzed. You will hear the results, hopefully, before the next meeting.

Perspectives of Theoretical Physics and Biology and Their Social Implications

Proc. 3rd int. Conf. From Theoretical Physics to Biology, Versailles 1971, pp. 434–435
(Karger, Basel 1973)

Introduction

S. L. SOBOLEV

Nous allons parler des perspectives du développement de la physique théorique de la biologie et de leurs implications pour toute l'humanité. De nombreuses questions se rattachent à ce développement. Maintenant que peut-être la plus grande découverte biologique a été faite, c'est-à-dire le code génétique, nous sommes dans la même position qu'une personne venue sur terre, d'une civilisation extra-terrestre placée devant une machine à écrire. Nous comprenons ce que sont les lettres et les machines à écrire, mais je pense que nous sommes encore loin de comprendre quelles sont les langues que parlent les habitants de la terre. Dans les perspectives du développement de la science, il faut tenir compte non seulement de la biologie, mais aussi d'autres sciences, la chimie surtout, la physique théorique et aussi les mathématiques, si vous me permettez de le dire. Pourquoi? Car il est des problèmes qui ne peuvent pas être traitées sans la collaboration de toutes les sciences. Nous avons déjà vu les exemples de travaux brillants, dans le domaine de la physiologie, au cours de ces journées, mais le contact entre ces sciences n'a pas été aussi général ni aussi productif que nous pouvons l'attendre dans le futur. C'est une direction dans laquelle, je crois, doit s'orienter la science moderne pour développer la biologie, compte tenu de tous les succès des mathématiques, de la physique théorique et de la chimie. Il y a d'autres domaines de collaboration: les problèmes généraux et l'application des sciences à la biologie. Plusieurs questions ont reçu leur réponse grâce aux mathématiques contemporaines, aux ordinateurs, à l'automation. L'informatique doit aider à développer chaque science, et surtout la biologie. Il existe donc deux voies de collaboration: poser et résoudre les problèmes généraux, et apporter la collaboration de la technique moderne physique et mathématique, pour résoudre, cas par cas, les problèmes limités. Un troisième type de problème est l'implication des nouveaux succès de la biologie. Les hommes disposent de nouveaux moyens beaucoup plus puissants qu'autrefois; par exemple l'énergie atomique dans les mains de l'humanité s'est trouvée être une arme extrêmement dangereuse. Je ne crois pas que les découvertes biologiques soient plus dangereuses, mais il faut craindre que tout le monde ne comprenne pas l'immense pouvoir contenu dans ces nouvelles découvertes. On peut changer même les qualités d'un homme, exercer une grande influence sur la nature qui nous entoure. Nous disposons désormais de moyens très grands et très puissants pour rendre la vie sur la terre beaucoup plus belle, beaucoup plus raisonnable. Nous pouvons éviter la grande pollution chimique de la nature. Chaque fois que l'humanité conquiert un nouveau pouvoir, le danger s'accroît, mais aussi l'espérance d'une vie plus belle, plus heureuse. Il est bien d'autres implications, par exemple sociales. Notre discussion d'aujourd'hui sera un pas très important.

Je donne la parole à M. Cournand sur le thème «Le code de la science».

Proc. 3rd int. Conf. From Theoretical Physics to Biology, Versailles 1971, pp. 436–450
(Karger, Basel 1973)

On the Codes of Science and Scientists

A. COURNAND

Columbia University, New York, N.Y.

My theme today is a problem which concerns both natural science—i.e., physics and biology—and sociology. At the outset may I note that the ideas which I should like to present are the result of my collaboration with a sociologist, HARRIET ZUCKERMAN. She was formerly the student of ROBERT MERTON, who more than anybody else in the United States has examined what might be called 'the social science of science'.

Scientists are compelled nowadays to explain the ways of science to their fellow men. In the past, as the result of an unending flow of achievements, most scientists came to think of themselves as independent of society, and to consider science as a self-validating enterprise, which was in society but not of it. More recently, this outlook has been modified. The impact of science on society has been so extensive and is increasing at such a pace that 'isolationism' is no longer tenable. Scientists have acquired a realistic view of the importance of their participation in social enterprises. This joining of the issue must necessarily lead to a clarification of the norms of operation, i.e., of the mores of science. A definition and an analysis of what may be termed the operational code of science will form the subject of this presentation.

Science can be defined—and here I am indebted to the philosopher-scientist KARL POPPER—as the society of past and present scientists, who are bound by traditions, concepts, methods, and techniques, and whose universal vocation is to comprehend the world and themselves. Science has been studied by philosophers, following the approach of the formal analysis of logical structure. It has been studied by historians, who seek to achieve the historical analysis of rational developments.

More recently, science has been studied by sociologists. They seek to explicate both its institutional character and the 'driving force' of its operations—what I might call its operational code. The main exponent of the view that science can be understood by reference to the driving force of its operations is ROBERT MERTON. He has defined that code as consisting of the cultural values and mores that govern the activities termed scientific. On the other hand, from the point of view of sociology, one has to consider the social structure, which provides the best additional context for development of knowledge. We must therefore first define the code of science as an institutional code, as opposed to what I shall speak about primarily, that is, the code of the individual scientist.

As an institutional code, science has four major imperatives: 1) universalism, or the principle that scientific work is to be assessed on no other criteria than its merit or its significance; 2) organized skepticism, i.e., the axiom that judgment of all scientific contributions should be suspended until the facts are at hand; 3) disinterestedness, i.e., commitments of scientists other than the extension of certified knowledge are not admissible; and 4) 'communism' (so termed by MERTON in the original edition of his book, but better referred to as 'communalism'). The scientific community has rights to knowledge produced by its members, who individually may not limit access to their products.

With regard to the individual scientist, we all know what his motivations are, and so I need not speak of them. I may note, however, that he must abide by the rules of operation of the enterprise. Also, as a member of society at large, he must anticipate the consequences for society of the activities of scientists.

Just as science the institution has been an object of study by the members of different disciplines, so too scientists have been studied by philosophers, historians, psychologists, behaviorists, and also sociologists. On the basis of the experience of at least one investigator, seven principles might be used to define the operational code of individual scientists: intellectual integrity and objectivity; recognition of priorities; tolerance; doubt of certitude; recognition of error; unselfish engagement; and sense of belonging. I first presented these seven principles several years ago in a colloquium; I shall now briefly characterize them.

1) Intellectual integrity and objectivity. Scientists are obliged to approach the natural world and their own investigation of it with as much objectivity and care as they can muster. And this, of course, to some extent leads to

2) Recognition of priority.

3) Tolerance. Since it is always possible for new ideas to seem absurd at first acquaintance, it is wise to be tolerant of them and to see whether their factual basis appears to fall within the boundaries of sound science. This is a principle that might be disputed. Science is also expressed through dissent so long as mutual respect is maintained.

4) Formulation of the principle of doubt of certitude followed a long, difficult tradition which holds that truth emerges from the confrontation of contraries. Scientists must approach what is generally considered certain with an ever-questioning mind. The tension so created is one of the mainsprings of the pursuit of knowledge.

5) Recognition of error. The systematic application of doubt is apt to reveal errors which may be recognized and acknowledged publicly.

6) Unselfish engagement. The scientist should be motivated by the desire to extend knowledge and not by the wish for personal gain or by the desire to foster the dominance of any one intellectual perspective.

7) Finally, the sense of belonging. Scientists should conceive of their work as being part of a larger enterprise and of themselves as joined to their scientific colleagues through collective contributions to this enterprise.

The points of view underlying these formulations are quite different, but not conflicting. So far as the institutional code is concerned, the imperatives which are transmitted by various examples and reinforced by sanctions become internalized by the scientist, thus fashioning his scientific conscience. With regard to the code of the individual scientist, the code of science serves as a guide in his research activity and in his association with his colleagues.

But both codes, the institutional and the individual, overlap in this respect: they are functional for the life of the scientific enterprise. I should say that they are utilitarian in the sense in which the American philosopher JOHN DEWEY used the term. That is, rather than having ethical and moral (intrinsic) value, these codes have instrumental value.

The rest of my talk will be devoted to a further analysis of these principles, and also to an effort to find out what are the possible transformations of the code of the scientist? What are the transformations that are already taking place? Where may these changes eventually lead?

To go a little further into the notion of universalism: it must be emphasized again that scientific merit is the only criterion. There are no other standards for evaluation of scientific work, or for admission to training or scientific assignments. Nationality, religion, social standing, prior research of quality, and even ideology should not enter or be introduced into questions of scientific worth. In other words, this approach must maximize the probability that both ideas and personnel are continually renewed.

With regard to objectivity and intellectual integrity, I have little to say about it except that in scientific work these qualities allow detection of relations not previously revealed, and they permit one to uncover weaknesses in accepted ideas.

The principle of doubt of certitude has been disputed by many. The phrase 'doubt of certitude' was coined by a Swiss essayist, DENIS DE ROUGEMONT, in a book that has been translated into English under the title *Man's Western Quest.* Perhaps the origin of this idea is Heraclitus' notion of flux, the thesis that the only constant in experience is the fact of ceaseless change. Or, leaping over 25 centuries, one might find a more contemporary formulation of this thought in the work of JOHN DEWEY, who asserted that the quest for certainty is an illusion that diverts man's attention from the possible and the practical. Or one might refer to the declaration of A. N. WHITEHEAD: there are two principles inherent in the nature of things, the spirit of change and the spirit of conservation, and there is nothing real without both. Or one can even evoke NIETZSCHE, who said that science arises from rationality, while remaining open to the irrational. Finally, I would say that the three stages of Socratic discourse to some extent express the doubt of certitude.

Now, organized skepticism. From the philosopher's point of view, organized skepticism is different from the doubt of certitude. As MERTON has described it, oranized skepticism refers to the suspension of acceptance of new scientific work until it has been subjected to a variety of institutionalized procedures for evaluation. MICHAEL POLANY, in his book *Personal Knowledge,* is completely opposed to the idea of doubt of certitude. A champion of authority, he thinks that doubt and belief are not the paths to discovery. Yet in the same book he recognized two types of scientists: those whom he called the 'literals', represented by LAUE, RUTHERFORD, and FRANCK, who deeply pursued ideas and their applications, and the revolutionaries, i.e., men who doubted the validity of accepted ideas, exemplified by EINSTEIN, PLANCK, and BOHR.

Now I come to POLANYI's notion of authority. He thinks that authority is necessary to favor work which is consistent with accepted ideas and is therefore more useful for further research. It also provides a defense against crackpots and producers of trivia. He argued that authority must be accepted even at the expense of putting aside good work. In fact, in his own experience as a scientist he presented an idea about adsorption which was

440 COURNAND

rejected, I think, by no lesser experts than EINSTEIN and BOHR. POLANYI declared that this negative judgment was correct since his idea was entirely new and was contrary to the general trend of thinking at the time. Furthermore, the passage of time was capable of indicating that his concept was correct.

This is all true. But, on the other hand, I cite another example, that of the physiologist ERNEST SCOTT. Professor SCOTT, in work leading to a thesis in 1911, had studied the effect of pancreatic extract on de-pancreatized dogs. This was approximately 10 years before the discovery of insulin. SCOTT reported that the pancreatic extract he had prepared was capable of keeping de-pancreatized dogs alive. The article which he submitted was first rejected, then editorially truncated, because of its deviation from accepted ideas, and thus the full text was not made available to the scientific community for many years[1]. Can this be considered an instance in which the exercise of the principle of authority had beneficial consequences?

That there is a tension between the principles of doubt of certitude and organized skepticism and authority need not be re-emphasized, and I pass on to the next principle, that of tolerance. Tolerance provides time to evaluate a new idea. It prevents immediate dissent or acceptance, thus helping to minimize conflicts between scientists. Clearly, the principle of tolerance (and also that of doubt of certitude) cannot consistently be applied at the same time as POLANYI's principle of authority. So far as the actions of scientists are concerned, it may be that tolerance is applied when there is a premium on obtaining new ideas, and authority, when the effective functioning of science has relative priority.

Concerning recognition of error, I wish to emphasize two points: 1) the basis of this principle is not that recognition of sins is desirable, but rather that it helps scientific work to advance. 2) There is another kind of error which is related to the overzealous acceptance of authority. One of the best historical examples is that of RICHARD LOWER, who was one of the three great physiologists in Oxford in the 17th century. He was a pupil of WILLIS and in a clash with an Irish doctor he took sides with WILLIS, asserting that the blood passing through the lungs does not change its color. Yet later on, in his book *De Corde,* he emphasized the fact that he had been in error. For those who are interested in history, I recommend this

[1]   In fact, not until DICKINSON W. RICHARDS recounted the story and presented the text in 'The effect of pancreatic extract on de-pancreatized dogs: Ernest L. Scott's Thesis of 1911', Persp. Biol. Med. *10:* 84 (1966–1967).

remarkable text. I do not think I have ever found a better way of expressing the necessity of recognizing error: to reaffirm the values of reason and experience.

Now I come to one of the most controversial elements in the code of the scientist, namely, unselfish engagement and disinterestedness. Historically, of course, the singleminded quest for knowledge can be traced far back. The astronomer EUDOXUS, as you know, said he would willingly 'burn to death like Phaeton were this the price for reaching the sun and learning its shape, size, and substance'. This is probably the spirit that animated the first man who went into space, and also the first to land on the moon.

Unselfish engagement is inculcated in the scientist by training and by education, which encourage one to value the intrinsic satisfaction associated with the work more than acclaim, rewards, or riches. Unselfish engagement, furthermore, compels acceptance of valid new ideas even if they supersede one's life work, rendering it technologically or intellectually obsolete. Unselfish engagement and disinterestedness must resist varied pressures, some political, some military, some economic. I would say that in the pursuit of knowledge, there is the danger of pressure to suppress information, particularly among scientists working for industry. I would take as an example the conflict that must exist among scientists working for the tobacco industry, which, as you know, promotes and sells a substance which pollutes the bronchi and leads to quite a number of unpleasant and dangerous developments.

What is really at issue? First is the notion that riches may legitimately result from the practice of the scientific vocation. I must say that in the past, we have had quite remarkable examples of people who were so disinterested that despite the extraordinary value—economic and military —of their discovery, they refused to accept patents or other non-scientific rewards. Thus the CURIES, who discovered radium, did not accept the idea that it be patented. More recently, cortisone was isolated by KENDALL, who refused to take commercial advantage of this discovery. Finally, I would say that G. H. SHULL did not take any advantage of his discoveries concerning hybridization of corn.

Then there is another danger, and that is the application to science of some of the mores prevalent in industry, sport, and spectacles in general, that is, publicity. The fact of publicity disturbs many. This is not to say that the scientist should not be available to the mass media; but he should be available to them for the purpose of education and not to publicize himself. With regard to this matter, I would say one word about the problem of recognition. Here I might cite what ROBERT MERTON has said concerning the 'ambivalence' of the scientist: 'He is modest, but he wants to be recognized.' There is a certain tension between those two possibilities, which I believe each of us has probably experienced. Recognition, of course, is a necessity for the scientist. He must be recognized by his colleagues; as a matter of fact, this is a guide in his work. He must also to some extent be recognized by the public, but not to the extent of recent years. I mention one example, which doubtless you all have in mind, that of the discovery of the double helix. In this connection I wish to emphasize the attitude of LINUS PAULING, who never made any public statement, except to recognize the value and the extraordinary importance of this discovery. There is a definite danger in the desire for reward, particularly as experienced or expressed by the young scientist. It is that of seeking 'the main chance', of taking on a problem because it is regarded as very important or because it is expected to lead to something of very great importance. Seeking 'the main chance' is of course not inherently invalid, provided that the results are scientifically sound. But it may lead to a number of dangers, among them the possibility of plagiary and of presentation of data that may not be absolutely satisfactory.

MERTON has emphasized the social consequences of unselfish engagement, or, more precisely, of disinterestedness, which he presented as an imperative of the institutional code of science. He said that 'the tenet of pure science and disinterestedness has helped to write its own epitaph'. I am not that pessimistic. But I would emphasize what to my mind appears to be the main danger of the opposition between the tenets of disinterestedness and of accepting rewards for scientific work, whatever they may be, and that is that it affects the cohesion among scientists.

Finally, I will speak of communalism, the sense of belonging. To belong enhances the sense of indebtedness to a common heritage. It enhances also the cumulative quality of scientific achievement. It favors free communication of ideas and helps to maintain a sense of purpose when the work is proceeding very slowly. But, of course, scientists have other commitments which are not always compatible one with another: family, nation, employer, and other parts of the social system. This aspect of the code of scientists will probably have to be modified, that is, in the relationship of the scientist to society, particularly as this relates to the wishes (or demands) of society.

I am sure that this can be accommodated, provided that the consequences are not inimical to the proper operation of science.

I would like particularly to mention the question of secrecy. Since the main, nay, the sole bond in science is the free flow of information, then secrecy by its very nature is inconsistent with the operation of the scientific enterprise. It is undeniable that some scientists adopt secrecy out of anxiety or fear of plagiarism. WARREN HAGSTROM, upon questioning 1700 academic scientists, found that one-half said that it was not safe to discuss their work with any, or nearly any, colleagues in other institutions. This has led, particularly in biology, to the formation of 'invisible colleges', i.e., societies in which all scientists except for immediate participants are excluded from attending reports and discussion of current research, and publication of findings outside the group is eschewed. Of course, the conditions that foster secrecy of this sort may be ones in which productive work is accomplished. But secrecy is nonetheless a condition which has to be looked at very closely. I could speak of the problem of secrecy in collective work, for example, the military. The circumstances in which secrecy develops, depend to a great extent on the social context[2]. In World War II, secrecy was considered essential in certain laboratories, and this was what the scientists who belonged to them really had at heart. There are indeed some obligations which go beyond those of science.

If I might offer some conclusions regarding the survival into the future of the culture of science as I have defined it, I would say that the moral, social, and ethical implications of scientific work are becoming increasingly important in decision-making. Obviously, priorities will have to be set, presumably in response to social pressures, if funds continue to be curtailed. With regard to personal gain, here the code of the scientist appears to be undergoing a transformation. Its outcome may well be a new ideology that relates social concerns and a new or different notion of the legitimate personal benefits to be derived from scientific work. In its impact, such an ideology would be comparable to the effect of the ideology of freedom on victims of persecution and authoritarianism.

At present there is considerable hostility to science. This was not obvious in this meeting, of course, where indeed I had a quite remarkable sense of collaboration and where most aspects of the code of the scientist as I have defined it were illustrated. But there is no doubt that the public and some government circles harbor great hostility to science. The hostility

2   It is interesting that many scientists who participated in the Manhattan Project are now, under different circumstances, strongly opposed to secrecy regarding the military applications of their work.

of the public is due, no doubt, to the 'balance of terror' which now prevails, but also, and perhaps more important, to the fact that on the one hand, science has developed in man the sense of his own insignificance, whereas on the other hand men of science, by acting as Prometheus, have spread the notion that man can control nature. In part, finally, a substantial portion of the public's resentment results from the fact that the values that scientists are trying to defend are not always understood.

But there is in the code of science, it seems to me, something which is fundamentally important, and which we can produce by means of the educational process. That is, that the code of the scientist, as I have defined it, favors dialogue as against suspension of efforts to communicate. At a time when in some situations recourse to violence is increasingly advocated, then one cannot help but think that if education emphasizes the values of the code of science, that is, intellectual integrity and objectivity, recognition of priorities, tolerance, doubt of certitude, recognition of error, unselfish engagement, and sense of belonging, then recourse to dialogue should prevail. Interestingly enough, we are starting, at least in the Western countries, to notice that there is a certain desire to confront ideology on the basis of that code, implicitly, not explicitly—to actually engage in dialogue on subjects that are divergent. And it is on that word of hope that I would like to end.

Author's address: Prof. A. COURNAND, Department of Medicine, Columbia University College of Physicians and Surgeons, 630 W. 168th St., New York, NY 10032 (USA)

Discussions

E. KATCHALSKI: Allow me, Professor COURNAND, to ask, do you feel that scientists should be permitted to pursue any research topic they like, or that society should—by appropriate means—restrict their activities to research projects that cannot lead to harmful results?

A. COURNAND: Regarding your question, I am aware that I am speaking to scientists; I will admit, therefore, that there is no boundary, i.e., that we are dealing with an *open* system. We have to take into account, however, the fact that we may have reached a time when the open system might become closed, because society is insisting on priorities in the problems attacked by scientists. This is the thesis which RENÉ DUBOS has presented so well.

E. KATCHALSKI: Should society be encouraged to regulate application of scientific discovery, abandoning the principle of free exploitation of scientific achievement?

A. COURNAND: I believe that you are speaking of the control of *technology*. I have been speaking mainly about scientists and science. I would only add that the scientist has the duty to anticipate the possible practical applications—both beneficial and destructive—of his work. This he might do as a consultant and not as one who makes decisions. There are, of course, a number of scientists who believe that science as an institution might regulate itself. In general, I believe that this will be done under external, social pressures, of which we are not the master.

E. KATCHALSKI: Would you encourage scientists to participate in social affairs and join decision-making bodies, such as those to be found within political parties and government?

A. COURNAND: To some extent, I think that if you take the scientist as an *ideal* man who is loyal only to science as an institution, he should be a consultant and not participate in decisions. However, there are good examples of scientists of renown who have played a role in politics. The names of two French scientists come to mind: PAUL BERT, who wrote a classic on the physiology of altitude, and E. BOREL, a mathematician who is known for his work on the theory of games (which he did in advance of VON NEUMANN).

U. PALMA: We have been participating to discussions about collective and cooperative phenomena for a week or so and I thought that the subject of today's discussions might have been approached from the same standpoint. That is, rather than seeing the scientist and his interaction with society, as a 'single particle' problem (that is, as a problem which mainly concerns the individual scientist), I am inclined to think of the problem as a collective one. By collective, I must specify, I mean pertaining to society as a whole, rather than to the minority group of scientists viewed as a (sub-)collectivity (perhaps more equal than others in a society of equals). I see instead that you use the 'single particle' approach. One reason for this could be a lack of hope that the collective approach could bring us anywhere. Another reason could be your being convinced that the present situation (allowing the 'single particle' approach only) is a good one. So my first question is, which of these two is your attitude?

A. COURNAND: Between what and what?

U. PALMA: The fact that you do not view the scientist as a member of a society of equals, giving some service and receiving something in return, just as anybody else, means that you are content with the present situation or that you do not hope there is much to be done in this direction? Before you answer my question, let me express my personal attitude. Apart from the hope we may, or may not, have in the collective approach, a think that we have already seen all the bad that the 'single particle' approach has brought about. Science was born as a struggle to substitute myth with logos and then the myth of the logos was created and further restricted to the myth of 'rationality', as opposed to 'reason'. For 20 centuries, from Anaxagora's expulsion from Athens, motivated by his teaching 'theories on celestial bodies', to Galileo's trial, the position of scientists as 'single particles', took them nowhere. Later, it was discovered that scientists could be exploited. They were given an ivory tower, but, alas, the tower was locked and the key was in somebody else's hands. Still, scientists kept feeling and behaving as a class. No wonder is, as a rule, they did not mature, did not let other people mature in human dimensions, did not care if and how their findings were interfering or used in connection with the most important problems of society. If this situation is going to go on, I am convinced it is not going to do any good, neither to science, nor to ourselves, let alone to society. As a class of the 'privileged-exploited', we have a twofold problem: to take immediately in our hands and to solve the most urgent problems that our findings have raised for society as a whole (and for this we must come out of the ivory tower, whatever this means) and to strive for a long-term cure and solution, in terms of a society of equals, where nobody is more equal than others.

A. COURNAND: I should like to emphasize that there was nothing in my presentation indicating that I am in favor of the scientist's remaining in his ivory tower. In fact, in the initial statement of my presentation, I said that isolationism is no longer tenable, and that scientists have acquired a realistic view of the importance of their participation in society. No longer do they see science as a self-regulating enterprise, that is in society but not of it.

G. LING: I cannot agree more with Prof. COURNAND's suggestion that the principles of the ethics of science should be made part of the educational program of the individuals of the society. One of the key processes in science, I believe, is debate. Without debate, errors in earlier views would be perpetuated and without dissenting views, there can be no debate. This is plain enough. Yet in some way, to the present group of highly successful scientists, the problem of minority view may seem a matter of insignificance. But, is this because there is no worthwhile minority view or because minority views are already fairly treated? Obviously, the answer depends on the person asked.

Thus, to a physicist, the minority view is hardly a problem because physics is a very mature science. Truly major revolutionary ideas are rare and if they do occur, are likely to be recognized as such. To a biologist, on the other hand, a totally different situation exists. Biology is in an infantile state of development. As such it is the science of the future. In other words, biology will need new and revolutionary concepts for a long time to come. Minority views, in this case are of vital importance.

In the corner of the editorial page of every issue of *Science* magazine, there is a statement that *Science* magazine is a forum for the presentation, for not only the majority view but also those of the minority and conflicting points of view. This is a very fine and pioneering statement.

In fact, however, regrettably there is no way to substantiate this aim because the way of selecting articles for publication, shared by many other journals, demands that the article be approved by a great majority of scientists in the same field of specialization with—overwhelming enthusiasm, and by definition, a minority view simply cannot get that kind of enthusiastic approval. More or less the same peer approval system is widely used as the basic guideline in the selection of research projects for funding.

We all know well that the ideas of neither DARWIN nor PLANCK, nor indeed many of our most illustrious scientists who have created our present day science, have ever been accepted immediately. However, if their views could survive as they did, why can't we anticipate more modern day scientists doing the same? The answer is obvious: times have changed. There were far fewer scientists then, and scientific journals did not reject a large percentage of submitted manuscripts (*Science* now rejects about two-thirds of submitted manuscripts). Scientists were also much less dependent on external financial support than they are now. Thus DARWIN and PLANCK could survive. It is doubtful whether future Darwins and Plancks will be as lucky unless contemporary scientists consider it their sacred duty to foster and nurture the views of the minority, not only in principle, but also in fact.

A. KATCHALSKY: I liked very much Prof. COURNAND's summary of ethics in science. I am not so sure, however, if his fine presentation covers the subject of our discussion today. The major issue of our discussion is the impact of science on society, and it seems that the self-recognition of scientists, and their understanding of those ethical principles which guide their scientific work, is not sufficient for present-day needs. During the last two and a half decades science has become such a powerful factor in shaping modern life, and has had such a deep impact on the structure of society, that the appreciation of scientific attitudes is not a sufficient basis for contemporary ethics.

Present-day science is not situated any more in the classical ivory tower, but through its impact on scientific technology has become a factor of change in the life of millions, without providing them with the modes of living required in a science-dominated world. It is, therefore, a meaningful problem, raised by the young generation, whether society should stop science and impose a moratorium on research—since its destructive influences may become more pronounced than its constructive effects. Indeed, even such 'harmless' sciences as molecular biology raise a wealth of moral problems which are thrown upon society without guidelines for correct behavior. For centuries, ethical solutions were based on theological reasoning and even the appeal to 'Laws of Nature' was based on the assumption that the natural laws express a divine will. But in the scientific world it is up to the scientists to search for guiding principles and not to rely on divine inspiration. Thus we read a short while ago that KHORANA has succeeded in synthesizing a gene, and there is little doubt that in the foreseeable future, genetic modification will become feasible. This possibility opens up, of course, a fertile field of genetic medicine and promises new ways of eradicating some of the horrible diseases of mankind. On the other hand, it brings us to ehical problems on embryological regulation, which neither theology nor legal philosophy could answer without the aid of a deeper scientific insight and without a profounder consideration of the value and the meaning of human life—of our responsibility toward future generations and of the meaning of right and wrong in a science-dominated society.

I. RABI: My remarks will be very brief and I will address myself to the part of the talk that refers to education. I am thoroughly in agreement with the remarks about the importance of science to society and the close connection. However, except through education, we cannot go very far in finding a way of dealing with changes in the culture and in society, now occurring through the growth of science and its applications. Basically, I feel that we must review our whole system of the education of the young so that they learn how to live with science, and to respect science profoundly enough to understand that science is for the present and future the background for society and of ehtics, and morality. Morality is a practical matter: it refers to action, and one cannot be practical now or in the future without a feeling for, an understanding for, and a deep respect for science. So we have to be agitators to the point where the misuse of science is a sin in the religious sense. Unless we move to reach that point, science may easily overwhelm us, as has been pointed out. Unfortunately in America and in other western countries, the movement is in the other direction. Interest in science goes down and science is regarded as an enemy, very often by the young who should embrace science and try to understand it. The fault lies with the general scientific community which failed to appreciate the larger context of science, and in their teaching of science, taught it somewhat in a way to show how clever scientists were, how glorious it all was, but failed to help either the young, or the public generally to find the connections between the vast scientific culture—this tremendous thing that has happened to humanity—and the meaning, responsibility and goals of humanity.

P. AUGER: Je voudrais dire un mot en ce qui concerne ce qu'a dit M. COURNAND sur l'éthique de l'homme de science, sur le code des sciences et surtout sur la façon de faire comprendre au public non scientifique ce que signifie vraiment ce code des sciences. Je crains que ce soit là une chose difficile, non pas chez les enfants, parce que je pense que l'éducation, en effet, doit comporter, devrait comporter, dès le plus jeune âge, une tentative de faire comprendre aux enfants, le plus tôt possible, ce qu'est la science, ce qu'elle peut donner, quel est son code éthique, quelles sont ses méthodes; je crois que ce serait une excellente chose de la faire à tous les enfants, et pas seulement à ceux qui se destinent à une carrière scientifique ou technologique.

Mais le problème le plus grave en ce moment, peut-être, immédiatement, est celui des adultes. A l'heure actuelle ce sont les adultes qui votent, ce sont ceux-là qui conduisent la société, et par conséquent c'est sur eux qu'on doit essayer d'agir, tout de suite, sans attendre. Il faut naturellement agir sur la jeunesse, mais pour un effet dans cinq, dix, quinze ans. Tandis que sur les adultes, il faut agir tout de suite pour des résultats immédiats. Et je crains que là-dessus, les hommes de science peut-être ne voient pas aussi clairement que certains d'entre eux et comme je le vois, la nécessité de faire quelque chose tout de suite et avec beaucoup de vigueur. En particulier, prenons la question qui a été soulevée par M. COURNAND, celle du doute. Dans le public des adultes, qui n'ont pas une grande formation scientifique, il règne une opinion générale, que les savants font, surtout dans la science appliquée, des choses merveilleuses, mais, au fond, ils savent très peu et les meilleurs d'entre eux disent qu'être vraiment savant ça veut dire surtout savoir que l'on ne sait pas, les choses qu'on ignore sont tellement nombreuses encore, qu'il n'est pas étonnant de voir des scientifiques changer d'avis, changer de théorie, douter d'eux-mêmes, douter de leurs collègues et retourner leur veste de temps en temps en faisant une théorique complètement nouvelle: NEWTON avait tort puisque EINSTEIN a montré qu'il y avait une autre théorie, qui était meilleure! Donc, on a l'impression dans le public, que, surtout du point de vue théorique, les hommes de science sont toujours prêts à changer d'avis et qu'au fond, ils ne savent pas très bien où ils en sont.

Je crois qu'il serait très important de faire connaître au public le sens véritable du doute scientifique, comme il a été exposé par M. COURNAND, c'est-à-dire que le doute est celui qui touche la théorie principalement, et non pas les expériences et que la science expérimentale, les faits scientifiques parfaitement établis le sont pour toujours, dans les conditions où ils ont été établis. C'est seulement l'interprétation théorique, le modèle intérieur que le scientifique construit pour représenter l'univers extérieur, c'est celui-là qui doit changer, et qui d'ailleurs, et cela a été dit très justement tout à l'heure, ne change que pour des conditions nouvelles. Il est toujours valable dans certaines approximations et c'est seulement lorsque les expériences sont meilleures qu'il faut changer la théorie et l'améliorer. Et là, bien entendu, il peut se produire de grandes révolutions. Mais que le public comprenne bien que ces révolutions sont normales pour la science, que c'est l'avancement régulier de la science, et que, par conséquent, il ne faut pas croire que tout à coup les savants deviendront par exemple des astrologues parce qu'ils auront découvert que les étoiles ont une action sur notre vie quotidienne. Beaucoup de personnes, dans le public, qui aiment beaucoup la science, croient que les choses sont tellement ouvertes, que l'on ignore tant de choses, que peut-être tout à coup on s'apercevra qu'effectivement la planète Mars fait que certaines personnes ont un esprit guerrier, batailleur, et que d'autres ne l'ont pas. Je crois que là, sur les adultes, il y aurait vraiment un travail très intéressant, très important à faire pour leur donner la compréhension véritable de ce que la science est, pourquoi elle change d'avis du point de vue des théories, pourquoi les savants communiquent entre eux de façon aussi continuelle, et ont besoin absolument de se rencontrer tout le temps pour se raconter leurs propres histoires. Ce besoin le public ne le comprend pas très bien non plus, il dit: «Ils n'ont qu'à rester dans leurs laboratoires.» Tout cela, je crois que vraiment il faut l'expliquer et que c'est seulement si on l'explique bien que le public pourra soutenir le mouvement scientifique de tout son cœur, et sans avoir trop d'inquiétude, au moins en ce qui concerne l'avancement de la science, si non pour ses applications.

L. ROSENFELD: With regard to the question of doubt I might perhaps say how theoretical physicists approach this idea. It is quite clear that an attitude of active doubt is hopeless. It doesn't lead to anything. There is one domain in which doubt is the actual method followed by scientists and which is necessary in order to get a broader view of unsolved problems: very often the problem is not solved because it is enclosed in too narrow a framework—a great discovery is just the breaking of such an artificial barrier—and doubt helps us to notice the barrier, which we can then try to break. But on the other hand, when one has recognized in a certain domain, that one has defined the limits of application of a theory, of a system of concepts—and has of course, verified by a sampling of experiments that the theory gives good results within these limitations—I would say that this theory is as certain as any human conception can be, and is valid for all times. Thus, the essential point in order to achieve certainty is to define the limits of applicability of a certain set of concepts. And it is at the fringe, where one tries to go beyond the known domain of applicability, that doubt is the attitude of mind which may lead us further outside that domain.

A. COURNAND: I would agree with you, since you partly agree with what I said. However, I may not have made myself clear in indicating that the doubt of certitude applies also to the scientist as he is working. I think that it is certain that we are continually criticizing ourselves—or we should—and we should continually exercise the faculty of criticism, which is to some extent the expression of the doubt of certitude. When we have a hypothesis, the experiment must allow us to modify that hypothesis.

T. L. BLUNDELL: One of the points that you mentioned in your very excellent talk was the effect of social position on the scientist; it seems to me that this emphasizes one of the biggest dilemmas for the scientist in society. Academic scientists who question scientific policy-making of their governments are often accused of not being in full possession of the facts on which the policies are based. They are accused of acting in public, on standards that are unscientific; standards that belittle the work of the scientific community. However, there is a difference between scientific policy and academic scientific research. Most military scientific policy is acted upon before academic scientists are allowed access to the facts. It seems to me that American military policy in Vietnam is a good example of this dilemma. It has been impossible for academic scientists to make a proper study of the effect of military policy on the ecology of Vietnam before voicing their suspicions. Would you therefore think that a scientist may have to adopt different standards in public than in the laboratory? And if so, what do you suggest should be the moral guidelines? Or, is this policy, this so-called unscientific stand that scientists are required to make, simply an extension of your philosophy of doubt?

A. LICHNEROWICZ: Juste quelques remarques pour dire mon accord avec COURNAND, KATCHALSKY et AUGER, et peut-être pour réaffirmer quelque chose. Les hommes de science forment d'abord une communauté scientifique, mondiale, dépassant tous gouvernements et toutes nations et se définissant en effet par une morale et par une éthique. C'est ce qui différencie un membre de la communauté scientifique d'un technologue, ou d'un technicien. Peut-être sommes-nous dans une mauvaise position pour avoir trahi, pour avoir mal plaidé notre dossier en prêchant trop souvent notre «utilité». La communauté scientifique vise des savoirs. Les pouvoirs viennent par surcroit et sont les garants qui nous permettent de tout remettre en question à chaque instant, sans rien perdre. Je voudrais que notre communauté scientifique s'affirme plus désintéressée qu'elle ne l'est jusqu'à présent. Je pense aussi, que le seul devoir impératif de cette communauté scientifique est le devoir d'information. Nous n'avons pas à nous transformer en technocrates. Nous avons un devoir d'information devant les citoyens du monde, et ce devoir doit être rempli avec la même imagination, la même intuition, mais aussi la même probité intellectuelle que celle avec laquelle nous faisons la science.

I. Prigogine: Je voudrais dire que dans toutes les remarques qui ont été faites il semble qu'on ait surtout parlé de crise, de difficultés entre société et science qui viennent, en quelque manière, de l'extérieur, mais je crois qu'un problème essentiel c'est l'évolution des sciences elles-mêmes qui a changé complètement leur situation propre. Ainsi, il y a vingt ou trente ans, on entendait très clairement ce qu'étaient les sciences fondamentales: la physique atomique, la relativité. Actuellement il est beaucoup plus difficile de savoir ce qui est fondamental, et les sciences sont beaucoup plus éparpillées. Dans ces conditions, l'unité de la vision scientifique a été perdue. Et quand nous parlons de crise, on peut se demander s'il ne s'agit pas en partie, simplement, d'une crise d'adaptation due à un développement rapide de nombreux domaines. Ainsi, on ne peut parler de crise au vu du développement des sciences économiques ou des sciences humaines en général. Nous ne devons pas nous limiter aux problèmes spécifiques de la physique ou même de la biologie. Je crois que dans aucun moment de l'histoire, les idées qui ont été à la base des sciences, d'abord mathématiques et physiques, n'ont autant pénétré d'autres activités humaines. Il y a là quelque chose comme un début de «rationalité» dans des domaines qui avant étaient purement empiriques.

Proc. 3rd int. Conf. From Theoretical Physics to Biology. Versailles 1971. pp. 451–467
(Karger. Basel 1973)

Social Implications of Biology

M. H. F. WILKINS

I generally find myself talking about social implications of science to an audience of students; it is an unusual privilege to be speaking to this distinguished, most erudite audience. I am embarrassed by the impression some of you have given me that you expect me to provide solutions to the problems facing us. My main hope is that the *audience* will suggest solutions, and what has been said for example by Professors RABI and KATCHALSKY, indicates that we may get good ideas. I am going to talk mainly about the important problem that Professor COURNAND referred to—current, anti-science attitudes and criticisms of science. I think this is a very important problem facing the scientific community. This type of attitude is widespread among our younger people; it affects the number and kinds of people entering the scientific community; it has important aspects from many angles. One narrow aspect, though it hits us very directly, is the funding of science. General attitudes towards science, social approval by the electorate, naturally affect the way governments spend their funds. While I was preparing this talk, a colleague brought me a newspaper cutting about an American meeting where they had had a 'trial of science'. The jury decided to condemn science and technology on the grounds that they had on balance not benefited humanity or raised the quality of human life. This is in striking contrast to the attitude which gave rise, after the Second World War, to enormous increase in funding of science. But I think the most profound importance, ultimately, of anti-science attitudes is their connection with the development of anti-rational attitudes generally; for example increasing interest in debased forms of astrology, in mysticism and taking hard drugs. I am not condemning mysticism in general but it has degenerate forms. Anti-science attitudes take various forms. First, there is the direct attack on science in general. One of the clearest exponents of this is THEODORE ROSZAK in his various writings[1]. Second, there are criticisms of science as it is today; for example, the demand by students and others for what is called relevant science. Further, there are various criticisms of science from those working in the general area of social responsibility in science. These various critical attitudes merge into another attitude which I think is hopeful and positive—the attitude of some of the younger artists today towards science, which is a very positive interest. The significant thing is that this new and positive attitude toward science is held by people who in fact share many of the views of THEODORE ROSZAK. This is, I think, an indication that the scientific community should not simply reject and react negatively to the ROSZAK views, but should partly accommodate them. There are sensible elements in them.

Consider the various aspects of ROSZAK's anti-scientific case. First there is the matter of plain mis-application of science—a very important area about which scientists will often agree with anti-scientists. This is basically anti-technology, not anti-science. There are the extremely important and urgent problems of science being used to develop new weapons. There is pollution and environmental deterioration. Second, there is the more subtle matter of the unpredicted side-effects of applying science. Here again we have pollution; also the population explosion, and all the various difficult moral and ethical problems which arise as new biology is applied in medicine. We would probably agree that, in principle, these two kinds of difficulties can be corrected by scientific planning and by political action. Relatively speaking, therefore, these are fairly straight-forward problems. But they are enormously important and urgent. If they are not dealt with, we may not survive. I think that the scientific community must not shirk the responsibility of going into the political area when it is necessary to do so on particular scientific issues, or on issues involving scientific elements. Although a great deal needs to be done here and it is of immense importance, the second main attack on science goes deeper. I think it is a more difficult one to answer. That is what ROSZAK calls de-humanization by science. He claims that objective thinking necessarily leads to lessening of moral sensibility. If you read him it is a bit difficult to find a clear statement to this effect. In his 'Making of a Counter Culture', I found I had to get into the appendix before I found this stated explicitly. But this is certainly the undercurrent and the implication of his writings. The

1   e.g. 'The Making of a Counter Culture', Faber, London 1970.

argument is that the need in scientists, to cultivate the ability not to be influenced by emotion and feeling, necessarily leads to alienation. He claims that scientists have an undesirable lust for knowledge, that curiosity is out of control. I have some sympathy with these views, but I think ROSZAK's final analysis goes off the rails. There is a quotation from KIERKEGAARD, which I think goes even beyond what ROSZAK had said about curiosity out of control ... this is a somewhat poetic statement not to be analyzed in too exact a way. He says 'knowledge is an attitude, a passion, actually an illicit attitude'. (I suspect that the word 'illicit' has been wrongly translated here from the Danish.) 'The compulsion to know is just like dipsomania, erotomania, homicidal mania in that it produces a character that is out of balance.' He doesn't say that the compulsion to know is equivalent to dipsomania, he says that it is similar in producing a character out of balance. 'It is not at all true that the scientist goes after truth. It goes after him. It is something he suffers from.' Some of you may find that statement exasperating or annoying, but I think KIERKEGAARD puts most clearly of all, the kind of objection anti-science people are raising. The attitude is that scientists are raping nature to discover its secrets. They don't really love nature. In this connection ROSZAK refers to the American Indians who will not cut down a tree for firewood but wait for the dead limbs to fall off and refuse to till the soil because they think this is hurting Mother Earth.

ROSZAK says, with regard to the organization of society, that science produces de-humanized technocratic bureaucracy. As examples of this he takes both the United States and the Soviet Union. He claims that it is intrinsic to the nature of science that it should produce this type of de-humanized society. In this type of society, he says, a basic attitude is that the problems which exist will be solved by scientific planning, by better management and by experts of various kinds. He says that in this society human attitudes are disinfected by cold reason, that thinking is regarded as superior to feeling. As MUMFORD says, you have mad rationality—which is epitomized in the nuclear bomb. Ultimately, ROSZAK claims that although science has replaced superstition by rational systems of thought, these rational systems are on balance not superior because they do not cater adequately for man's spiritual needs. He claims that reason constrains the human spirit whereas things like poetry and so on liberate the spirit.

There are a number of obvious answers to this anti-science case, these answers being commonly given. First, one says that science is the only means we have of providing a way of avoiding starvation, drudgery and premature death for the majority of people. In underdeveloped countries, for example, people don't doubt the social value of science. But on the other hand we have to face the fact that in technologically advanced countries the very success and power of science and technology has certainly given rise to negative effects as well as to positive. The general line commonly coming from scientists is that all this misapplication of science in technology is due to the fact that the politics are wrong but that science itself is all right. They imply that one can in fact clearly separate science from politics. Now my own view is that the politics is wrong and that a great deal needs to be done in that direction. Also, politics and science are so closely interrelated that they can barely be separated; moreover, insofar as science can be separated from politics, science itself needs to be changed. Another main answer commonly given to the anti-science critics is that apart from useful beneficial applications science gives an overall enlightenment. This is obviously a very important point; but I think that this idea needs developing further than it is at present. It is sometimes said that modern research laboratories correspond to the cathedrals of the Middle Ages. If one considers the positive attitude of some of the younger artists towards science today this particular idea is not quite so silly as it might a first seem. But I think that these answers to the anti-science case need a great deal of working on and I am going to make some suggestions about how thinking might go on this subject. What I shall to say is not original, or at least if it does have originality this probably derives from DAVID BOHM, the theoretical physicists. What I am really interested in is in your views on what I am going to say and on the problems generally.

First, I think in answering the anti-science attitude we need to develop a deeper understanding of the nature of science itself. This is a first step in social responsibility in science generally. The majority of scientists don't spend much time in thinking about the nature of science—they just get on and do it. Those people who do study the philosophy of science are very often not really working scientists. Scientist don't pay much attention to them, and in many cases quite rightly because these people have had little direct experience of doing science and don't understand its nature. There really is something anomalous, that the scientific community should be working away, producing all this knowledge which is having such a great impact on society today without, in my view, really trying to understand what they're doing. To use an analogy, I think that scientists are really rather like racehorses: they are very good at going somewhere very fast, but they don't really know what the race is all about. Some of them might say that it's not their job to think about that, their job is to win the race. But I think that inasmuch as the very essence of science is that it is an inquiry, it does seem rather odd that scientists shouldn't inquire more about the overall nature of their activities and what it means in a broad sense to man, to his culture and to his future. Basically, I suppose you would all agree that science is an inquiry, that you get ideas and you test these ideas by experience. But I don't think there is any way of defining uniquely the scientific method, or defining science. The scientific approach is applied in all academic fields outside science subjects themselves. Going further, one can reasonably argue that the whole of life itself consists of inquiry, having ideas and testing one's ideas against one's experience. I don't think there is any unique method of science; this is probably what MEDAWAR meant when he spoke of science as the 'art of the soluble'. Science consists basically of finding certain areas in which the rational approaches pay off very rapidly, where one can quantify and systematize; its nature is as much determined by this selecting process as by the method itself.

It is important that it is not only in science and other academic pursuits that one has a large rational element. The rational element is essential to all those areas with which people like ROSZAK are so much concerned, such as poetry, fine arts and religion. Unless there is a strong rational element in these, they have little form or meaning. There is no clear way in my mind to distinguish science from other human activities, no clear way to define its method. The nature of science depends on its context. ROSZAK—I keep referring to him but I mean anti-science people generally —are wrong in attacking reason and science as such. They have to accept the presence of reason as essential to all human activities. The important thing is to search for the desirable form which science should take if we are to avoid the negative effects which it has. In avoiding the misuse of science, I think an important and essential step is to change the nature of science today.

Science takes a particular form today. In the past it has had various forms. For example, in primitive societies, classification of natural objects was a form of science. Science, religion and magic were combined together in alchemy and in Egyptian astronomy. In the 17th century our present forms of science began to emerge clearly. Recently we have seen the science of Lysenko and the 'Aryan science' of Nazi Germany. You may say that these last two aren't really science, but I think that is avoiding the very question I am raising. If one considers 17th century science, one sees that at that time there was no clear distinction between pure and applied science and between science and philosophy. All that was excluded from the consideration of science were moral issues; e.g. when the Royal Society set itself up. Professor HODGKIN mentioned to me that there were good reasons then, because of religious and political differences, that moral issues should have been excluded.

As science developed during 300 years and has refined its nature, one of the main tendencies has been towards increasing specialization. It was very necessary to focus attention on particular aspects of science in order to make rapid progress, e.g. we have physics, chemistry, biology and, further, the division of biology into botany, zoology, anatomy and physiology and so on. Within science itself it has been increasingly realized in recent years that the reverse of this specialization is now desirable for the development of science. One therefore sees, for example, the setting up of Schools of Life Sciences in which botany and zoology and other aspects are no longer treated as separate. We need a wider perspective for the development of science itself; this present meeting of physicists and biologists is a further example of the recognition of this need.

Pursuing these thoughts further I would like to discuss here ideas developed by DAVID BOHM about fragmentation of thought which I think are very basic in the whole area of social implications and social responsibility of science. Specialization and focusing of attention may be very efficient but the very success of this leads to reluctance to think otherwise—to mental blockage of wider types of thinking. As a result, fragmentation of thought has developed. This means that the area of specialization is no longer regarded as part of a whole but is treated as a separate thing complete within itself. Inasmuch as this is untrue—nothing is an island that has no interconnections with other things—this fragmentation of thought must give rise to contradictions and difficulties. I believe that what we need now, corresponding to the approach of treating biological sciences as a whole and bringing them into relation with the physical sciences, is a broader science in which human needs and moral values are brought into science to become part of scientific inquiry. Just what this means in practice of course is not very easy to get clear.

The concept of pure science derives from fragmented thinking. It involves the separation of science and human needs. The development of the concept has caused science to lend itself to misuse. And in this respect I think that ROSZAK is partly right in saying that the form which science takes today does tend inevitably to lead to undesirable applications of science and to the undesirable types of society which we have. But I don't believe this applies to science in general; instead, science can be developed in new, unfragmented ways so that it will not necessarily lend itself to misuse. We need to seek to rebuild science in a broader form: at the moment it is too narrow. After 300 years of refining the nature of science, I think we are not properly aware of the type of thinking we have built up. We are—to use a rather horrible phrase—prisoners of our own expertise. And, as an example, with due respect to Professor MacKay, I think his suggestion earlier, that we should quantify 'doubt' is an example of the way in which we think in terms of fixed frameworks which work well only when applied to the limited problems which science has traditionally set itself to solve. As long as science remains in this limited and fragmented form we will continue to accumulate very serious social problems. You may well ask how are we going to find this broader science which is going to help? It is slow and difficult work thinking about this, but I have a few thoughts I would like to put forward.

First of all I think we should reject the whole concept of pure science. Most distinguished scientists today (e.g. Medawar, Bernal) agree in rejecting this concept, although their political views may differ. Adopting this attitude leads one to think, as many of you would agree, that we want some readjustment of priorities within scientific research. The general demand for what is called relevant research has got some sense in it. We ought to spend a bigger proportion of our effort in studying not what one might call science itself but trying to tackle the very difficult problems and intellectual challenge of finding how, in the world today, we are to get the applications we wish and how we can avoid the undesirable.

Secondly, I think much more work needs to be done in trying to understand the nature of science in relation to our culture as a whole. For example, we need to increase the funding and extent of activity in areas of study relating social, historical, economic and political and other factors to science. We need to study more the way science influences the social system and also the way social factors determine science. I think it will be fruitful to explore the relations between art and science. The reason for this is that the basic problem today is to relate science to moral values. As I said, moral values were cut out of science at the time of the formation of the Royal Society. We need to go back and somehow build them back into science itself. This is where the exercise of studying art and science may be valuable, because in art one has predominantly what one might call feeling as well as thinking, whereas in science one has predominantly thinking, though there is some feeling too. To some of you this must sound dreadfully woolly or academically esoteric, but I think it may provide a glimmering of some new kind of thinking. Studying the relations between art and science one finds that the distinctions become unclear. The distinction is probably quite as much in the area of study rather than the method actually used. Today many artists regard art as merely an inquiry which leads to some statement about man in the universe; it does not lead to the production of art objects. The artists use their personal feelings to cast light on universal problems. Similarly there is an increasing tendency for science to be regarded not as giving final truths, but as giving simply aspects of a truth, that science is an open-ended inquiry,—an inquiry which is guided much more than people have realized in the past by man's social and philosophical attitudes. But I think that to trace interconnections between art and science as they exist now is probably not very significant, because they are both fragmented activities; both are separated from living as a whole. What is more significant is to seek their underlying unity; thus we may hope to develop new forms of broader inquiry which embrace both science and art. The purpose of this exercise is, as I said, to try to reunite science with moral values. Art in a wider sense necessarily embraces religion and deals with moral and ethical problems.

One may see a little more clearly what I am trying to get at when one considers these things in the educational context. Take children of 12 or 14 in school. Normally there is the science class, the art class, the history class and so on: they are taught these subjects. But there is a lot of scope to teach without subject boundaries. This is already done at lower age levels in the primary schools in Britain, but is little applied at higher ages. Take a subject like water. Why couldn't one just go into a room with the children and say, 'We're going to learn about water. We're going to deal with the physics, chemistry, biology of water, its physiological importance . . . leading on to public health and economics, water transport and water supply, water conservation, agriculture and so on; water in art, sport, literature, mythology and religion'. As a result, we would not have the science, art or history class, but would study things as a whole in a comprehensive unitary approach. Learning in this way should help young people to develop new ways of thinking which are barely possible for people educated to be either scientists or artists. When such children grow up they should not regard science as something discrete and separable from politics. They would see that inevitably work at the laboratory bench interacted with politics and was a social activity which had immense consequences for the rest of society. As a result one might hope that those who became scientists would feel more responsibility for matters of science and society.

While I am talking about philosophy and art and so on some of you may get impatient. You may say that we are at the barricades... for example the United States is involved in a war, there are race problems, pollution, acute lack of freedom for scientists in some countries. These are pressing problems which we have to do something about now. If we allow nuclear bombs to proliferate we may all be blown up before we have any time to develop new education and new hybrid areas embodying science and art for example. I have much sympathy with this point of view. It is certainly very important for scientists to take direct and, where necessary, political action on the problems facing them. Apart from the direct benefits of trying to solve pressing problems, I think that taking part in the attempt to solve these problems will itself be educational to scientists and will help them to see more clearly the nature of science itself.

In exploring these matters we need to adopt a general exploratory experimental approach. To find how to proceed we need to try all sorts of things. That is the justification for my discussion of art and science. Some of you might consider that discussion rather irrelevant, but I feel these things may ultimately have profound significance in relation to social problems generally. We can do a great deal by developing thought within the scientific community itself. We can also hope to gain much by contacts with other specialists such as poets, philosophers, religious thinkers, historians and economists. But all these specialists are like scientists in that they have their own mental blockages because their thinking is fragmented. How do we get out of this difficulty? I think it may be useful to take the old idea, like that of TOLSTOY, that we need to go out to the people, that in the ordinary mass of people there is a wisdom which the specialized thinkers have lost. We can do with all the help we can get because these are very difficult problems to contend with. There is of course a great deal of popular science today, on the radio and television, in newspapers and in books. But in all this there is a remoteness between the scientist and ordinary people. ARTHUR GALSTON, after his recent visit to China, said that in China there is direct contact between every research laboratory and some factory. I don't know what this entails, but I think it is an idea worth exploring. We can of course begin this direct contact with ordinary people in our own laboratories. Some years ago the administrative officer in our lab, who was not a scientist, pointed out that the girls who do our typing, the secretaries, the mechanics in our workshops, the porters and the storekeepers did not feel very much a part of our laboratory. The research scientists themselves knew what they were doing but they had their own community: the others

would like to know more about it. They came to work every day but did
not really know why. They would like to have talks to explain what the
work is all about. What does it mean to ordinary people for scientists to
be doing research? I was dismayed to find the extent to which the non-
scientific people had no direct contact. The majority of the workshop
mechanics in our lab had never once looked down a light microscope.
Several people in our lab claimed they had never even *seen* an electron
microscope, and we have half a dozen. When they walked down the corridor
the door didn't happen to be open or if it was open they wouldn't have
known it was an electron microscope they saw. In response to this I gave
some talks and we had some discussion. It was quite fruitful. Some of my
senior colleagues had thought that giving a talk of this nature would be
inherently too difficult and would fail. It is not easy to do, but we now
have several colleagues giving talks and discussions and showing how the
apparatus works. The purpose of this exercise is not only to educate the
non-scientific people but also to educate the scientists who do the explaining. Discussing with ordinary people does help us to get out of the fixed
intellectual framework used in our community and to get a better feeling
for what science means to society at large. If it is the essence of science
to be an inquiry, then surely we should be trying to find out more.

I realise that the ideas I have expressed may not be clear and that the
same may be true of the way I have expressed them. But really the purpose
of my talk was to find what ideas the audience here would have. How can
scientists become more effective in dealing with the pressing problems that
face them, in meeting the anti-science attitudes and the criticism of science?
Should we or can we change the nature of science itself?

Discussion

L. COOPER: I agree so much with Professor WILKINS' clear and moving statement that
I cannot refrain from adding a small elaboration. We do best, I believe, if we regard the
scientific method not as something that is particular and unique but more perhaps as an
extension or refinement of what is normally called common sense—that vast, but commonplace
endeavor engaged in by all humans, children and adults, in which one attempts to introduce
concepts to organize experience: what distinguishes science from other activities is perhaps
the variety of experience that is considered.

If one regards science in this way, among other things, it makes it much easier to address
students, especially those who are not interested primarily in science, since one is attempting
to convince them that what the scientist does is no mystery, but an application of what they
do every day to some particular realm of experience.

Consider as an example, a subject mentioned by Prof. WILKINS, the relation between science and art. I think there is a deep relation. But it is not to be found in a comparison of techniques or in the use of science to analyze or improve the technique of the artist (although this might be useful—how to make better pigments, for example) but rather in the similarity of the underlying endeavor—the attempt of scientist and the artist to organize a certain realm of experience.

NEWTON regarded a variety of planetary and other experience with moving bodies; he introduced what are now called Newton's Laws to organize this experience into a vast structure. Some believe that Newton's Laws are eternal truths—a reality of nature, but it's hard to take that point of view in view of the competing 'truths', general relativity and quantum theory. However, true or not, NEWTON's organization shaped people's mind for 200 years afterward. So much that they often didn't look at the phenomena; they looked first at how NEWTON had put them together. They were living in a Newtonian World—but of course that world was in their heads.

When the Impressionists first began to put paint on canvas they were greeted with ridicule; it was said, not too far from this room, that these men couldn't see properly and certainly couldn't paint. But we believe, I presume, that the lens and the retina in Monet's or Renoir's eyeballs were not that different from the optical apparatus possessed by David. They chose, and it was a voluntary choice, to (shall we say) look at things differently. What Monet gave us was not the façade of a cathedral but—clearly—his perception, his way of regarding that façade. It was this perception, so different from that of David (not to mention Gustave Moreau) that was shocking. But nowadays, if one looks at a calendar or a magazine of a more popular kind one will as likely as not find an Impressionist type composition. The *avant garde* perception has once again become banal. Which reveals the extent to which the Impressionist has imposed his vision of light and color on the rest of us.

It seems to me that in this most important aspect—where the artist or the scientist imposes his order on experience and presents us with the furniture of our intellectual residence, they are most alike. Whether that order is 'true' or 'false' is perhaps not the most relevant consideration—as long as it is 'true' enough to enable us to function. If the artist/scientist is powerful enough, his vision dominates the imagination—future generations live in the world he has created. The musician, we might say, teaches us what to listen for and how to regard sound, the painter—what to look at and how to regard light, and the scientist what to think about and how to regard nature.

A. KATCHALSKY: The major issue facing mankind is the need for a new attitude to treat the global problems of the world. During millenia man could regard the globe as an open system, which could absorb any fall-out thrown beyond the national boundaries. Scientific technology and population explosion have changed the picture drastically. Today the globe became a single, closed system, and its survival depends on the human capacity to maintain a well balanced international equilibrium. The outcry on pollution and extermination of wild-life is only one aspect of the detrimental effect of pushing the local classical materialistic attitudes towards a world which has to be controlled by an overall policy.

Such a global attitude is not contradictory to the cultural or national allegiance—as belonging to a family or being devoted to your city is not contradictory to your being a member of a national group. This is slowly being recognized not only by intellectuals but even by broad-minded political leaders. To bring about a global approach which has an operational validity and is not only a set of slogans and empty declarations, *internationalization* must undergo a process of *internalization*. That such an internalization is possible and can be brought about by a conscious educational effort can be derived from a classical study of JEAN PIAGET carried out already 30 years ago:

PIAGET and his coworkers asked young Genevoise children to depict the relation of Geneva to Switzerland. All of them drew two circles: one circle for Geneva and another circle for Switzerland; the circles did not touch and there was no relation between the two. When the same question was posed to older children, they still drew two circles, but now the circle of Geneva was put within the circle representing Switzerland. Thus the children retained their identity as Genevoise, but underwent a process of internalization as Swiss citizens. Even older children, however, did not know how to put Switzerland on the bigger circle representing the globe, and it was clear that the process of becoming a citizen of the world was never accomplished.

To prepare citizens of the world—who could handle the problems of modern mankind—a major effort of scientists and psychologists, teachers and political leaders is required, and it is up to us to mobilize the good-will of all people whose help is needed, and who could provide the moving power for this international effort.

Only within the framework of a global approach is there a hope to bridge the gap between the developed and under-developed countries; within this framework there is a hope to find a way to control poverty and to develop deserted areas; and with this approach the impact of science may become a positive asset for the future of man.

P. AUGER: I just wanted to say a word about the problems Dr. KATCHALSKY has mentioned: the problem of the underdeveloped countries. I want to mention that there is now something in the making which is called the International Science Foundation and which is a tentative action by scientists in order to help the underdeveloped countries to organize their own science. The International Science Foundation is supposed to try to pinpoint in the different developing countries the teams and the research scientists, the young ones if possible, who really can give some promise to be active, to be really capable of organizing good science in their own countries, and to help them. To give financial, technical help and to make it possible for them to travel to other countries and also to give them the possibilities to get the books and the views they should have. The International Science Foundation should try to work outside of direct governmental activities. The help which is given by the United Nations to developing nations always goes completely through local governmental authorities. This, of course, is one of the difficulties. These authorities do not know exactly, where the real scientists are in their own country. We think that the international scientific community would be able to find out who is really worth-while helping in the developing countries: young scientists and teams of young scientists. Of course this cannot be done outside the knowledge of the different governments; but not *through* the governments. The choice must be made by the scientists and not by local authorities. I just wanted to mention that because it is one of the ways through which the scientific community can help the developing countries.

H. SPEKREYSE: I am afraid I shall be a bit down to earth after what we have heard. I should like to say that I found myself in deep resonance with most of what Prof. WILKINS said and perhaps I can add some information about two points that he made. The first place, the breaking down of the myth of pure science. He mentioned the attempts to get in closer contact with the technicians in the laboratory. This is being done, in the laboratory of High Energy Physics, in which we have a very substantial contribution from a number of unskilled workers—usually girls or women, who do the scanning of the films that are studied and then interpreted by the physicists. We were not animated by any moral inspiration in deciding that it was necessary to instruct those unskilled workers in the meaning of the work they were doing. It was found that this increased their efficiency and reliability. So from a purely practical approach one is led (as an after effect of the consequence), to beneficial consequences of the kind you suggest. The second point you mentioned is the problem of education. That is a long-term problem and there are, of course, more pressing ones, but after all, the mess that we are making now has to be cleaned-up by the next generation and so it is not useless to do whatever we can to help them. At the primary level, you mentioned some developments in Britain. I might point out that this kind of global education, breaking the artificial barriers between various parts of science, was introduced and practised by pioneers in Holland and in Belgium 50 years ago or more, 70 years ago. I am alluding to Mrs. Montessori, who is Italian of course, but who lived in Holland; Dr. Decrolis in Belgium. They were guided by just the principle you mentioned and those schools have really produced. One can see the difference in the children coming from those schools and those who have gone through the traditional routine. I have my own experience with that from Holland. I could immediately distinguish, when being on an examination commission, the voice coming from those schools and from the others, by the independence of judgement that they displayed. Here we have positive experience that this should be continued. However, when I tried to advocate the extension of those schools, I immediately noticed a hostility on the part of the government, because of course, this is a more difficult kind of education and the people in power are not prepared to make any effort, any concrete effort or any sacrifice, in order to introduce changes that are very necessary. Now, what can we do as scientists? In isolation, or even in a very distinguished group as this one, we can only complain about the situation and point to the direction in which we see a possibility of progress, but the real problem is that of practical action. Practical action can only succeed by the weight of numbers. Now this unfortunate distinction between pure and so-called upright sciences has produced a separation, a segregation of scientists which are but a tiny minority who have the leisure to think about those problems. The great mass of applied scientists is not privileged in that respect. However, I should point out that even here, there is an organization which has been in existence about 25 years, which has, at least potentially, the element of number and that is the World Federation of Scientific Workers, created in 1945 on the initiative of the British Association of Scientific Workers. Now this federation is not perfect, of course, but it is a starting point from which one could achieve concrete results, by real action, practical action based not only on the influence of authority, but also of number, which is quite essential.

A. COURNAND: At this point, I shall add nothing further to my presentation, but I should like to comment on what Professor WILKINS has so remarkably set forth: the role of the scientist in relation to the world, and the necessity, if I may say so, of recreating a natural philosophy. The great physiologist SHERRINGTON, at the end of his life, wrote about JEAN FERNEL, who in the sixteenth century was the exponent of a natural philosophy. As a physician, he emphasized the importance for the doctor, of establishing dialogue with his patients, of learning about their trades and vocations, their training, their culture. He indicated the need for the physician to develop his sense of the human, and to have some conception of the problems to which natural philosophy addressed itself.

We know, of course, that HIPPOCRATES, 25 centuries ago, in certain of his aphorisms stressed that the physician should not only understand disease, help the sick person, and provide for the needs of his family; he should also try to understand himself and nature. I recommend the admirable article written by my colleague, DICKINSON RICHARDS, 'Hippocrates and History', in which the de-humanization of science is severely criticized.

Recently, a new philosophy, that of Prospective, has been elaborated by GASTON BERGER and his colleagues. Prospective, I believe, has remarkable relevance to the concerns that Professor WILKINS has evoked. Prospective has set itself the problem of evolving ways of dealing with the future—more exactly, using the words of PAUL VALÉRY, of helping man to adapt himself 'to what has never been'. Seeing the future as free to be shaped by man, Prospective has taken on three principal tasks: 1) developing images of various possible future situations that may be desirable, to provide guidance in making decisions in regard to planning; 2) helping man to adapt to a world that is subject to continual but rapid transformation; and 3) preparing man to act.

It seems to me that this philosophy applies remarkably well to our concern as scientists to become part of the world and to construct a new philosophy applicable to this problem.

K. MENDELSSOHN: I would like to take up the very first point that Prof. WILKINS made and that is the question of the attitude of anti-science, of ROGER and so on. What does a world without science really look like? As an experimentalist, I have had the experience. I spent three weeks among the Ashanti in the rain forest of Ghana. That world has no science, it has no bureaucracy, it is very quiet. It doesn't pollute, it doesn't engage in the worst pollution of all which consists in polluting the globe with *Homo sapiens* because the intestinal parasites of both men and the anopheles mosquito see to it. What does this world look like? Even at the age of 65 where one values rest and relaxation, I would find it somewhat intellectually frustrating. It has certain advantages, but not all of them. Now it is just as well to remember that 500 years ago the great civilizations of China and India were rich compared with Europe. This position has been reversed, simply because at about that time we developed a philosophy based on experimentation and quantitative relations that endowed us with the means of fairly accurate prediction and, therefore, of extreme economy of effort. We call it science. The only trouble with it all is that we really don't know what brought about the renaissance. One is slowly driven to the assumption that science is simply a phase in the development of *Homo sapiens.* Now perhaps this conference should have shown us better than anything else that we are the results of the processes of life, puppets on a stage, performing a play whose plot we don't know and which we must just go on following. Perhaps this is the answer to ROGER. I would go further than Prof. WILKINS who said that mankind in its development should follow reason. I would suspect that mankind in its development *must* follow reason. Why? We don't know. Because we simply don't know the intricacies of life; we don't know its development. Once you accept the fact, however, that science is part of our organic development then it doesn't really matter how much we exercise and in which direction we exercise our free will. One would suspect that we must follow reason whether we like it or not.

D. HODGKIN: I have found much of this discussion extremely interesting and hopeful in the sense that we can look forward to a future in which we slowly transform science and society, and there are many experiments going on in different parts of the world attempting to do just this at the moment. Since Professor WILKINS brought us back news through Professor ARTHUR GALSTON from China, I found myself remembering talking to the Chinese Chargé d'Affaires in London who was saying very much the same kind of thing: science itself is changing in China. On the other hand, and in other directions we have a sense of time running out, in the development of the war in Indochina, for example, where a tremendous use of modern technology and of scientific ingenuity is being more than wasted, being used for sheer destruction of a whole countryside. Somehow I feel as though we should be thinking of positive ways in which not only the scientific community, of course, but everyone should get some control of the direction to which we all are concerned to move here today.

M. H. F. WILKINS: I couldn't agree more with what Professor HODGKIN said about the time running out and the need to face the very urgent problems. I also agree very much that political change is very important. My own position is that I am a socialist, basically. If any of my general discussion distracted people from the urgency of our practical problems then I would regret this. But what I feel is that, if we try to deal with the practical, urgent problems facing us and if we make the political changes that many of us think ought to be made, we may find ourselves afterwards facing further problems which may not be all that different from those we have now. The scientific community needs to get down to the underlying philosophical problems which are involved. There is one important thing that those of us who are in universities and other educational institutions can do. We can try to find new ways of teaching science.

J. NINIO: I don't feel guilty about being a scientist. It seems to me that to put too much emphasis on the responsibility of the scientist—of course it is a generous feeling—is not very commendable because, in a certain way, we would be assuming the sins of the society. It is a Christian attitude. But if we do so, it means that we, as scientists, are influenced by non-scientific thinking, by irrational thinking. We are speaking of Society in a general way. I am not convinced that an adequate description of the world as it is today would be in terms of a Society with a capital 'S', and not in terms of two kinds of systems struggling each against the other. I am not convinced that on close scrutinity, socialism and capitalism are really identical, as would appear in some of the talks here. Perhaps we can study the text of the founders, or of the great thinkers of one system or the other, and see what is the underlying attitude of one economic system as contrasted with that of the other, in the long run and the short run—towards science. Which system favors rational thinking and which system tries to go against thinking? I take my own country: it is perfectly clear whether or not the government allows—say astrology and anti-scientific thinking, to be presented on radio and television. I won't give the answer, because I don't want to get into a political debate, but it seems obvious that the answer is essentially here.

I. RABI: Mr. Chairman, a few brief remarks. I also agree with the previous speaker in that I am not at all ashamed of being a scientist. I don't believe that science de-humanizes; on the contrary, it brings out the highest human faculties. Science brings people together; science helps us organize the world. It seems that some of us have been brainwashed by the critics, and we beat our breasts and say that we are de-humanizers, which is arrant nonsense. However, the advance of science as noted by Prof. KATCHALSKY has taught us now that we are part of the system of the globe. Our systems of morality according to which we have been brought up, are individual moralities. You are responsible for your own actions. We know perfectly well that this is no longer adequate as we have seen developing concerns about France with respect to Algeria, the United States with respect to Vietnam. This storm of feeling and indeed a common feeling of community morality is in a certain sense one of community guilt. This concern has not yet been spelled out. The philosophic basis for this concern does not yet exist because our moralities are founded on individual morality, although we see some developments. The Nuremburg trials were one such in the trials occurring in the United States now, for My Lai and so on; we see the beginnings of new structures, perhaps new legal structures. We need an underpinning for those things, we lack an appropriate rationale. We need to construct a future morality, but not out of vague feelings, because science, particularly biological science, has gone under the skin of man and begins to understand the deeper, hidden motives of our actions. I won't even call it motives; it may be the wrong term. The future morality must have scientific basis from which we can make broader moral judgements with regard to the whole of humanity because the old categories, whatever they may be, family, city, state, country, have ultimately to become subordinated to some form which I cannot see, to a deeper feeling for mankind and its goals. We can no longer talk of the destiny of a particular country, which was so successfull for a while here in France, we have to talk in some sense of the whole world. I do not mean to say that all nations are equal as they are, so to speak, in the United Nations. We do not begin to have the proper categories. I have the feeling that we need a new generation which has to be educated to see the world in a newer light as demanded by the developments of science. It is not an accident, so to speak, that in the Bible, the children of Israel had to wander for 40 years to forget that they had been slaves.

A. LICHNEROWICZ: Je voudrais me borner à deux remarques fort simples. Je suis d'accord avec l'analyse du monde fermé de notre ami KATCHALSKY, mais nous nous trouvons placés devant un problème qui n'est pas celui de la responsabilité de la communauté scientifique, mais qui importe: c'est celui que je traduirai en parlant de l'unité de la science et de la diversité des cultures. Il nous faut trouver, avec maturité, la manière de contrôler un monde qui respecte la diversité de cultures dont chacune est porteuse de valeurs humaines. La seconde est qu'il y a, particulièrement dans les domaines écologiques, économiques, quelques quiproquo qui ne sont pas toujours innocents. Nous analysons par exemple des économies très différentes de la nôtre avec nos standards et ces analyses sont toujours très implicitement *normatives*. Cela n'a pas grand sens dans l'histoire, nous l'avons appris, de pouvoir comparer véritablement, quantitativement, rationnellement, les taux de développement d'un siècle à l'autre. Cela n'a pas non plus grand sens de traduire en dollars le revenu par tête correspondant à certaines cultures. Nous devons lutter contre ces déformations qui sont non innocentes à tous les points de vue.

A. COURNAND: There is one group of future scientists who haven't been mentioned in this discussion so far. That is the group of students now in our schools. It is quite important for us to establish seminars, as I have done in my own university, where we are able to find out what the young think about the future world in which they are going to live, about the future of their own professions and vocations—scientific, medical, or other. I believe that the development of such seminars where faculty members participate with and learn to engage in dialogue with the young is of considerable importance. I would like to emphasize that the future belongs to the young people, and that the old must contribute to this belief.

P. AUGER: Just one word to answer remarks about the future of mankind. Science is really,part of the development of the human race: normal development. Any action on babies from the time of birth on, is the conditioning of the child, changing it into a citizen of the present world. We have to make this change very carefully, and I feel that it is not being done so at present. It should start straightaway at the very first hour after birth. It should be studied scientifically, much more carefully, how to organize the lives of children from the start, to make them better citizens of the present society.

G. L. Sobolev: Avant de clore la séance, je dois livrer ma propre opinion. Rappelons-nous l'époque de la Renaissance. Les savants, les artistes, les peintres étaient alors très fiers de ce qu'ils faisaient, ils étaient absolument sûrs que leur œuvre était bonne. Il faut que nous nous remettions en mémoire l'esprit de la Renaissance; il faut que nous soyons fiers de ce que nous faisons maintenant; aucun doute, aucun scepticisme ne doit se développer: l'œuvre de la science constitue un progrès pour l'humanité. La masse de toute l'humanité continue sa marche et nous sommes les représentants de nos ancêtres qui ont commencé cette marche à l'époque de la Renaissance déjà.